Thermodynamics of Atmospheres and Oceans

This is Volume 65 in the
INTERNATIONAL GEOPHYSICS SERIES
A series of monographs and textbooks
Edited by JAMES R. HOLTON

A complete list of books in this series appears at the end of this volume.

Thermodynamics of Atmospheres and Oceans

Judith A. Curry and Peter J. Webster
PROGRAM IN ATMOSPHERIC AND OCEANIC SCIENCES
DEPARTMENT OF AEROSPACE ENGINEERING
UNIVERSITY OF COLORADO
BOULDER, COLORADO
USA

ACADEMIC PRESS
San Diego London Boston
New York Sydney Tokyo Toronto

Academic Press
24–28 Oval Road, London NW1 7DX, UK
http://www.hbuk.co.uk/ap/

Academic Press
a division of Harcourt Brace & Company
525 B Street, Suite 1900, San Diego, California 92101-4495, USA
http://www.apnet.com.

ISBN 0-12-199570-4

A catalogue record for this book is available from the British Library

Printed and bound by Antony Rowe Ltd, Eastbourne
Transferred to digital printing 2005

Contents

Part I   Basic Concepts

Chapter 1   Composition, Structure, and State

Chapter 2   First and Second Laws of Thermodynamics

Chapter 7 Static Stability of the Atmosphere and Ocean

Chapter 8 Cloud Characteristics and Processes

Chapter 9 Ocean Surface Exchanges of Heat and Freshwater

Chapter 10 Sea Ice, Snow and Glaciers

Chapter 11   Thermohaline Processes in the Ocean

Part III   Special Topics

Chapter 12   Global Energy and Entropy Balances

Chapter 13   Thermodynamic Feedbacks in the Climate System

Chapter 14 Planetary Atmospheres

Appendices

Preface

Thermodynamic processes are inherent to the atmosphere and oceans. Winds and ocean currents owe their existence to the thermodynamic imbalances that arise from the differential heating of the Earth's surface and air by the sun. Thermodynamic processes on Earth are especially interesting and challenging because of the proximity of its climate to the triple point of water. Changes of phase of water in the atmosphere result in the formation of clouds and precipitation. Associated with the changes of phase are the release of latent heat and modifications to the atmospheric radiative transfer. Freezing and melting of seawater in high latitudes influences profoundly the exchange of radiant, sensible, and latent heat between the atmosphere and oceans. Accounting for heat exchanges within the atmosphere and ocean (and *between* the atmosphere and ocean) is essential in any predictive model of the ocean and/or atmosphere. Thermodynamic feedback processes in the atmosphere and ocean are critical to understanding the overall stability of the Earth's climate and climate change.

Typically, atmospheric thermodynamics receives no more than one or two chapters in textbooks on dynamic meteorology, atmospheric physics, or cloud physics. Two modern textbooks have been dedicated entirely to aspects of atmospheric thermodynamics: *Atmospheric Thermodynamics* by Iribarne and Godson and *Atmospheric Thermodynamics* by Bohren and Albrecht. The thermodynamics of the ocean receives less emphasis in oceanography texts, although this subject receives more substantial treatment in a few chemical oceanography texts. Sea ice is barely mentioned in any of the modern textbooks on oceanography. The present text expands the subject of atmospheric and oceanic thermodynamics considerably beyond these prior treatments, integrating the treatment of thermodynamics of the atmosphere, ocean, and cryosphere.

The book *Atmosphere-Ocean Dynamics* by Gill presented for the first time a unified treatment of the large-scale dynamics of the atmosphere and ocean. In the same spirit, *Thermodynamics of Atmospheres and Oceans* presents a systematic and unifying approach to the thermodynamics of the atmosphere and ocean and establishes the interrelationship between these subjects. There is commonalty between the atmosphere and ocean of many of the important thermodynamic processes, and simultaneous consideration of the thermodynamics of the atmosphere and ocean enriches our understanding of both. A compelling reason for the unified treatment of atmospheric and oceanic thermodynamics is the importance of air-sea interactions in topics ranging from the forecasting of severe weather (e.g., hurricanes) to understanding climate variability on time scales ranging from the interannual to the millennial and longer. A major motivation for writing this text arises from our involvement over the past decades in the World Climate Research Programme and associated U.S. programs, and the need for a unified treatment of processes involved in energy and water exchange in the atmosphere, ocean and cryosphere, over a broad range of time scales.

Thermodynamics of Atmospheres and Oceans has evolved from a course of the same title that has been taught at the University of Colorado since 1993, in a class with both graduate students and senior undergraduates. This book is intended to serve as the text of an introductory graduate course for students in atmospheric science and/or oceanography. The text is targeted at students with diverse backgrounds in physics, chemistry, mathematics or engineering. Undergraduates at the senior level with a similar background should also find this to be a suitable text. Researchers will find the book useful, not only for its systematic treatment of thermodynamics of the atmosphere, ocean and cryosphere, but also because of the many diagrams and formulas and the tables in the Appendices.

Part I of the text develops the basic concepts of classical thermodynamics. Statistical thermodynamics is presented in a heuristic manner, to develop a conceptual understanding of various thermodynamic state functions and processes. The selection of topics is focused towards establishing the foundations for specific thermodynamic applications dealt with in subsequent chapters. In Chapter 1, the composition and vertical thermodynamic structure of the atmosphere and ocean are described. The equations of state for the atmosphere and ocean are developed, including the state in a gravitational field. Chapter 2 reviews the first and second laws of thermodynamics, including applications to ideal gases. Chapter 3 introduces time-dependent thermodynamics and heat exchange processes. Chapter 4 focuses on the thermodynamics of water, including phase equilibria. The physical chemistry of water solutions is described, with specific applications to understanding the colligative properties of seawater and sea ice. The nucleation and diffusive growth of condensed phases are described in Chapter 5, with specific applications to cloud particles and sea ice.

Part II addresses applications of the basic concepts developed in Part I to specific processes in the atmosphere and ocean, including cloud formation, thermohaline processes in the ocean, and heat transfer in sea ice. Examples are provided from two regions that represent the Earth's thermodynamic extremes, specifically the tropical ocean "warm pools" and the Arctic Ocean. Chapter 6 describes moist atmospheric processes and Chapter 8 relates the physical characteristics of the clouds to these processes. Chapter 7 addresses the statics and stability of the atmosphere and ocean, while Chapter 9 introduces air/sea interactions in the context of the ocean surface buoyancy flux. Chapter 10 describes the thermodynamic processes associated with the annual cycle of growth and ablation of sea ice and also the thermodynamics of seasonal snow and glacier ice. Chapter 11 applies the concepts of thermodynamics and statics to ocean thermohaline processes.

Part III addresses the role of thermodynamics in the global climate and in planetary atmospheres. The interfaces of thermodynamics with radiative transfer and large-scale dynamics are introduced. Chapter 12 presents a global view of atmospheric and oceanic thermodynamics, including global scale heat transports and the hydrological cycle in the global ocean and atmosphere. Chapter 13 extends the applications of Part II to address thermodynamic feedbacks in the climate system, including those that involve clouds, snow and sea ice, and the ocean thermohaline circulation. Applications of thermodynamics to planetary atmospheres are presented in Chapter 14, providing a broader perspective and context of the Earth thermodynamic properties and processes.

The notes at the end of each chapter list some relevant books and review articles that provide the reader with a starting point for further investigation. Individual citations are made within the text only to cite specific data that we use in figures and tables, figures that we have obtained copyright permission to reproduce, and equations or empirical relations that are not readily derived from principles in the text or are relatively recent and not in common use. No attempt has been made to provide a comprehensive bibliography on the subject.

Problems at the end of the chapter are designed to enforce the principles developed in each chapter as well as to extend the students' appreciation of the subject matter. The problems range from practical to didactic in nature, and from simple to challenging. Some of the problems are best solved numerically, although these problems are not so indicated in the text, since we believe that graduate students need practical experience in determining the best method to solve a problem. Answers to selected problems are provided after the appendices.

The presentation emphasizes understanding in the context of fundamental physical and chemical principles while at the same time providing useful tools for the practicing scientist or engineer. Hence the treatment is deductive as well as descriptive. Elegance in formal derivations is sacrificed in favor of developing simple conceptual models, including many schematic diagrams. The mathematics of thermodynamics, particularly with differentials and subscripted partial differentials, is cumbersome and remains the subject of considerable debate. Because of the complex applications of Maxwell's relations that arise particularly in the context of the ocean, for the sake of clarity we do not make the simplifications to the traditional mathematics and notation proposed by some authors. Combining atmospheric and oceanic sciences presents a substantial challenge to devising a system of notation. The two fields often use different notation for the same variables, or the same notation for different variables. We have made every attempt to develop a logical notation scheme that is consistent with each of the fields and nonduplicative. Unfortunately there aren't enough letters and typefaces to meet completely both of these requirements. Appendix A describes the notation in detail.

Students whom have taken the course *Thermodynamics of Atmospheres and Oceans* at the University of Colorado have taken a prior course in thermodynamics (e.g., engineering thermodynamics, statistical thermodynamics, physical chemistry). Because of the students' previous background in thermodynamics, we have been able to cover in a one-semester class all of the material in Parts I and II plus Section 12.1, with one chapter selected from Part III. Since many atmospheric science departments teach an undergraduate course in atmospheric thermodynamics or a graduate course to students with little previous background in thermodynamics, this book has been designed so that it can also be used for such courses. An undergraduate semester course in atmospheric thermodynamics might cover Chapters 1, 2, 3, 4, 6, and 7 in depth, eliminating Sections 1.2, 1.8, 1.9, 2.11, and 4.6 that address specific topics in oceanography. Hence there is scope for considerable flexibility in the use of this book as a text.

Judith A. Curry
Peter J. Webster

Boulder, Colorado
August, 1998

Acknowledgements

Numerous colleagues contributed to the text in various ways. We owe particular thanks to the following individuals who reviewed most of the chapters in the book: Dr. James Holton, Dr. Branko Kosovic, Dr. James Pinto, and Dr. Scott Doney. We are also grateful to colleagues who reviewed individual chapters: Julie Schramm, Dr. William Rossow, Dr. Geoffrey Considine, Dr. Chris Torrence, Dr. Gary Maykut, Dr. Jeffrey Weiss, and Dr. Andrew Moore. Students in the class of Fall 1997 provided considerable assistance in improving the presentation of the material, identifying errors, and solving the homework problems. Original figures were provided by Julie Schramm, Dr. Konrad Steffen, and Dr. Richard Moritz. Last but certainly not least, we would like to thank Karyn Moore for her excellent technical assistance in preparing the manuscript, including preparation of the figures, editing, and text formatting.

Finally, we would like to acknowledge the continuing support of our research from the National Science Foundation, National Oceanic and Atmospheric Administration, Department of Energy, and National Aeronautics and Space Administration, as well as the University of Colorado, which made it possible for us to write this book.

Judith A. Curry
Peter J. Webster

Boulder, Colorado
August, 1998

Publisher's Credits

The following publishers have kindly granted permission to reprint or adapt materials in the following figures. We have attempted to reach the senior authors to obtain their permission as well; we apologize to those whom we have not reached, for whatever reason. Complete citations of the sources are found in the bibliography at the end of the book.

3.7:   From *An Introduction to Boundary Layer Meteorology* by R.B. Stull. Copyright 1988 by Kluwer Academic Publishers. Reprinted with kind permission from Kluwer Academic Publishers.

4.2:   Reprinted with permission from *Ice Physics* by P.V. Hobbs. Copyright 1974 by Clarendon Press.

4.9:   Reprinted with permission from *Arctic Sea Ice*. Copyright 1958 by National Academy Press.

5.2:   Reprinted with permission from *Elements of Cloud Physics* by H.R. Byers. Copyright 1965 by University of Chicago Press.

5.6:   Reprinted with permission from *A Short Course in Cloud Physics* by R.R. Rogers and M.K. Yau. Copyright 1989 by Pergamon Press.

5.9:   From *Microphysics of Clouds and Precipitation* by H.R. Pruppacher and J.D. Klett. Copyright 1997 by Kluwer Academic Publishers. Reprinted with kind permission from Kluwer Academic Publishers.

5.11:   Reprinted with permission from *The Growth and Decay of Ice* by G.S.H. Lock. Copyright 1990 by Cambridge University Press.

5.12:   Reprinted with permission from *Sea Ice Biota*, Copyright 1985 CRC Press, Boca Raton, FL.

5.13   Reprinted with permission from *Sea Ice Biota*, Copyright 1985 CRC Press, Boca Raton, FL.

5.14:   From Sinha, 1977. Reprinted with permission from the International Glaciological Society.

6.4:   From *Clouds and Storms* by F.H. Ludlam. Copyright 1980 by Pennsylvania State University Press. Reproduced with permission of the Pennsylvania State University Press.

6.5:   From *Principles of Atmospheric Physics and Chemistry* by R.E. Goody. Copyright 1995 by Oxford University Press, Inc. Used by permission of Oxford University Press, Inc.

8.3:   Used with permission from *The Physics of Rainclouds* by N.H. Fletcher. Copyright 1962 by Cambridge University Press.

8.4:   From Klett and Davis, 1973. Reprinted with permission of the American Meteorological Society.

12.2:   Kiehl and Trenberth, 1997. Reprinted with permission by the American Meteorological Society

12.3:   Reprinted with permission from *Global Physical Climatology* by D.L. Hartmann. Copyright 1994 by Academic Press.

12.4:   Reprinted with permission from *Global Physical Climatology* by D.L. Hartmann. Copyright 1994 by Academic Press.

12.5:   Zhang and Rossow, 1997. Reprinted with permission of the American Meteorological Society.

12.7:   From *Monsoons*, J.S. Fein and P.L. Stephens, eds., Copyright 1987 by John Wiley & Sons. Reprinted by permission of John Wiley & Sons.

12.8:   Webster et al., 1998. Reprinted with permission of the American Geophysical Union.

12.9:   Stephens and O'Brien, 1993. Reprinted with permission of the Royal Meteorological Society.

13.3:   Schlesinger, M.E., 1986. Reprinted with permission of Springer-Verlag.

13.4:   Spencer and Braswell, 1997. Reprinted with permission of the American Meteorological Society.

13.5:   Curry et al., 1995. Reprinted with permission of the American Meteorological Society.

13.9:   Reprinted with permission from *Energy and Water Cycles in the Climate System*, E. Raschke and D. Jacob, eds. Copyright 1993 by Springer-Verlag.

14.7:   Reprinted with permission from *Physics and Chemistry of the Solar System* by J.S. Lewis. Copyright 1995 by Academic Press.

14.4:   Webster, P. J., 1977, The Low-latitude circulation of mars. ICARUS, **30**, 626-664.

14.13: Reprinted with permission from *Physics and Chemistry of the Solar System* by J.S. Lewis. Copyright 1995 by Academic Press.

14.14: Reprinted with permission from *Physics and Chemistry of the Solar System* by J.S. Lewis. Copyright 1995 by Academic Press.

14.15: Reprinted with permission from *Physics and Chemistry of the Solar System* by J.S. Lewis. Copyright 1995 by Academic Press.

Chapter 1 | Composition, Structure, and State

Thermodynamics deals broadly with the conservation and conversion of various forms of energy, and the relationships between energy and the changes in properties of matter. The concepts of classical thermodynamics were developed from observations of the macroscopic properties of physical, chemical, and biological systems. In the early part of the 20th century, it became apparent that the empirically-derived laws of thermodynamics are deducible by application of classical and quantum mechanical principles to atoms; this is known as *statistical thermodynamics* or *statistical mechanics*. In applying thermodynamics to the atmosphere and ocean in this book, we draw upon classical thermodynamics to describe what happens in a thermodynamic process, while using the molecular view of statistical thermodynamics to increase our understanding of why.

Why do we need an understanding of thermodynamics to study the atmosphere and ocean? Some of the reasons include:

a) The forces that drive the motions of the atmosphere and ocean are created by differential heating of the Earth's surface and atmosphere by the sun. Because of the Earth's spherical shape and axial tilt, the tropics receive more energy than the poles. Furthermore, the heat capacities of water, land, and air are very different, as are the efficiencies at which they absorb solar radiation. Differential heating spanning a wide range of spatial scales creates thermodynamic imbalances, which in turn create winds and ocean currents as the atmosphere/ocean system attempts to return to thermodynamic equilibrium.

b) Changes of phase of water in the atmosphere result in the formation of clouds and precipitation. Associated with the formation of clouds and precipitation are the release of latent heat and modifications to atmospheric radiative transfer. Freezing and melting of seawater in high latitudes influences profoundly the manner in which heat is exchanged between the atmosphere and ocean.

c) Accounting for heat exchanges in the atmosphere and ocean is essential in any predictive model of the ocean and/or atmosphere, for any space or time scale that is considered.

d) Thermodynamic feedbacks in the atmosphere and ocean are critical to understanding climate change. For example, increasing the concentration of carbon dioxide in the atmosphere tends to heat the planet. However, changes in the amount and phase of water in the atmosphere and at the Earth's surface caused by this warming may enhance or mitigate the warming.

Although other motivations can be provided, these motivations determine the focus of this book.

To begin our discussion of thermodynamics, some definitions are needed. A *thermodynamic system* is a definite quantity of matter which can exchange energy with its surroundings by performing mechanical work or by transferring heat across the boundary. A system may be *open* or *closed*, depending on whether or not it exchanges matter with its environment. A system is said to be *isolated* if it does not exchange any kind of energy with its environment. The *environment* comprises the surroundings of the thermodynamic system. The thermodynamic systems addressed in this book are portions of air or seawater undergoing transformations, both of which are open systems. For the sake of simplicity, we will sometimes treat portions of the atmosphere and ocean as closed systems, which is a reasonable approximation if the volumes are large enough to neglect exchanges with the surroundings or if the surroundings have the same properties as the system.

A thermodynamic state variable is a quantity that specifies the thermodynamic state of a substance (e.g., temperature). For a closed system, the mass and chemical composition define the system itself; the rest of the properties define its *state*. For a homogeneous system of constant composition, there are three variables that describe the state of the substance, only two of which are independent. These variables are the pressure p, the volume V, and the temperature T. If any two of the three thermodynamic variables are known, then the value of the third will be fixed, because the variables are related in a definite way. Thus for a homogeneous system we have the following *equation of state* relating the three variables:

$$f(p,V,T) = 0$$

For an ideal gas, we have the familiar equation of state,

$$pV = nR^*T$$

where n is the number of moles and R^* is the universal gas constant.

Thermodynamic variables and the functions derived from these variables are called *extensive* if they depend on mass (e.g., volume, internal energy), and *intensive* if they do not depend on mass and can be defined for every point of the system (e.g., temperature, density). We shall, with some exceptions (such as temperature), use capital letters for extensive properties and lower case letters for intensive variables. Intensive variables are particularly advantageous in studying atmospheres and oceans since they make keeping track of the number of moles, mass, etc. unnecessary in these large thermodynamic systems.

Central to the description of a thermodynamic system is its chemical composition and physical state. In the remainder of this chapter the composition, structure, and state of the atmosphere and ocean are described.

1.1 Composition of the Atmosphere

The Earth's atmosphere consists of a mixture of gases, water in the liquid and solid states, and other solid particles that are very small in size. The Earth's atmosphere up to about 110 km (the *homosphere*) is well mixed by turbulent air motions, and the composition and concentration of the passive constituent gases (i.e., those that do not undergo phase changes or extensive chemical reactions) is fairly constant with height. Above the homosphere, the composition of the *heterosphere* (or *exosphere*) is subject to diffusive stratification by the molecular weight of the gases and strong chemical and photochemical alterations. The focus here is on the composition of the homosphere.

The concentrations of the major gaseous constituents in the homosphere are shown in Table 1.1. It is seen that N_2 and O_2 constitute approximately 99% of the volume and mass of the homosphere. The concentrations of the gases in Table 1.1 are relatively constant throughout the homosophere, with the exception of water vapor. There are many other gases present in trace amounts in the homosphere (besides those shown in Table 1.1) that are of importance in atmospheric chemistry and radiative transfer, including ozone, methane, nitrous oxide, sulfur dioxide.

Water vapor constitutes 0 to 4% of the atmospheric concentration of gases, the exact amount varying with time and location. Figure 1.1 shows the distribution of water vapor in the atmosphere and how it varies with height and latitude. The maximum concentration near the surface of the Earth indicates that the surface is the principal source of atmospheric water vapor. The general decrease in water vapor concentration above the Earth's surface arises from condensation that occurs in clouds. Water vapor is the most important gas in the atmosphere from a thermodynamic point of view because of its radiative properties as well as its ability to condense under atmospheric conditions. Water is the only substance in the atmosphere that occurs naturally in all three phases. Condensed water in the atmosphere consists of

Table 1.1 Main gaseous constituents of air, relative to the percent composition of dry air.

Constituent	Formula	Molecular weight	% by volume	% by mass
Nitrogen	N_2	28.016	78.08	75.51
Oxygen	O_2	31.999	20.95	23.14
Argon	Ar	39.948	0.93	1.28
Carbon dioxide	CO_2	44.010	0.03	0.05
Water vapor	H_2O	18.005	0–4	

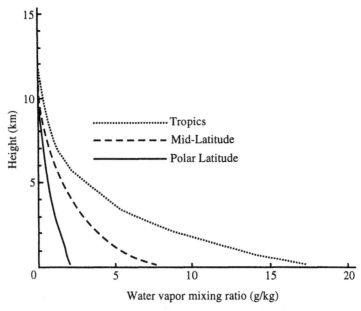

Figure 1.1 Atmospheric water vapor mixing ratio profiles (annual mean). Atmospheric water vapor decreases with both height and latitude. (Data from Oort, 1983.)

suspended cloud particles and hydrometeors. Cloud particles are either liquid (water droplets) or solid (ice crystals). *Hydrometeors* are bodies of liquid or solid water such as raindrops and snow that are falling through the air.

In addition to the gaseous constituents and condensed water particles, the atmospheric composition includes *aerosol* particles that are suspended in the air. Atmospheric aerosol particles are composed of dust, sea-salt particles, soil particles, volcanic debris, pollen, by-products of combustion, and other small particles that arise from chemical reactions within the atmosphere. Concentrations of atmospheric aerosol particles range from values on the order of 10^3 cm^{-3} over the oceans, to 10^4 cm^{-3} over rural land areas, to 10^5 cm^{-3} or higher in cities. Aerosol number density generally decreases with height above the Earth's surface, where most of the aerosol and gaseous precursors originate. Atmospheric aerosol particles commonly have sizes in the range between 0.1 and 10 μm. Concentrations of aerosol fall off sharply with increasing size. Aerosol sizes are limited on the upper end by gravitational fallout and on the lower end by aggregation processes. Scavenging by precipitation accounts for about 80–90% of the mass of aerosol removed from the atmosphere. The primary importance of aerosols to atmospheric thermodynamics is that a select group of them are crucial for cloud formation. The atmospheric aerosol also influences radiative transfer within the atmosphere.

1.2 Composition of the Ocean

The principal characteristic that differentiates seawater from pure water is the presence of dissolved salts, or its *salinity*. For purposes here, a *salt* is defined as a substance that forms ions in a water solution. Dissolved salts comprise about 3.5% of seawater by weight. Table 1.2 lists the elements that comprise 99.9% of the dissolved constituents of seawater. Together, sodium and chloride account for approximately 90% of the ocean salinity. Although the total amount of dissolved salts may vary with time and location, the fractional contribution of the major ions to total salinity of the ocean remains approximately constant.

Salinity was originally conceived as a measure of the mass of dissolved salts in a mass of seawater. However, it is difficult to determine the salt content by drying seawater and weighing the remaining salts, since chemical changes occur at the heats necessary to accomplish the drying. A complete chemical analysis of seawater is too time consuming to be conducted routinely. A more practical method of estimating salinity is to infer the salinity from the electrical conductivity of seawater. Although pure water is a poor conductor of electricity, the presence of ions allows water to carry an electric current, and the conductivity of seawater is proportional to its salinity.

By international convention, the practical salinity of a sample of seawater is defined in terms of the ratio of the conductivity of seawater to the conductivity of a standard KCl (potassium chloride) solution with concentration of 32.4356 $g\,kg^{-1}$ at 15°C and 1 atm pressure. Conductivity depends not only on salinity, but also on temperature and pressure. Therefore conductivity cannot be interpreted unambiguously as the total dissolved salts in a seawater sample. The unit of salinity determined in this manner is the *practical salinity unit* (psu), which is nearly equivalent to the total mass of dissolved solids in parts per thousand (‰).

Table 1.2 Average concentrations of the principal ions in seawater (total 34.482‰).

Ion	Formula	‰ by weight
Chloride	Cl^-	18.900
Sodium	Na^+	10.556
Sulfate	SO_4^{--}	2.649
Magnesium	Mg^{++}	1.272
Calcium	Ca^{++}	0.400
Potassium	K^+	0.380
Bicarbonate	HCO_3^-	0.140
Bromide	Br^-	0.065
Borate	$H_2BO_3^-$	0.026
Strontium	Sr^{++}	0.013
Fluoride	F^-	0.001

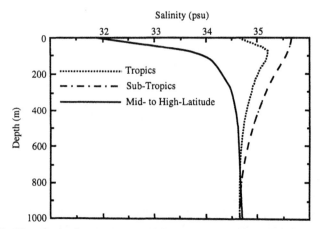

Figure 1.2 Profiles of annual mean ocean salinity in the upper ocean. High salinities near the surface are observed in subtropical latitudes because evaporation of fresh water from the ocean exceeds precipitation in these regions. In latitudes where precipitation exceeds evaporation, and hence the flux of fresh water into the ocean is positive, salinities are relatively low. In the deep ocean, salinity varies little with latitude, since evaporation and precipitation—the primary mechanisms influencing salinity—act at the surface. (Data from Levitus, 1982.)

The average salinity of the world ocean is 34.7 psu. Salinity in the open ocean ranges from about 33 to 38 psu. Higher values of salinity occur in regions of high evaporation such as the Mediterranean and Red Seas, where salinity values reach as high as 39 and 41 psu, respectively. Profiles of annual and zonal mean salinities are given in Figure 1.2. The salinity is large in the subtropical latitudes because evaporation exceeds precipitation and leaves the water enriched with salt. Salinities are low in the tropics and the mid-latitudes, where precipitation is high. The salinity of the North Atlantic Ocean averages 37.3 psu, compared with the North Pacific Ocean salinity of 35.5 psu.

In addition to dissolved salts, seawater also has dissolved gases (e.g., oxygen, carbon dioxide, and sulphur dioxide) and a variety of suspended particles (e.g., soil, atmospheric aerosol, and biogenic particulate matter).

1.3 Pressure

Pressure is defined as force per unit area, $p = \mathscr{F}/A$. The principal force contributing to pressure in the atmosphere and ocean is the gravitational force.[1] The mass per unit

---

[1] Bannon *et al.* (1997) estimate that the surface pressure is a factor of 0.25% less than the weight per unit area of a resting atmosphere, because of lateral pressure forces associated with a curved surface geometry. Vertical accelerations can also contribute to the pressure in a column.

area of the atmosphere is approximately 10^4 kg m^{-2}, and since the the acceleration due to gravity is about 10 m s^{-2}, the surface atmospheric pressure is about $p_0 = 10^5$ Pa. Since the mass of the world ocean is about 270 times the mass of the atmosphere, pressures in the ocean are substantially greater than those in the atmosphere. The pressure at any point in the ocean is the sum of the atmospheric pressure plus the weight of the ocean in a column above the point per unit area.

In SI units, the pascal (Pa) is the unit of pressure, where 1 Pa = 1 N m^{-2}, and N is a newton. Alternative units of pressure include:

bar (bar):	1 bar	=	10^5 Pa
millibar (mb):	1 mb	=	10^2 Pa
torricelli (torr):	1 torr	=	133.322 Pa
atmosphere (atm):	1 atm	=	1.01325 bar
		=	760 torr
		=	1.01325×10^5 Pa

While Pa is the preferred unit of pressure, torr is a unit commonly used by atmospheric chemists and chemical oceanographers. The unit mb is frequently used by meteorologists, and oceanographers commonly use decibars (db). The preferred pressure unit for meteorology is hPa (10^2 Pa), which is equivalent to mb.

The vertical variation of pressure in the atmosphere is shown in Figure 1.3 to decrease almost exponentially with height, from a mean sea-level pressure of 1013.25 hPa. Approximately 90% of the weight of the atmosphere lies below 15 km. The

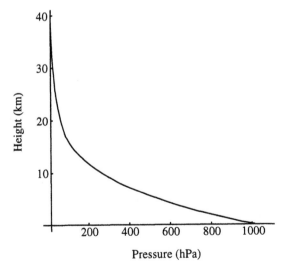

Figure 1.3 Variation of atmospheric pressure with height (U.S. Standard Atmosphere, 1976).

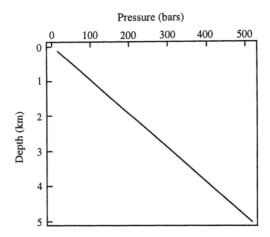

Figure 1.4 Vertical variation of pressure with depth in the ocean.

vertical variation of pressure with depth in the ocean is shown in Figure 1.4 to be approximately linear. In dealing with the pressure in the oceans, the atmospheric pressure is subtracted and the pressure at the sea surface is entered as zero. An increase of 10 m in depth in the ocean corresponds to an increase of 10^5 Pa, which is approximately 1 atm. Thus the pressure at a depth of 1 km in the ocean is equivalent to approximately 100 atm.

1.4 Density

Because of the large volumes characterizing the atmosphere and ocean, an intensive volume is desired. Such a volume, v, referred to as the *specific volume*, is given by $v = V/m$ so that the units of v are m^3 kg^{-1}. The specific volume is the inverse of the *density*, ρ, which has units of kg m^{-3}.

Figure 1.5 shows the vertical variation of density with height in the atmosphere. Density decreases with height nearly exponentially, which is related to the pressure decrease (Figure 1.3). A typical value of surface air density is 1.3 kg m^{-3}. The *mean free path* of molecules, which is determined by the frequency of intermolecular collisions, is inversely proportional to density. The mean free path increases exponentially from a value of about 10^{-7} m at the surface to the order of 1 m at 100 km.

Liquid water is almost three orders of magnitude more dense than air. Hence, the interface between the atmosphere and ocean is very stable. Since liquid water is nearly

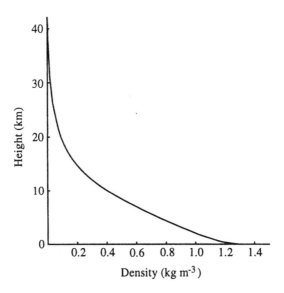

Figure 1.5 Vertical variation of density with height in the atmosphere. Density, like pressure, decreases nearly exponentially with height (U.S. Standard Atmosphere, 1976).

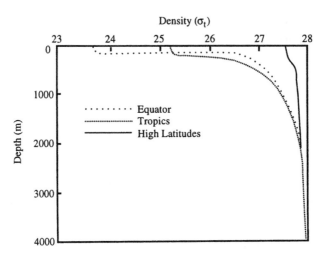

Figure 1.6 Density profiles in the ocean. Near the equator and throughout the tropics, the density increases rapidly with depth below a shallow surface layer of nearly constant density. This is due to the rapid cooling with depth in the ocean in these regions (see Figure 1.8). At very high latitudes, salinity is more important than temperature in regulating the density, and thus the σ_t profile near the surface is very different from the low-latitude profiles (see Figure 1.2). (Data from Levitus, 1982.)

incompressible, density changes in the ocean are relatively small. The density of seawater is a function not only of temperature and pressure, but also of salinity. Because of the small variations of density in the ocean, we employ the notation σ_t as a shorthand way of expressing the density in terms of its departure from a reference value, ρ_0:

$$\sigma_t = \rho - \rho_0 \qquad (1.1)$$

The reference value is $\rho_0 = 1000$ kg m^{-3}, which is the density of pure water at 4°C and 1 atm. A density value $\rho = 1025$ kg m^{-3} thus corresponds to $\sigma_t = 25$ kg m^{-3}.

Depth profiles of σ_t for different latitudes are given in Figure 1.6. The shallow layer of near-constant density just below the ocean surface is called the *ocean mixed layer*. Regions in which density changes sharply with depth are known as *pycnoclines*.

1.5 Temperature

An important property of temperature is that whenever two bodies are separately brought to be in equilibrium with a third body, these two bodies are then found to be in equilibrium with each other. This property is called the *zeroth law of thermodynamics*. The zeroth law tells us how to obtain a universal temperature scale: choose a particular system as a standard, select one of that system's possible temperatures and assign it a numerical value, and the same numerical value can be assigned to any other system in thermal equilibrium with the standard system.

A *thermometer* is an example of such a standard system. In order to assign a number to the temperature of a system, we need to define a temperature scale. This is done by choosing a thermometric substance and a thermometric property X of this substance which bears a one-to-one relation to its possible thermal states. The use of a thermometer allows us to specify temperature by using an arbitrary relation such as $T = aX + b$, which requires the choice of two well-defined thermal states as fixed points, to determine the constants a and b. The Celsius temperature scale assigns the fixed points so that 0°C is the temperature of melting ice and 100°C is the temperature of boiling water. An ideal gas thermometer can be used to determine an absolute temperature scale by making the temperature proportional to the pressure exerted by a sample of gas of low density held in a container at fixed volume. By defining the coldest possible temperature at zero, and using the unit of temperature in the ideal gas scale of temperature as equal in magnitude to the unit of the Celsius scale, the triple point of water is 273.16. Thus the ideal gas temperature scale can be regarded as an absolute temperature scale, where temperatures cannot be negative.

The modern absolute temperature scale, which is independent of the nature of the thermodynamic system used as a thermometer, is called *Kelvin* (K) and is defined by setting the temperature of the triple point of water at exactly 273.15. This choice establishes the unit of temperature so that there are 100 units between the freezing point and the boiling point of water, consistent with the Celsius scale. An absolute temperature scale such as the Kelvin is the appropriate temperature scale to use for thermodynamic calculations.[2] However, both the Celsius and Fahrenheit scales are commonly used to describe the weather, and the Celsius scale is widely used in ocean-ography. Scale conversions between the Kelvin, Celsius, and Fahrenheit temperature scales are:

$$K = 273.15 + {}^\circ C$$

$$^\circ C = \frac{5}{9} ({}^\circ F - 32)$$

$$^\circ F = \left(\frac{9}{5} {}^\circ C \right) + 32$$

The vertical temperature structure of the atmosphere below 50 km is given in Figure 1.7. In the lower atmosphere, below approximately 10–15 km, temperature decreases with height except in the polar regions where surface temperatures are very cold. At a height ranging from 8 km in the polar winter to 17 km in the tropics, an inflection point is seen in the temperature profile, called the *tropopause*. The atmosphere below the tropopause is called the *troposphere*. Above the troposphere (up to about 50 km) is the *stratosphere*.

The atmospheric temperature *lapse rate*, Γ, is defined to be minus the rate of temperature change with height:

$$\Gamma = -\frac{\partial T}{\partial z} \tag{1.2}$$

If the atmosphere has temperature that increases with height ($\Gamma < 0$), we call this layer a *temperature inversion*. If $\Gamma = 0$ (i.e., zero temperature change with height), the layer is called *isothermal*. Temperature inversions are seen in Figure 1.7 near the surface in the polar winter profile where the surface is very cold, and in the stratosphere, where ozone absorbs solar radiation. The average lapse rate in the troposphere is $\Gamma = 6.5°C$ km^{-1}.

---

[2] This is easily illustrated by substituting a temperature such as –30°C into the ideal gas law; this would result in either a pressure or volume that is negative!

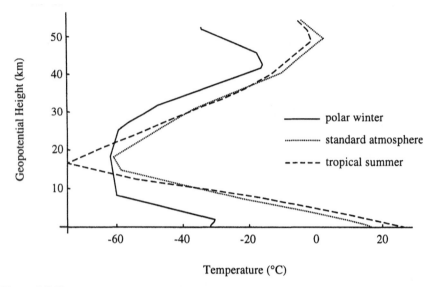

Figure 1.7 Vertical temperature structure in the atmosphere below about 50 km. Temperature decreases with height in the troposphere, except for the polar winter, where surface temperatures are very low, causing a temperature inversion near the surface (U.S. Standard Atmosphere, 1976).

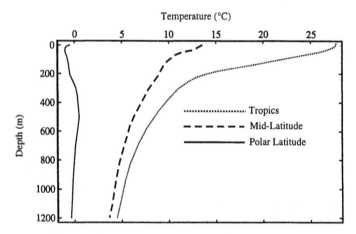

Figure 1.8 Variation of temperature with depth in the upper ocean. The vertical temperature gradient in the upper ocean is largest in the tropics, where surface water is warm. The latitudinal variation of temperature with depth is large in the upper ocean, where surface effects are important. (Data from Levitus, 1982.)

The distribution of temperature with depth in the ocean is shown in Figure 1.8. In sub-polar regions, large variations in temperature occur in the upper 100 m or so (the ocean mixed layer), due to fluctuations in forcing from the atmosphere. Between about 300 m and 1 km, the temperature decreases rapidly with depth. This region of steep temperature gradient is the permanent *thermocline*, beneath which there is virtually no seasonal variation, and the temperature decreases gradually to between 0 and 3°C. A seasonal thermocline often develops above the permanent thermocline. Strong latitudinal variations are seen particularly in the upper ocean. In the high latitudes there is no permanent thermocline because the temperature of the surface water is very cold, often covered by sea ice.

1.6 Kinetic-Molecular Model of the Ideal Gas

The preceding sections have described pressure and temperature based upon experimental observations. To increase our understanding of thermodynamic systems, it is useful to examine the thermodynamic variables from a microscopic point of view, using a simple model that treats a gas as a collection of moving molecules.

Central to the microscopic view of matter is *Avogadro's law*, which states that equal volumes of different gases at the same pressure and temperature contain equal numbers of molecules. Avogadro's law is used for determining the relative masses of the various atoms, using as a reference the carbon–12 nucleus which has been assigned a molecular mass of 12. A quantity of any substance whose mass in grams is equal to its molecular mass is called a *mole*. The volume occupied by a mole of gas at standard atmospheric pressure and 0°C is 22.4 liters, and is the same for all gases. The number of molecules contained by a mole is therefore a constant and is called *Avogadro's number* ($N_0 = 6.02 \times 10^{23}$ mole^{-1}).

Empirical evidence that all matter consists of a very large number of molecules has led to the development of several statistical theories, including the *kinetic theory of gases*. According to the kinetic hypothesis, the individual molecules that constitute matter are continually in motion, regardless of whether or not the matter as a whole is moving. These individual motions take place randomly in all directions and with a variety of speeds so that, as far as gross motion is concerned, the contributions of individual molecules tend to cancel. From the macroscopic viewpoint, however, the molecular motion has two major consequences:

1. the impacts of the moving molecules contribute to the pressure exerted by the material on its surroundings; and

2. the kinetic energy of the molecules contributes to the internal energy of the material (which we will later show is related to temperature).

In applying kinetic theory to the atmosphere, consider the case of a dilute gas, where the gas molecules are sufficiently far apart so that they are independent and exert no forces on one another, except during occasional collisions between two molecules. In any collisions in a dilute gas, we assume that the translational momentum and kinetic energy of the molecules are conserved. We are in fact equating a dilute gas with an ideal gas. A simple theory of dilute gases based on classical mechanics has been developed that accounts quite accurately for the behavior of the permanent gases under atmospheric pressures and temperatures.

Consider N molecules in a cubical box of side a and volume V (Figure 1.9), where the length a is large relative to the molecular diameters. The molecules are separated by distances large compared to their own diameters. Between collisions, molecules move in straight lines with constant speed. It is also assumed that the distribution of molecular velocities is the same in all directions, so that the average molecular velocity is zero (*velocity* is a vector; average molecular *speed* is not zero).

Since pressure is defined as force per unit area, the pressure on the walls of the box by the gas arises from collisions of molecules with the walls. Hence the pressure depends on:

1. the speed of the molecules;
2. the mass of the molecules; and
3. the frequency of the molecular impacts.

The greater the number of molecules, the greater the number of collisions with the wall per second, so that

$$p \propto N$$

It is also reasonable to suppose that, if the box were somehow made smaller without letting any of the molecules escape, the pressure would increase because each molecule would travel a shorter distance before colliding with a wall and thus hit a wall more frequently. So a smaller volume increases the pressure:

$$p \propto \frac{1}{V}$$

If the temperature of the gas rises, the molecules will move faster, hitting the walls of the flask more often and with greater force. Hence we see that pressure increases when temperature increases:

$$p \propto T$$

By combining these three relations, we obtain

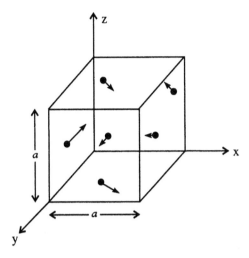

Figure 1.9 In the kinetic-molecular model of an ideal gas, individual molecules move randomly in all directions, colliding with each other and with the walls of their container. The pressure exerted by the molecules on the walls of the container is determined by the speed of the molecules, the mass of the molecules, and the frequency of the impacts.

$$p \propto \frac{NT}{V}$$

which shows the proportionalities that we recognize from the ideal gas law.

We would now like to calculate the pressure in terms of molecular quantities, relative to Figure 1.9. The velocity u of any molecule may be resolved into components u_x, u_y, and u_z, parallel to the three orthogonal axes x, y, z, so that its magnitude is given by

$$u^2 = u_x^2 + u_y^2 + u_z^2$$

Collisions between a molecule and the walls are assumed to be perfectly elastic; that is, the angle of incidence equals the angle of reflection and the velocity changes in direction but not in magnitude. At each collision with a wall that is perpendicular to x, the velocity component u_x changes sign from $+u_x$ to $-u_x$ or vice versa; the momentum component of the molecule accordingly changes from $+mu_x$ to $-mu_x$, where m is the mass of the molecule. The magnitude of the change in momentum is therefore $2m|u_x|$, where $|u_x|$ denotes the absolute value of u_x. The number of collisions in unit time with two walls perpendicular to x is equal to $|u_x|/a$, where a is the dimension of the wall and also the separation between the two walls. Thus the change in the x-component of momentum in unit time is $2mu_x(u_x/a) = 2mu_x^2/a$.

If there are N molecules in the box all moving with velocity u_x, the change of momentum in unit time becomes $2(Nmu_x^2/a)$. This rate of change of momentum is simply the force exerted by the molecules colliding against the two container walls normal to x whose area is $2a^2$. Since pressure is defined as the force normal to unit area, we have

$$p = \frac{2Nmu_x^2}{2a^2 \cdot a} = \frac{Nmu_x^2}{V} \tag{1.3}$$

In the atmosphere below 100 km, even in small volumes there is a countless number of molecules, all of which are moving randomly. To describe this chaotic motion, we consider the statistical properties of a very large number of molecules. Since the individual molecules are moving randomly with different velocities, we can introduce a velocity distribution. The distribution of a single velocity component u_x can be represented by $f(u_x)$ where

$$\int_0^\infty f(u_x)\, du_x = 1$$

If the directions of the molecular motions are random, then $f(u_x) = f(u_y) = f(u_z)$. The average velocity of the molecules is therefore zero. The mean of the square of the x-component of the velocity is given by

$$\overline{u_x^2} = \int_{-\infty}^\infty u_x^2 f(u_x)\, du_x = \frac{1}{N}\int_0^N u_x^2\, dN$$

Because we have assumed that there is no preferred direction of velocity, we have

$$\overline{u_x^2} = \overline{u_y^2} = \overline{u_z^2} = \frac{1}{3}\,\overline{u^2} \tag{1.4}$$

where the quantity $\overline{u^2}$ is called the *mean square speed* of the molecules.

Incorporating (1.4) into (1.3), the expression for pressure becomes

$$p = \frac{Nm\overline{u^2}}{3V} \tag{1.5}$$

The average translational kinetic energy of the molecules is $\mathcal{E}_k = (1/2)Nm\overline{u^2}$. Hence (1.5) can be written as

$$pV = \frac{2}{3} \mathcal{E}_k \qquad (1.6)$$

According to kinetic theory, the transformation of mechanical work into heat is simply a degradation of large-scale motion into molecular-scale motion. Thus an increase in the temperature of a body is equivalent to an increase in the average translational kinetic energy of its constituent molecules. We may express this mathematically by saying that the temperature is a function of $\mathcal{E}_k$ alone, so that $T = f(\mathcal{E}_k)$. The exact relationship between the kinetic energy and temperature depends on the scale of temperature. For the Kelvin scale of temperature the constant of proportionality is $(1/2)k$, where $k = 1.38 \times 10^{-23}$ J K^{-1} is the *Boltzmann constant* and $\mathcal{E}_k = (1/2)kT$. For N molecules moving in all three directions, we have

$$\mathcal{E}_k = \frac{3}{2} NkT \qquad (1.7)$$

or

$$T = \frac{2}{3} \frac{\mathcal{E}_k}{Nk} \qquad (1.7a)$$

Temperature is therefore proportional to the average translational kinetic energy of the molecules. The kinetic theory interpretation of absolute zero temperature is thus the complete cessation of all molecular motion: the zero point of kinetic energy. It should be noted that this picture has been changed somewhat by quantum theory, which requires a small residual energy even at absolute zero.

Substitution of (1.7) into (1.6) gives the *ideal gas law*

$$pV = NkT \qquad (1.8)$$

In practice, since the number of molecules is so large, we use moles instead, where $n = N/N_0$ is the number of moles of gas. We then define the *universal gas constant*, R^*, to be $R^* = N_0 k = 8.314$ J mole^{-1} K^{-1} so that the ideal gas law can be written as

$$pV = nR^*T \qquad (1.9)$$

By applying Newton's laws of motion to moving gas molecules, we have derived a relationship between pressure and temperature that is consistent with the relations derived from experimental observations of the variations of pressure, temperature, and volume.

1.7 Equation of State for Air

Except when water vapor is near condensation, air is observed to obey the ideal gas law. The ideal gas law (1.9) is written in extensive form, since the volume V, and number of moles n, are extensive variables. When applying the ideal gas law to the atmosphere, it is convenient to write the equation in terms of intensive variables. This is accomplished by dividing both sides of (1.9) by mass, m, yielding

$$p\frac{V}{m} = \frac{n}{m} R^* T \qquad (1.10)$$

Using the definition of *molecular weight*, $M = m/n$, and the definition of specific volume, (1.10) can be written as

$$pv = \frac{R^*}{M} T \qquad (1.11)$$

A *specific gas constant*, R, may be defined as $R = R^*/M$, so that (1.11) becomes

$$pv = RT \qquad (1.12)$$

Strictly speaking, air does not have a molecular weight, since it is a mixture of gases and there is no such thing as an "air molecule." However, it is possible to assign an apparent molecular weight to air, since air as a mixture is observed to behave like an ideal gas. To apply the ideal gas law to the mixture of atmospheric gases, consider first the mixture of "dry-air" gases, excluding for now the variable constituent water vapor. To understand the behavior of a mixture of gases, we employ *Dalton's law of partial pressures*. Dalton's law states that the total pressure exerted by a mixture of gases is equal to the sum of the partial pressures that would be exerted by each constituent alone if it filled the entire volume at the temperature of the mixture. That is,

$$p = \sum_j p_j \qquad (1.13)$$

where p is the total pressure and the p_j are the partial pressures.

Dalton's law implies that each gas individually obeys the ideal gas law and that the ideal gas law (1.12) for a mixture of gases can be written using (1.13) as

$$V \sum_j p_j = T \sum_j m_j R_j$$

where we have used $v = V/m$. We can now define a mean specific gas constant as

$$\overline{R} = \frac{\sum_j m_j R_j}{m} \qquad (1.14)$$

The equation of state for the mixture of dry-air gases can therefore be written in intensive form as

$$pv = R_d T \qquad (1.15)$$

where R_d is the specific gas constant for dry air. Using Table 1.1 and (1.14), a value for R_d is determined to be 287.104 J K^{-1} kg^{-1}. The mean molecular weight of the mixture is

$$\overline{M} = \frac{\sum_i n_i M_i}{n} = \frac{m}{n} \qquad (1.16)$$

The mean molecular weight for dry-air gases, M_d, is determined to be 28.96 g mole^{-1}.

The equation of state for air is complicated by the presence of water vapor, which has a variable amount in the atmosphere (Table 1.1). Assuming that the water vapor is not near condensation, the ideal gas law may be used and we have

$$e = \rho_v R_v T \qquad (1.17)$$

where the notation e is commonly used to denote the partial pressure of water vapor and the subscript v denotes the vapor. The specific gas constant for water vapor is $R_v = R^*/M_v = 461.51$ J K^{-1} kg^{-1}. In a mixture of dry air and water vapor (*moist air*), the equation of state is

$$p = p_d + e = \left(\rho_d R_d + \rho_v R_v\right) T \qquad (1.18)$$

The subscript d denotes the dry-air value, and the absence of a subscript denotes the value for the mixture of dry air plus water vapor.

The specific gas constant for moist air is determined from (1.14) to be

$$R = \frac{m_d R_d + m_v R_v}{m_d + m_v} \qquad (1.19)$$

where m_d and m_v are the mass of dry air and water vapor, respectively, and $m = m_d + m_v$.

An intensive variable, the *specific humidity*, q_v, is defined as

$$q_v = \frac{m_v}{m_v + m_d} \tag{1.20}$$

so that the specific gas constant for moist air can be written as

$$R = \left(1 - q_v\right) R_d + q_v R_v \tag{1.21}$$

Using the definition of the specific gas constant, the specific gas constant for water vapor, R_v, may be written in terms of R_d

$$R_v = \frac{M_d}{M_v} R_d = \varepsilon^{-1} R_d \tag{1.22}$$

where $\varepsilon = M_v/M_d = 18/29 = 0.622$. The specific gas constant for moist air may then be written as

$$R = R_d\left[1 + q_v\left(\tfrac{1}{\varepsilon} - 1\right)\right] = R_d\left(1 + 0.608\, q_v\right) \tag{1.23}$$

Incorporating (1.23) into (1.18), the equation of state for moist air becomes

$$pv = R_d\left(1 + 0.608\, q_v\right) T \tag{1.24}$$

It is awkward to have a variable gas constant, so it is the convention among meteorologists to make the humidity adjustment to the temperature rather than to the gas constant. Thus we define a *virtual temperature*, T_v

$$T_v = \left(1 + 0.608\, q_v\right) T \tag{1.25}$$

so that the ideal gas law for moist air becomes

$$pv = R_d T_v \tag{1.26}$$

The virtual temperature may be interpreted as the temperature of dry air having the same values of p and v as the moist air under consideration. Since q_v seldom exceeds 0.02, the virtual temperature correction rarely exceeds more than 2 or 3°C; however, it is shown in Chapter 7 that the small virtual temperature correction has an important effect on buoyancy and hence vertical motions in the atmosphere.

1.8 Equation of State for Seawater

For a one-component fluid such as pure water, density is a function only of temperature and pressure. Since seawater is a multi-component fluid owing to its dissolved salts, its density is a function of temperature, pressure, and salinity: $\rho = \rho(T,p,s)$. Seawater density is observed to increase with increasing pressure and salinity, but decrease with increasing temperature.

An accepted theory for the density of pure water, analogous to the kinetic theory of ideal gases, does not exist. Therefore, an empirically-determined equation of state is used for seawater. An internationally agreed-upon equation of state (UNESCO, 1981) fits the available ocean density measurements to high accuracy. This equation has the form

$$\rho = \rho(T,p,s) = \frac{\rho(T,0,s)}{1 - K_T(T,s,p)} \tag{1.27}$$

where $K_T(T,s,p)$ is the *mean bulk modulus,* which is inversely proportional to the compressibility (see Section 1.9). Each quantity on the right-hand side of (1.27), except pressure, is expressed as a polynomial series in s and T, expanded about values for zero salinity and a pressure of 1 bar. The density at the surface pressure ($p = 0$) is given by the polynomial form

$$\rho(T,0,s) = A + Bs + Cs^{3/2} + Ds^2 \tag{1.28}$$

The mean bulk modulus is given by

$$K_T(T,s,p) = E + Fs + Gs^{3/2} + (H + Is + Js^{3/2})p + (M + Ns)p^2 \tag{1.29}$$

The coefficients A, B, ... N in (1.28) and (1.29) are polynomials up to fifth degree in temperature (Table 1.3). In Table 1.3 and (1.27)–(1.29), the temperature is specified in °C, the pressure in bars, the salinity in psu, and density is $m^3\,kg^{-1}$. This equation of state is accurate to within a standard error of approximately 0.009 kg m^{-3} over the entire oceanic pressure range.

For seawater at standard atmospheric pressure, a contour plot of ρ is given in Figure 1.10 as a function of temperature and salinity. Values of constant density are called *isopycnals.* Near the freezing point, the density of seawater is relatively insensitive to temperature variations and small salinity differences can play a major role in density variations.

An expression for the temperature of maximum density of seawater, T_ρ, can be obtained by differentiating with respect to temperature the equation of state for seawater.

Table 1.3 Coefficients A, B, ..., N for the equation of state for seawater (UNESCO, 1981).

T^n	A	B	C
T^0	9.99842594×10^2	8.24493×10^{-1}	-5.72466×10^{-3}
T^1	6.793952×10^{-2}	-4.0899×10^{-3}	1.0227×10^{-4}
T^2	-9.095290×10^{-3}	7.6438×10^{-5}	-1.6546×10^{-6}
T^3	1.001685×10^{-4}	-8.2467×10^{-7}	
T^4	-1.120083×10^{-6}	5.3875×10^{-9}	
T^5	6.536332×10^{-9}		

	D	E	F
T^0	4.8314×10^{-4}	1.965221×10^4	5.46746×10^1
T^1		1.484206×10^2	-6.03459×10^{-1}
T^2		-2.327105	1.09987×10^{-2}
T^3		1.360477×10^{-2}	-6.1670×10^{-5}
T^4		-5.155288×10^{-5}	

	G	H	I
T^0	7.944×10^{-2}	3.239908	2.2838×10^{-3}
T^1	1.6483×10^{-2}	1.43713×10^{-3}	-1.0981×10^{-5}
T^2	-5.3009×10^{-4}	1.16092×10^{-4}	-1.6078×10^{-6}
T^3		-5.77905×10^{-7}	

	J	M	N
T^0	1.91075×10^{-4}	8.50935×10^{-5}	-9.9348×10^{-7}
T^1		-6.12293×10^{-6}	2.0816×10^{-8}
T^2		5.2787×10^{-8}	9.1697×10^{-10}

A simple empirical expression for T_ρ is given by (Neumann and Pierson, 1966)

$$T_\rho = 3.98 - 0.200\,s - 0.0011\,s^2 \tag{1.30}$$

where T_ρ is in °C and s is in psu. It is seen that the temperature of maximum density of seawater, T_ρ, varies with salinity. For pure water, $T_\rho = 3.98$°C, while T_ρ decreases below 0°C for $s > 17$ psu.

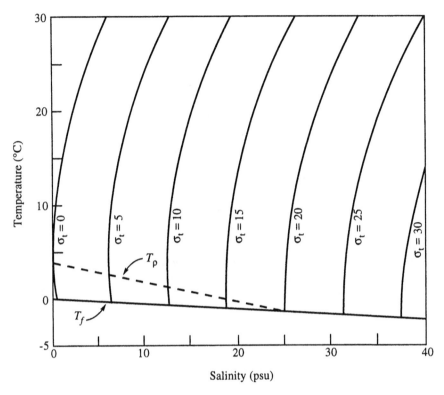

Figure 1.10 The density of seawater as a function of temperature and salinity for $p = 1013$ hPa. An increase in temperature at constant salinity may increase or decrease the density, depending on the salinity and the initial temperature. An increase in the salinity at constant temperature acts to increase the density, regardless of the temperature and initial salinity. Notice that near the freezing point (T_f line), the density is relatively insensitive to temperature changes.

1.9 Compressibility and Expansion Coefficients

The ideal gas law can be shown graphically on a three-dimensional surface, described by the variables p, ρ, and T (Figure 1.11). Because of the complexity of the equation of state for the ocean, an analogous diagram would require four dimensions (to include salinity). Lines of constant pressure (isobars), contant temperature (isotherms), and constant density (isopycnals) are projected onto the surface, showing the relationships among the variables. For example, the slope of an isobar gives the rate of change of volume with temperature at the constant pressure chosen. The slopes of the isolines in Figure 1.11 can be interpreted in the following way.

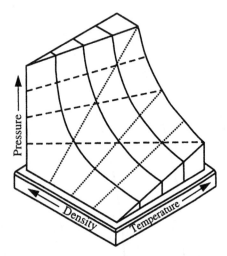

Figure 1.11 Graphical representation of the ideal gas law. The solid lines in the figure are isotherms, showing the relationship between p and ρ when T is constant. The dashed lines are isobars, showing the relationship between ρ and T when p is constant. The dotted lines are isopycnals, showing the relationship between p and T when ρ is constant.

The fractional rate of decrease of ρ with T at constant pressure is called the *coefficient of thermal expansion*, α:

$$\alpha = -\rho^{-1}\left(\frac{\partial \rho}{\partial T}\right)_p \tag{1.31a}$$

where the negative sign allows α to be positive for an ideal gas and over most of the temperature range for seawater. For an ideal gas, α is easily evaluated to be equal to T^{-1}.

In a similar way, the slope of an *isothermal* (constant temperature) curve gives the variation of volume with pressure at constant temperature. We define γ, the *compressibility* of a substance, as

$$\gamma = \rho^{-1}\left(\frac{\partial \rho}{\partial p}\right)_T \tag{1.31b}$$

For an ideal gas, $\gamma = p^{-1}$. The compressibility of seawater decreases as temperature, pressure, and salinity increase.

Since the density of seawater is influenced by salinity, we define the *coefficient for saline contraction, β*, as

$$\beta = \rho^{-1}\left(\frac{\partial \rho}{\partial s}\right)_{p,T} \tag{1.31c}$$

Values of α and β are given in Table 1.4 for the ocean at the sea surface ($p = 0$).

Useful linear approximations (accurate to within 5%; following Krauss and Businger, 1994) for temperatures above 5°C and at atmospheric pressure have the form

$$\alpha \approx 77.5 + 8.70\,T \qquad \text{and} \qquad \beta \approx 779.1 - 1.66\,T$$

where T is in °C, α has units $10^6\ °C^{-1}$, and β has units $10^6\ psu^{-1}$. As pressure increases, α increases and β decreases (see Appendix F). At temperatures below 5°C and for salinities below about 28 psu, the coefficient of thermal expansion is negative, indicating an decrease in density with decreasing temperature. This anomalous behavior of water at low temperatures and salinities gives rise to a density maximum at a temperature that is higher than the freezing point of pure water for $s > 17$ psu.

Table 1.4a. Coefficient of thermal expansion, α ($10^{-6}\ °C^{-1}$), as a function of temperature and salinity at atmospheric pressure (sea surface). (Data from UNESCO, 1981.)

s (psu) \ T(°C)	−2	0	5	10	15	20	25	30
0	−105	−67	17	88	151	207	257	303
10	−65	−30	47	113	170	222	270	314
20	−27	5	75	135	188	237	281	324
25	−10	21	88	146	197	244	287	329
30	7	36	101	157	206	251	292	332
35	25	53	113	167	214	257	297	334
40	38	65	126	177	222	263	301	337

Table 1.4b. Coefficient of saline contraction, β ($10^{-6}\ psu^{-1}$), as a function of temperature for $s = 35$ psu at $p = 0$. (Data from UNESCO, 1981.)

T (°C)	−2	0	5	10	15	20	25	30
β	795	788	774	762	752	744	737	732

1.10 Hydrostatic Equilibrium

Thus far we have examined the thermodynamic state of individual masses of air and seawater. Here we consider the state of the atmosphere and ocean in the presence of a gravitational field, particularly the height dependence of pressure, temperature, and density. The strength of the gravitational field, which depends primarily on the mass of the planet, is a central determinant of the mass of the atmosphere and ocean. Although both the atmosphere and ocean are bound to the earth by gravity, the ocean has a finite depth, while the atmosphere does not have a top and blends slowly into interplanetary space. The reason for this difference is that the atmosphere is compressible while the ocean is nearly incompressible, since the density of the atmosphere varies with pressure whereas the density of the ocean hardly varies at all.

The vertical variations of pressure in the atmosphere and ocean are observed to be much larger than either the horizontal or temporal variations. The decrease in pressure with height in the atmosphere, and the increase of pressure with depth in the ocean gives rise to a *vertical pressure gradient force*, $\mathscr{F}_p$

$$\mathscr{F}_p = -\frac{1}{\rho}\frac{\partial p}{\partial z} \tag{1.32}$$

where z is depth/height from the surface and ρ is the density.

The vertical pressure gradient force results in a vertical acceleration in the direction of decreasing pressure (upwards). The vertical pressure gradient force is generally in very close balance with the downward force due to gravitational attraction. This is called *hydrostatic balance*, and is written as

$$g = -\frac{1}{\rho}\frac{\partial p}{\partial z} \tag{1.33}$$

where g is the acceleration due to the Earth's gravity. The hydrostatic balance is applicable to most situations in the atmosphere and ocean, exceptions arising in the presence of large vertical accelerations such as are associated with thunderstorms.

Equation (1.33) can be integrated to determine a relationship between pressure and depth or height:

$$-\int dp = \int \rho g\, dz \tag{1.34}$$

To integrate (1.34), it is commonly assumed that g is constant; however, g varies with distance from the Earth's center and also with latitude because of the nonsphericity of the Earth.

To account for these variations in g, the *geopotential* ϕ is often introduced

$$\phi(z) = \int_0^z g \, dz \qquad (1.35)$$

where the geopotential at sea level $\phi(0)$ is taken to be zero by convention. ϕ is the gravitational potential energy per unit mass, with units $J\,kg^{-1}$. Using the geopotential, we may write an alternative and equivalent statement of the hydrostatic balance:

$$dp = -\rho \, d\phi$$

The *geopotential height*, Z, can be defined for application to the atmosphere as

$$Z = \frac{\phi(z)}{g_0} = \frac{1}{g_0} \int_0^z g \, dz \qquad (1.36a)$$

where $g_0 = 9.8$ m s^{-2} is the globally averaged acceleration due to gravity at the Earth's surface. The force of gravity is thus perpendicular to surfaces of constant ϕ, while not exactly perpendicular to surfaces of constant z. Geopotential height is used as a vertical coordinate in many atmospheric applications. In the lower atmosphere, Z is very nearly equal to z; at a distance of $z = 10$ km above the Earth's surface at 40°N, $g = 9.771$ m s^{-2} and $Z = 9.986$ km. In oceanographic applications, the *dynamic depth*, D, is used analogously to the geopotential height in the atmosphere

$$D = \frac{\phi(z)}{g_0} = \frac{1}{g_0} \int_{p_0}^p v \, dp \qquad (1.36b)$$

where $D(p_0) = 0$ is assumed by convention and the *dynamic meter*, dm, is the common unit of dynamic depth. The pressure change is usually expressed in decibars (db, where 1 db = 100 mb), since a pressure of 1 db is equivalent to a change of dynamic depth of about 1 dm.

Since the focus of this text is on the ocean and lower atmosphere, we will assume that $g = g_0$ is a constant, which simplifies the integration and evaluation of (1.34). However, integration of (1.34) also requires some assumption about the vertical variation of the density, ρ.

Because seawater is nearly incompressible, density is nearly constant within the ocean, and thus there is a nearly linear relationship of pressure with depth (Figure 1.4). In the ocean, the following integrated form of the hydrostatic equation is used:

$$p(-z) = p_0 + \rho g z \qquad (1.37a)$$

where p_0 is the atmospheric pressure. If we assume that $\rho = 1036$ kg m^{-3} and $g = 9.8$ m s^{-2}, we may then write

$$p(-z) = p_a + 10153\, z \qquad (1.37b)$$

where z is in meters and p is in Pa. Because 10^5 Pa $= 1$ bar, it is easily seen that the oceanic pressure increases at approximately 1 db per meter of depth. Ocean pressures given in db are numerically equivalent to the depth in meters to within 1–2%. However, if seawater were actually incompressible, the sea level would rise by more than 30 m, because the hydrostatic pressure in the deep ocean is so great.

Because air is compressible and density decreases with height in the atmosphere (Figure 1.5), integration of (1.34) for the atmosphere is more complicated than for the ocean. However, useful insights can be derived from examining an idealized *homogeneous atmosphere*, where density is assumed constant. Consideration of a homogeneous atmosphere with finite surface pressure implies a finite total height for the atmosphere, which is called the *scale height H.* Assuming that density is constant, we can integrate (1.34) from sea level, where the pressure is p_0, to a height H, where the pressure is zero, to obtain

$$p_0 = \rho g H \qquad (1.38)$$

The height of the homogeneous atmosphere (often referred to as the *scale height*) is therefore

$$H = \frac{p_0}{\rho g} = \frac{R_d T_0}{g} \qquad (1.39)$$

where T_0 is the surface temperature and H can be evaluated from the surface temperature and known constants to be approximately 8 km. From the ideal gas law, it is easily inferred that temperature must decrease with height in the homogeneous atmosphere. The lapse rate of the homogeneous atmosphere is obtained by differentiating the ideal gas law with respect to z, holding density constant

$$\frac{\partial p}{\partial z} = \rho R_d \frac{\partial T}{\partial z} \qquad (1.40)$$

Combining (1.40) with the hydrostatic equation (1.33) leads to the result

$$\Gamma = -\frac{\partial T}{\partial z} = \frac{g}{R_d} = 34.1\,°\text{C km}^{-1} \qquad (1.41)$$

Thus the lapse rate of a homogeneous atmosphere is constant and about six times larger than the lapse rate normally observed in the atmosphere (which is $\Gamma \approx 6.5°C$ km^{-1}). The lapse rate for the homogeneous atmosphere is referred to as the *autoconvective lapse rate* for the following reason: if the lapse rate exceeds the autoconvective value, it is implied that the lower air is less dense than the air above, causing the atmosphere to overturn and the spontaneous initiation of convection. Values of the atmospheric lapse rate as large as the autoconvective value are observed over desert surfaces in summer when the solar heating is high; however, lapse rates in the atmosphere typically do not exceed $\Gamma \approx 10°C$ km^{-1}.

Further insight is gained by examining the characteristics of yet another idealized atmosphere, called the *isothermal atmosphere*. After substitution of the ideal gas law for density, we can write the hydrostatic equation in the following form:

$$\partial p = -\frac{pg}{R_d T} \partial z \tag{1.42}$$

This equation is easily integrated for a constant temperature from sea level ($z = 0$, $p = p_0$) to some arbitrary height z

$$\int_{p_0}^{p} \frac{dp}{p} = -\frac{g}{R_d T} \int_{0}^{z} dz \tag{1.43}$$

or

$$\ln \frac{p}{p_0} = -\frac{gz}{R_d T} \tag{1.44a}$$

Taking antilogs and using $H = RT/g$, we have

$$p = p_0 \exp\left(-\frac{z}{H}\right) \tag{1.44b}$$

Thus pressure decreases exponentially with height in an isothermal atmosphere, and there is no definite upper boundary to this atmosphere. Note that when $z = H$, the pressure is $1/e$ of its surface value. The isothermal atmosphere resembles the real atmosphere more closely than does the homogeneous atmosphere; however, (1.44b) is not applicable to the real atmosphere except when applied over a shallow layer above the ground.

Many meteorological applications require an accurate relationship between atmospheric pressure and height, which necessitates considering the variation of temperature with height. These applications include: determination of the elevation at which the observations of pressure, temperature, and humidity are obtained from balloons carrying radiosondes; conversion between pressure and height as a vertical coordinate in numerical models of the atmosphere; reduction of surface pressure to sea–level pressure over land; and determination of the thickness between pressure levels. The vertical variations of the temperature profile can be accounted for by integrating (1.42) in a piecewise manner, between height levels that are close enough so that a mean atmospheric temperature in the layer can be defined. Thus we have

$$\int_{z_1}^{z_2} g \, dz = - \int_{p_1}^{p_2} \frac{R_d T_v}{p} \, dp$$

Assuming that $\overline{T}_v$ is constant within the layer, we can integrate to obtain

$$z_2 - z_1 = - \frac{R_d \overline{T}_v}{g} \ln \frac{p_2}{p_1} \tag{1.45}$$

or

$$p_2 = p_1 \exp\left[\frac{g}{R_d \overline{T}_v} (z_1 - z_2) \right] \tag{1.46}$$

Equation (1.45) is referred to as the *hypsometric equation*. From (1.45), it is seen that the *thickness* $\Delta z = z_2 - z_1$ of a layer bounded by two isobaric surfaces is proportional to the average virtual temperature of the layer ($\overline{T}_v$). Figure 1.12 shows the variation with latitude of the relative thickness of isothermal atmospheric layers. Since temperature decreases with latitude away from the equator, the distance between two isobaric surfaces decreases from equator to pole.

An additional application of the hydrostatic equation to the atmosphere is integration under the assumption of a constant lapse rate. Assuming that temperature varies linearly with height with a lapse rate Γ, we have

$$T = T_0 - \Gamma z \tag{1.47}$$

Substituting (1.47) into (1.42) yields

$$\frac{dp}{p} = - \frac{g}{R_d} \frac{dz}{T_0 - \Gamma z}$$

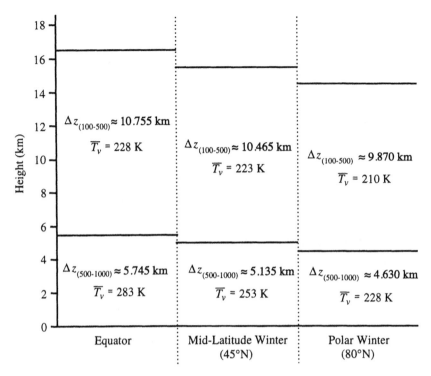

Figure 1.12 Representation of the thicknesses of the 1000–500 hPa and 500–100 hPa layers and their variation with latitude. The thickness of the layer between two isobaric surfaces is determined by the mean virtual temperature in the layer, according to (1.45), resulting in layers of decreasing thickness from equator to pole.

This equation is easily integrated between the limits $(z = 0, p = p_0)$ and (z, p) to obtain

$$\ln \frac{p}{p_0} = \frac{g}{R_d \Gamma} \ln \left(\frac{T_0 - \Gamma z}{T_0} \right)$$

or

$$p = p_0 \left(\frac{T}{T_0} \right)^{g/R_d \Gamma} \tag{1.48}$$

Note that the exponent in (1.48) is equal to the ratio of the autoconvective lapse rate (1.41) to the actual lapse rate.

Notes

Singh (1995) *Composition, Chemistry, and Climate of the Atmosphere* gives an extensive description of the composition of the atmospheric gases and aerosol.

Millero and Sohn (1992) *Chemical Oceanography* provides an extensive discussion of the ocean composition and equation of state.

Problems

1. If 10^6 molecules are required to ensure a statistically uniform distribution of velocities in all directions, what is the minimum volume for which the atmospheric state can be defined at standard atmospheric conditions?

2. A molecule of N_2 travels with a speed of 515 m s^{-1}.
a) What is its temperature?
b) What is its kinetic energy?
c) If there were 1000 such molecules in a 1 cm^3 box, what would the pressure be inside the box?

3. A new power-generating plant is designed for a large city. The plant will consume about 6000 metric tons of coal per day. Assume that the coal contains 2% sulfur, which produces SO_2 on burning. The area of the city is 100 km^2 and the temperature is 17°C. If all the SO_2 produced in a day were evenly distributed over the surface of the city to a height of 200 m:
a) Calculate the constituent density of SO_2 in the air you breathe in mg m^{-3} (the limit set by the U.S. Environmental Protection Agency is 80 mg m^{-3}).
b) Calculate the partial pressure of SO_2.

4. Using the data listed in Table 1.2, consider the following questions:
a) In seawater, the proportion by weight of negative ions greatly exceeds that of positive ions. Why does seawater not carry a net negative charge?
b) What is the ratio of potassium concentration to total salinity? What would the potassium concentration be if the salinity rose to 36 psu? If it fell to 33 psu?
c) Speculate on some processes that might change the salinity in a given sample of seawater in the upper 100 m or so of the ocean surface.

5. A spherical weather balloon has a diameter of 10 m when filled with helium gas at a pressure of 106 kPa. What is the total mass the balloon can lift at 27°C?

6. Diving bells provide a method of transporting divers while under pressure and of supplying breathing gas to the diver. Diving bells may have an open bottom (i.e., a "wet" bell) or closed. Consider a diving bell with volume of 1 m^3 that is lowered into the ocean from a ship. At the surface, the temperature is 20°C; at 30 m depth, the temperature is 4°C.

a) If the diving bell is open, calculate the volume of the air space in the bell at depths of 10, 20, and 30 m.

b) If the diving bell is closed, calculate the pressure on the bell when it is at a depth of 30 m.

7. Evaluate the thickness of a layer of atmosphere between 800 and 900 mb with average temperature 300 K and specific humidity 20 g kg^{-1}. Compare the thickness determined with the virtual temperature versus that determined without the virtual temperature correction.

8. Derive a formula for the dependence of density upon height in a hydrostatic atmosphere of constant lapse rate of temperature, Γ ($T = T_0 - \Gamma z$).

9. At what height above sea level does the partial pressure of oxygen fall to one-half of its sea-level value if the temperature is assumed to be a constant 300 K?

10. Consider a hydrostatic atmosphere where pressure varies with height in the following way:

$$p(z) = \frac{p_0}{1 + \left(\dfrac{z}{H}\right)^2}$$

a) Determine the corresponding variation of temperature with height (valid for $z > 0$).

b) Evaluate the height at which $\Gamma = 10°C$ km^{-1}.

11. Consider a hydrostatic atmosphere with constant lapse rate.

a) Derive an expression for the variation of height with pressure, $z(p)$, in terms of the surface pressure p_0, surface temperature T_0, and lapse rate Γ. This equation forms the baisis for the calibration of aircraft pressure altimeters, where $T_0 = 288$ K, $p_0 = 1013.25$ hPa, and $\Gamma = 6.5°C$ km^{-1} (U.S. Standard Atmosphere).

b) An aircraft flying at pressure of 850 mb is preparing to land. Calculate the height above the surface (assume the surface is at sea level) that the aircraft is flying, using the altimeter correction for the standard atmosphere.

c) On February 3, 1989, sea-level pressure reached a North American record of
    1078 hPa. Surface temperature reached a minimum value of 217 K. The verti-
    cal temperature profile in the lower atmosphere was nearly isothermal. For an
    aircraft flying at a pressure of 850 mb above a surface that is at sea level, estimate
    the error in the altimeter reading that would be made under these conditions.
    (Note: The U.S. Federal Aviation Administration banned night and instrument
    flights in Fairbanks, AK, because altimeters could not be accurately calibrated to
    give altitude readings.)

Chapter 2 | The First and Second Laws of Thermodynamics

The classical physics principle of the conservation of mechanical energy states that while energy may manifest itself in a variety of forms (e.g., kinetic energy, gravitational potential energy), the sum of all different forms of energy in any particular system is fixed. Energy can be transformed from one type to another, but total energy can be neither created nor destroyed. Thermodynamics extends the principle of conservation of energy to include heat.

The *first law of thermodynamics* arose from a series of experiments first carried out in the 19th century. These experiments demonstrated that work can be converted into heat and that the expenditure of a fixed amount of work always produces the same amount of heat. The first law of thermodynamics places no limitations on the transformation between heat and work. As long as energy is conserved, these transformation processes do not violate the first law of thermodynamics.

The *second law of thermodynamics* limits both the amount and the direction of heat transfer. According to the second law, 1) a given amount of heat cannot be totally converted into work, thus limiting the amount of heat transfer; and 2) the spontaneous flow of heat must be from a body with a higher temperature to one with a lower temperature, thus stipulating the direction of heat transfer.

2.1 Work

When a force of magnitude $\mathscr{F}$ is applied to a mass which consequently moves through a distance dx, the mechanical work done is

$$dW = -\mathscr{F}\cos\theta\,dx \qquad (2.1)$$

where θ is the angle between the displacement dx and the applied force. Only the component of the displacement along the force enters the computation of work. There is no universal sign convention for work, so we adopt the following convention: work done *on* a system is positive; work done *by* a system is negative. It makes no difference which convention is adopted as long as it is used consistently.

An important kind of work in thermodynamics is the work systems do when they expand or contract against an opposing pressure. *Expansion work* is defined as

$$dW = -\mathscr{F}dx = -pA\,dx = -p\,dV$$

where $Adx = dV$ is the differential volume change associated with the work done against the external pressure, p. The specific work, $w = W/m$, is an intensive variable, independent of mass, and thus

$$dw = -p\,dv \qquad (2.2)$$

There are numerous examples of expansion work in the atmosphere (Figure 2.1), wherein a parcel of air rises in the atmosphere and its pressure decreases and volume increases. Some processes that cause air to rise are:
a) orographic lifting;
b) frontal lifting;
c) low-level convergence;
d) buoyant rising of warm air; and
e) mechanical mixing.
Analogous processes occur in the ocean. Work of expansion also occurs in the change of phase of water from liquid to gas and from liquid to ice.

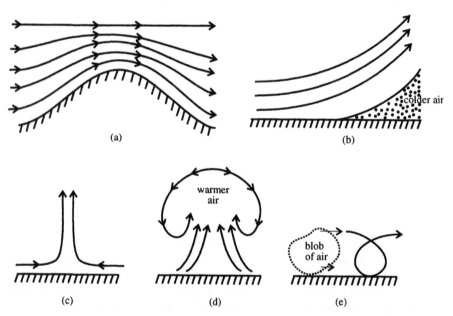

Figure 2.1 Rising motion occurs in the atmosphere due to (a) orographic lifting, (b) frontal lifting, (c) low-level convergence, (d) buoyant rising of warm air, and (e) mechanical mixing. Expansion work is done by an air parcel when it rises.

For a finite expansion or compression from v_1 to v_2, work is determined by integrating (2.2):

$$w = -\int_{v_1}^{v_2} p \, dv \tag{2.3}$$

The expansion from v_1 to v_2 is illustrated in Figure 2.2a by the top curve (A to B). The work done in this expansion is represented geometrically by the area under the curve. The area, and thus the work done, depends on the specific path followed during the expansion. For example, the temperature may remain constant or may vary during the expansion, resulting in different expansion paths. In fact, there is an infinite number of curves connecting the initial state v_1 to the final state v_2. If the system is compressed back to v_1 via a different process, net work will be done even though the system has returned to its initial state, as indicated by the shaded area between the two curves in Figure 2.2b.

Cyclical processes have the same initial and final states. A *cycle*, therefore, is a transformation that brings the system back to its initial state. The total work done in a cyclical process depends on the path, and is not necessarily zero. The work done by

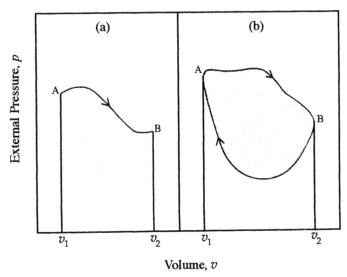

Volume, v

Figure 2.2 (a) The amount of work done in the expansion from v_1 to v_2 is equal to the area under the curve. In (b), the system is compressed back to v_1 via a different process. Even though the system has returned to its initial state, net work has been done, as indicated by the shaded area between the two curves.

a system in going from one state to another is a function of the path between the states. Therefore, generally

$$\oint dw \neq 0$$

To evaluate (2.3), the path of the expansion must be specified. Therefore, work is not an *exact differential* since dw cannot be obtained by differentiating a function of the state of the system alone, knowing only the initial and final states.

2.2 Heat

Heat is an extensive measure of the energy transferred between a system and its surroundings when there is a temperature difference between them. When two systems are placed in thermal contact, energy flows spontaneously from one system to the other. This energy flow can occur by various mechanisms, such as the transfer of vibrational energy between one solid and another whose surfaces are in contact, or the exchange of electromagnetic radiation. Such a spontaneous movement of energy is called a *heat flow*. It can be shown experimentally that if equal masses of water, one at 100°F and the other at 150°F, are mixed, then the resulting temperature is midway between the two extremes, or 125°F. If the same mass of warm mercury is used in place of the warm mass of water, however, the resulting final temperature is not midway between the two extremes, but rather 115°F, indicating that water has a greater "capacity" for heating than does mercury. That is, it takes more heat to raise the temperature of a given mass of water by one unit than it does to raise the temperature of the same mass of mercury by the same amount.

When two bodies with different temperatures, T_1 and T_2, are brought into contact with each other, the temperature difference eventually disappears, and the final temperature, T' is intermediate between the two initial temperatures. Experiments show that this heat transfer is governed by the following formula:

$$c_2 m_2 (T' - T_2) + c_1 m_1 (T' - T_1) = 0$$

where c is the *specific heat capacity*, which depends on the physical state and chemical composition of the substance. The amount of heat ΔQ lost by the warmer body is equal in magnitude to the amount of heat gained by the cooler body, so that

$$\Delta Q = c_1 m_1 (T_1 - T') = c_2 m_2 (T' - T_2) \tag{2.4}$$

The final equilibrium temperature is thus

$$T' = \frac{c_2 m_2 T_2 + c_1 m_1 T_1}{c_2 m_2 + c_1 m_1}$$

In differential form, the equation for heat (2.4) is

$$dQ = mc\, dT \tag{2.5}$$

The differential dQ is not exact since

$$\oint dQ \neq 0$$

To integrate dQ, one must know how the pressure and volume change during the transformation and if any phase changes occur during the transformation (e.g., gas to liquid).

Experiments have shown that the specific heat capacity is itself a function of temperature and is defined in terms of the differential heat flow and temperature change as

$$c = \frac{dq}{dT} \tag{2.6}$$

where $q = Q/m$ is the intensive heat.

Heat transfer processes in the atmosphere and ocean include radiation, molecular conduction, and the release of latent heat in phase changes (see Chapter 3).

2.3 First Law

The first law of thermodynamics is an extension of the principle of conservation of mechanical energy. We can use the conservation principle to define a function U called the *internal energy*. When an increment of heat dQ is added to a system, the energy may be used either to increase the speed of the molecules (i.e., to increase the temperature of the system), to create motion internal to each molecule (e.g., rotation and vibration), or to overcome the forces of attraction between the molecules (e.g., change of state from liquid to vapor), all of which contribute to the internal energy of the system. The internal energy of a system can increase when heat enters the system from the surroundings, and/or when work is done on the system by the surroundings.

If we take dU to denote an increment of internal energy, then

$$dU = dQ + dW \tag{2.7}$$

This statement is the differential form of the first law of thermodynamics. The intensive differential form of the first law of thermodynamics is written as

$$du = dq + dw \tag{2.8}$$

From the law of conservation of energy, the total energy of the system plus its environment must be constant. That is, the total energy change in the system plus its environment is zero:

$$0 = \Delta U_{syst} + \Delta U_{env}$$

What happens in a cyclical process?

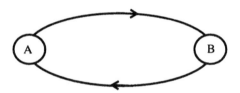

In a cyclical process, $\Delta U_{syst}(\text{A} \rightarrow \text{B} \rightarrow \text{A}) = 0$, since otherwise we would be creating energy. Therefore,

$$0 = \Delta U_{syst}(\text{A} \rightarrow \text{B} \rightarrow \text{A}) = \oint dU$$

and ΔU depends only on the initial and final states but not on the path followed between them. The first law thus states that although dQ and dW are not exact differentials, their sum $dU = dQ + dW$ is an exact differential and thus a thermodynamic state variable.

An exact differential $d\xi$ has the following properties:

1. The integral of $d\xi$ about a closed path is equal to zero ($\oint d\xi = 0$).
2. For $\xi(x,y)$, we have $d\xi = (\partial \xi / \partial x)\, dx + (\partial \xi / \partial y)\, dy$ where x and y are *independent variables* of the system and the subscripts x and y on the partial derivatives indicate which variable is held constant in the differentiation.

3. If the exact differential is written as $d\xi = Mdx + Ndy$, we obtain *Euler's relation:*[1]

$$\frac{\partial M}{\partial y} = \frac{\partial N}{\partial x} \tag{2.9}$$

If an experiment is conducted under conditions of constant volume, the first law of thermodynamics (2.7) becomes

$$dU = dQ \tag{2.10}$$

since we are allowing only for the possibility of expansion work ($dW = -p\,dV$) and no other type of work, and since no expansion work is done during a process carried out at constant volume ($dV = 0$). The change in heat at constant volume gives us an experimental measure of ΔU for any process involving the same initial and final states.

What happens when we do an experiment at constant pressure, and no work is done except expansion work? Consider the changes accompanying a process at constant pressure:

$$\Delta U = U_2 - U_1 = Q_p + W = Q_p - \int_{V_1}^{V_2} p\,dV = Q_p - p(V_2 - V_1) \tag{2.11}$$

where Q_p denotes heating at constant pressure, and no work other than expansion work is done. We can rearrange (2.11) to obtain

$$(U_2 + pV_2) - (U_1 + pV_1) = Q_p$$

It is convenient to define a new function called the *enthalpy*, H, by

$$H = U + pV \tag{2.12}$$

so that

$$\Delta H = H_2 - H_1 = (U_2 + p_2 V_2) - (U_1 + p_1 V_1) = Q_p$$

---

[1] We depart here from the tradition in thermodynamics where it is customary to enclose partial derivatives in parentheses and append subscripts to denote the variable(s) held constant in the differentiation, e.g., $\partial M / \partial y \equiv (\partial M / \partial y)_x$. The subscripts on the partial differential are usually not required mathematically, and their use serves to make the equations unneccessarily cumbersome. In those cases where omitting subscripts and parentheses may cause confusion, they are retained.

Since $H = H(U, p, V)$ and $U, p,$ and V are all state functions, H is also a state function; this is another way of saying that dH is an exact differential. In differential form,

$$dH = dU + p \, dV + V \, dp = dQ + V \, dp \qquad (2.13)$$

and in intensive form

$$dh = dq + v \, dp \qquad (2.14)$$

where $h = H/m$. From (2.14), it is clear that when we allow an expansion at constant pressure $(dp = 0)$, we obtain an experimental measure of a state property, enthalpy. Equations (2.13) and (2.14) are equivalent forms of the first law of thermodynamics to (2.7) and (2.8). The enthalpy form of the first law is advantageous when considering constant-pressure processes.

Since u and h are state functions, we can write

$$u = u(p,v,T) = u(v,T)$$
$$h = h(p,v,T) = h(p,T)$$

Although u and h are functions of three variables (p, v, T), an equation of state allows us to eliminate one of the three variables. Since u and h are exact differentials, we can expand du and dh as follows:

$$du = \left(\frac{\partial u}{\partial T}\right) dT + \left(\frac{\partial u}{\partial v}\right) dv$$

$$dh = \left(\frac{\partial h}{\partial T}\right) dT + \left(\frac{\partial h}{\partial p}\right) dp$$

At constant volume $dv = 0$ and $du = dq_v$, which leads to

$$du = \left(\frac{\partial u}{\partial T}\right) dT = dq_v$$

In a constant-pressure process, $dp = 0$, and

$$dh = \left(\frac{\partial h}{\partial T}\right) dT = dq_p$$

where q_v and q_p refer to constant-volume and constant-pressure heating, respectively. From the definition of specific heat (2.6), we can write

$$c_v = \frac{dq_v}{dT} = \frac{\partial u}{\partial T} \tag{2.15a}$$

and

$$c_p = \frac{dq_p}{dT} = \frac{\partial h}{\partial T} \tag{2.15b}$$

where c_v and c_p are defined, respectively, as the *specific heat at constant volume* and the *specific heat at constant pressure*. We may thus write

$$du = c_v \, dT + \left(\frac{\partial u}{\partial v}\right)_T dv$$

$$dh = c_p \, dT + \left(\frac{\partial h}{\partial p}\right)_T dp$$

For an ideal gas, it has been shown experimentally that $(\partial u/\partial v)_T = 0$, so that internal energy is a function only of temperature for an ideal gas, i.e., $u = u(T)$. It can also be shown that $(\partial h/\partial p)_T = 0$ and $h = h(T)$. This implies that for ideal gases

$$du = c_v \, dT \tag{2.16}$$
$$dh = c_p \, dT$$

How does c_v differ from c_p quantitatively? In a constant-pressure process, some of the added heat must be expended in doing work on the surroundings, while in a constant-volume process, all of the heat is devoted to raising the temperature of the substance. Therefore it takes more heat per unit temperature rise at constant pressure than at constant volume, and $c_p > c_v$. The difference between c_p and c_v can be evaluated from

$$c_p - c_v = \left(\frac{\partial h}{\partial T}\right)_p - \left(\frac{\partial u}{\partial T}\right)_v$$

Using the definition of enthalpy, $h = u + pv$, we can write

$$c_p - c_v = \left(\frac{\partial u}{\partial T}\right)_p + p\left(\frac{\partial v}{\partial T}\right)_p - \left(\frac{\partial u}{\partial T}\right)_v \tag{2.17a}$$

Expanding the differential $du(v,T)$ as

$$du = \left(\frac{\partial u}{\partial T}\right)_v dT + \left(\frac{\partial u}{\partial v}\right)_T dv$$

and dividing by dT while requiring constant pressure, we obtain

$$\left(\frac{\partial u}{\partial T}\right)_p = \left(\frac{\partial u}{\partial T}\right)_v + \left(\frac{\partial u}{\partial v}\right)_T \left(\frac{\partial v}{\partial T}\right)_p$$

We can now write (2.17a) as

$$c_p - c_v = \left(\frac{\partial u}{\partial v}\right)_T \left(\frac{\partial v}{\partial T}\right)_p + p \left(\frac{\partial v}{\partial T}\right)_p \tag{2.17b}$$

For an ideal gas, $(\partial u/\partial v) = 0$ and $p(\partial v/\partial T)_p = R$, so (2.17b) can be evaluated to be

$$c_p - c_v = R \tag{2.17c}$$

where R is the specific gas constant.   Hence for an ideal gas, the magnitude of the difference between the two specific heat capacities is simply the specific gas constant.

2.4 Applications of the First Law to Ideal Gases

We now apply the first law of thermodynamics to ideal gases, which is useful in the interpretation of thermodynamic processes in the atmosphere. The thermodynamic characteristics of an ideal gas have been shown to be:

1. The equation of state is $pv = RT$.
2. The internal energy is a function of its temperature alone
    ($du = c_v\, dT$; $dh = c_p\, dT$).
3. The specific heats are related by $c_p - c_v = R$.

The first law of thermodynamics for an ideal gas is thus written as

$$c_v\, dT = dq - p\, dv \tag{2.18a}$$

$$c_p\, dT = dq + v\, dp \tag{2.18b}$$

in internal energy (2.18a) and enthalpy (2.18b) forms.

Consider the isothermal ($dT = 0$) expansion of an ideal gas. Because internal energy is a function only of temperature, the internal energy of the gas is unchanged in an isothermal expansion. The first law of thermodynamics (2.18a) for an isothermal expansion may therefore be written as

$$dq = p\,dv$$

assuming that the only work done is expansion work. In the isothermal expansion of an ideal gas, the system does work, and the energy from this work comes from the environment and enters the system as heat. Since work is not an exact differential, we cannot integrate the right-hand side of the equation until we specify a path. As seen from Figure 2.2b, an infinite number of paths can be specified. Here we consider the path of an isothermal reversible expansion. A reversible path is one connecting intermediate states, all of which are equilibrium states. Exact conditions for reversible processes and how they differ from irreversible processes are described in Section 2.5. For now, we consider a reversible path where the equation of state is exactly satisfied during all stages of the expansion. Therefore, p may be evaluated using the ideal gas law, and the equation becomes

$$dq = RT\frac{dv}{v}$$

Integrating from v_1 to v_2 yields

$$\Delta q = RT\ln\!\left(\frac{v_2}{v_1}\right) = RT\ln\!\left(\frac{p_1}{p_2}\right)$$

The solution states, for example, that the amount of heat required to expand a gas from 10^6 Pa to 10^5 Pa is the same as that required to expand from 10^5 Pa to 10^4 Pa.

For a constant-volume process ($dv = 0$), the first law (2.18a) may be written as

$$du = dq$$

From the definition of internal energy for an ideal gas, $du = c_v dT$, the amount of heat required to raise the temperature of the gas from T_1 to T_2 at constant volume is

$$\Delta q = c_v\,(T_2 - T_1)$$

For a constant-pressure process ($dp = 0$), it is advantageous to use the first law in enthalpy form (2.18b), so that the first law for a constant-pressure process becomes

$$dh = dq$$

From the definition of enthalpy for an ideal gas, $dh = c_p dT$, the amount of heat required to raise the temperature of the gas from T_1 to T_2 at constant pressure is

$$\Delta q = c_p \left(T_2 - T_1 \right)$$

The constant-volume and constant-pressure results may be anticipated from the definitions of specific heat in Section 2.3.

An *adiabatic* process is one in which no heat is exchanged between the system and its environment, so that $dq = 0$. The first law for a reversible adiabatic process may thus be written as

$$du = dw$$

An adiabatic compression increases the internal energy of the system. The first law (2.18a,b) for an adiabatic expansion of an ideal gas is thus written

$$c_v \, dT = - p \, dv \tag{2.19a}$$

$$c_p \, dT = v \, dp \tag{2.19b}$$

Considering a reversible adiabatic expansion for an ideal gas, we have from (2.19a) and the equation of state (1.12)

$$c_v \frac{dT}{T} = - R \frac{dv}{v}$$

which may be integrated between an initial and final state (assuming that c_v is constant) to give

$$c_v \ln\left(\frac{T_2}{T_1}\right) = - R \ln\left(\frac{v_2}{v_1}\right)$$

so that

$$\frac{T_2}{T_1} = \left(\frac{v_1}{v_2}\right)^{R/c_v} \tag{2.20}$$

During an adiabatic expansion of a gas, the temperature decreases. In the reverse process (adiabatic compression), work is done on the gas and the temperature increases. Using the ideal gas law and the relationship $c_p - c_v = R$, we may write (2.20) in the following equivalent forms:

$$\frac{p_2}{p_1} = \left(\frac{v_1}{v_2}\right)^{c_p/c_v} \tag{2.21}$$

$$\frac{T_2}{T_1} = \left(\frac{p_2}{p_1}\right)^{R/c_p} \tag{2.22}$$

Equations (2.20), (2.21), and (2.22) are commonly referred to as *Poisson's equations*. It is noted here that (2.22) may also be derived directly by starting from the enthalpy form of the first law (2.19b).

Figure 2.3 compares an isothermal expansion with a reversible adiabatic expansion on a p, V diagram. It is seen that a given pressure decrease produces a smaller volume increase in the adiabatic case relative to the isothermal case, because the temperature also decreases during the adiabatic expansion.

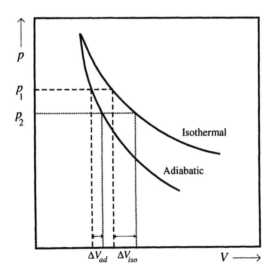

Figure 2.3 Isothermal expansion compared with a reversible adiabatic expansion. For a given drop in pressure, $\Delta V_{iso} > \Delta V_{ad}$, since during the adiabatic expansion, the temperature also decreases.

2.5 Entropy

Before discussing entropy, we first consider the difference between reversible and irreversible processes. In all thermodynamic processes, the changes that occur in the environment must be considered in conjunction with the changes that occur in the thermodynamic system. A *reversible process* is one in which the system is in an equilibrium state throughout the process. Thus the system passes at an infinitesimal rate through a continuous succession of balanced states that are infinitesimally different from each other. In such a scenario, the process can be reversed, and the system and its environment will return to the initial state. *Irreversible processes* proceed at finite rates: if the system is restored to its initial state, the environment will have changed from its initial state. The term "irreversible" does not mean that a system cannot return to its original state, but that the system plus its environment cannot be thus restored.

A comparison between reversible and irreversible atmospheric processes is illustrated in Figure 2.4. If a mass of moist air rises adiabatically and then descends adiabatically to the initial pressure level, the final temperature and mixing ratio of the air will be equal to the initial values and the process is thus reversible. However, if clouds form during the ascent and some of the cloud water rains out, then the air mass when brought down to the initial pressure will have a higher temperature and lower specific humidity than the initial values. Precipitation is an example of an irreversible process. If the rain falls to the ground and does not evaporate in the sub-cloud layer, then the total water content of the atmosphere decreases irreversibly and the temperature of the atmosphere increases irreversibly.

Consider the first law of thermodynamics in enthalpy form (2.18b) for a reversible process:

$$dq = c_p \, dT - v \, dp$$

Reversible heating is an abstract concept, whereby heating of a system occurs infinitesimally slowly through contact with an infinite heat reservoir. For the reversible expansion of an ideal gas, we may substitute for the specific volume from the equation of state and divide by temperature

$$\frac{dq}{T} = c_p \frac{dT}{T} - R \frac{dp}{p} = c_p \, d(\ln T) - R \, d(\ln p) \tag{2.23}$$

The two terms on the right-hand side of (2.23) are by definition exact differentials, and their sum must also be an exact differential. Therefore dq/T is an exact differential, i.e.,

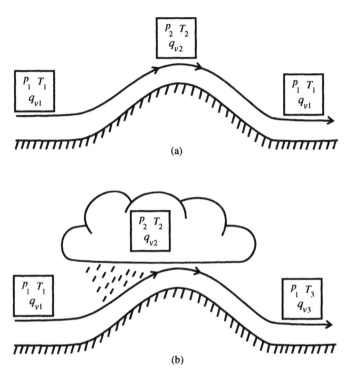

Figure 2.4 Comparison of a reversible and an irreversible process in the atmosphere. In (a), moist air initially at pressure p_1, and having temperature T_1 and specific humidity q_{v1}, rises adiabatically to the top of a mountain. It then descends adiabatically on the other side to the initial pressure. Because the process of passing over the mountain was done reversibly and adiabatically, the temperature and specific humidity are restored to their inital values, and the process is thus reversible. In (b), clouds form as the mass of moist air rises, and some of the cloud water rains out. When the mass of air descends on the other side to its initial pressure, its specific humidity is lower and its temperature is higher than the original values, and the process is thus irreversible: the total water content of the atmosphere decreases irreversibly and the atmosphere is warmed irreversibly.

$$\oint \left(\frac{dq}{T} \right)_{rev} = 0 \qquad (2.24)$$

where the subscript *rev* emphasizes that this relationship holds only for a reversible process. Dividing heat by temperature converts the inexact differential dq into an exact differential. We can now define a new thermodynamic state function, the *entropy*, η, with units J K^{-1} kg^{-1}, to be

$$d\eta = \left(\frac{dq}{T}\right)_{rev} \qquad (2.25a)$$

It is important to remember that entropy is defined so that the change in entropy from one state to another is associated with a reversible process connecting the two states.

When a change in entropy between two given states occurs via an irreversible process, the change in entropy is exactly the same as for a reversible process: this is because entropy is a state variable and $d\eta$ is an exact differential, which means that integration of $d\eta$ does not depend on the path of integration. Although the change in entropy is exactly the same for reversible and irreversible processes that have the same initial and final states, the integral of dq/T is not the same for reversible and irreversible processes. In fact,

$$\Delta\eta > \int \left(\frac{dq}{T}\right)_{irrev} \qquad (2.25b)$$

where the subscript *irrev* indicates an irreversible process. This suggests that to accomplish a given change in entropy (or state) by an irreversible process, more heat is required than when a reversible process is involved. This implies that reversible processes are more efficient than irreversible processes.

Entropy changes for an ideal gas in a reversible process can be determined from (2.18a) and (2.25a):

$$d\eta = c_v\, d(\ln T) + R\, d(\ln v) \qquad (2.26a)$$

or alternatively from (2.18b) and (2.25a):

$$d\eta = c_p\, d(\ln T) - R\, d(\ln p) \qquad (2.26b)$$

The entropy change for isobaric heating is thus

$$\Delta\eta = c_p \ln\left(\frac{T_2}{T_1}\right)$$

and for isothermal processes

$$\Delta\eta = R \ln\left(\frac{v_2}{v_1}\right) = R \ln\left(\frac{p_1}{p_2}\right)$$

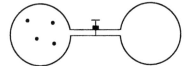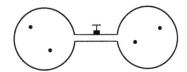

Figure 2.5 Expansion of an ideal gas illustrating the relationship between entropy and probability. Initially, four molecules of the gas are placed in the left bulb, and the right bulb is empty. When the stopcock is opened, the volume doubles, and the molecules are distributed between the left bulb and the right bulb. In this process, the number of possible configurations of molecules, and hence the entropy of the system, has increased.

As introduced above, entropy has arisen from purely mathematical considerations. Entropy can be interpreted physically in the context of statistical mechanics. The relationship between entropy and probabilities is illustrated using a simple example. Consider the ideal gas expansion shown in Figure 2.5. Two isolated bulbs, each of volume V, are connected by a stopcock. Initially, four molecules of the gas are placed in the left bulb, and the right bulb is empty. The stopcock is opened and the volume doubles (this is an example of an irreversible adiabatic expansion). The change in entropy from the intial (*init*) to final (*fin*) state is

$$\Delta \eta = Nk \ln\left(\frac{V_{fin}}{V_{init}}\right)$$

where N is the number of molecules and k is the Boltzmann constant (gas constant per molecule). Since $N = 4$ and $(V_{fin}/V_{init}) = 2$, we may write

$$\Delta \eta = 4k \ln 2 = k \ln 2^4$$

The entropy change is thus proportional to $\ln 2^4 = \ln 16$.

In the final state, the molecules are distributed between the left bulb and the right bulb. Table 2.1 lists the numbers and probabilities of the possible configurations of the final distribution of molecules. There are 16 ways of arranging the four molecules between the two bulbs in the final state. There is only one configuration for the initial state: all four molecules in the left bulb. The ratio of the final to the initial probability, P_{fin}/P_{init} and the final to the initial number of possible configurations, C_{fin}/C_{init} are

$$\frac{P_{fin}}{P_{init}} = \frac{C_{fin}}{C_{init}} = 16 = 2^4$$

Table 2.1 Ways of arranging four molecules in two bulbs of equal volume.

# in left bulb	# in right bulb	# of ways to achieve configuration, C	Probability of the configuration, P
0	4	1	1/16
1	3	4	4/16
2	2	6	6/16
3	1	4	4/16
4	0	1	1/16
		Total: 16	1

This suggests that we can associate entropies with probabilities, or numbers of possible configurations.

The equilibrium state of the four molecules distributed in two bulbs is more random than four molecules in one bulb, since we are less definite about the location of the molecules in the more random (or disordered) state. More rigorous developments of this relationship can be done in the context of quantum mechanics. However, the present example suffices to associate entropy with randomness. The natural path of all processes is from order to randomness. Entropy in an isolated system will tend to increase as the probability spreads out over the possible states and the system approaches equilibrium.

2.6 Second Law

The second law of thermodynamics forbids certain processes, even some in which energy is conserved. The second law of thermodynamics may be stated in several different ways, which appear to be different in content but can be shown to be logically equivalent.

The entropy statement of the second law is:

> *There exists an additive function of state known as the equilibrium entropy, which can never decrease in a thermally isolated system.*

In other words, a thermally isolated system cannot spontaneously regain order which has been lost. The second law may be applied to a system and its surroundings to determine the total entropy change $\Delta \eta_{tot}$

$$\Delta \eta_{tot} \geq 0$$

which is known as *Clausius' inequality*. For a reversible process we cannot have $\Delta\eta_{tot} > 0$, since we would have $\Delta\eta_{tot} < 0$ upon reversing the process, which would violate Clausius' inequality. Therefore, $\Delta\eta_{tot} = 0$ for all reversible changes. For the special case of a reversible adiabatic process, the entropy change is zero in the system, $\Delta\eta_{syst} = 0$. Reversible adiabatic processes are therefore *isentropic*. Using the definition of entropy in (2.25), we may write Clausius' inequality as

$$\oint \frac{dq}{T} \leq 0 \tag{2.27}$$

where the equal sign holds for a completely reversible process.

The temperature or Clausius statement of the second law is:

No process exists in which heat is transferred from a colder body to a less cold body while the constraints on the bodies and the state of the rest of the world are unchanged.

A quantitative statement of this principle in terms of entropy can be made as follows. Consider a process that transfers heat between two bodies A and B, leaving the surroundings and the constraints on the bodies unchanged. After a small heat transfer,

$$d\eta_{tot} = d\eta_A + d\eta_B$$

This can be expanded for a constant-volume process as

$$d\eta_{tot} = \left(\frac{\partial\eta_A}{\partial u_A}\right)du_A + \left(\frac{\partial\eta_B}{\partial u_B}\right)du_B$$

If the heat transfer is denoted by $dq_A = du_A = -du_B$ we have

$$\eta_{tot} = \left(\frac{\partial\eta_A}{\partial u_A} - \frac{\partial\eta_B}{\partial u_B}\right)dq_A \geq 0 \tag{2.28}$$

We now define a quantity, T, the *absolute thermodynamic temperature*, as

$$\frac{1}{T} = \frac{\partial\eta}{\partial u}$$

We may therefore write (2.28) in terms of T as

$$\left(\frac{1}{T_A} - \frac{1}{T_B}\right) dq_A \geq 0 \qquad (2.29)$$

This equation shows that dq_A cannot be positive if $1/T_B > 1/T_A$. It follows that the thermodynamic temperature alone determines the direction of heat transfer between bodies and that the heat transfer proceeds from warm to cold. The absolute thermodynamic temperature can be shown to be proportional to the gas scale temperature (Section 1.5) by evaluating the entropy change of an ideal gas over a cyclic process. Equality between these two temperatures is achieved by choosing the value 273.15 K for the reference state (the Kelvin scale).

The third statement of the second law is the heat engine or Kelvin statement. This statement derives its name from the problem that originally stimulated the formulation of the second law: the efficiency of a heat engine, a device that turns heat abstracted from a heat source into work. The heat engine statement of the second law is:

No process exists in which heat is extracted from a source at a single temperature and converted entirely into useful work, leaving the rest of the world unchanged.

This statement tells us that a heat engine cannot have an efficiency of 100%. Part of the heat absorbed must be rejected to a heat sink. The second law implies a certain degree of unavailability of heat for the production of work. If all of the heat were converted into work, the total entropy would decrease, which is not physically possible.

The simplest possible heat engine is a device which works in a cycle, and in one cycle takes heat q_1 from a source at a high temperature T_1, converts part of the heat into useful work, w, and rejects waste heat q_2 to a heat sink at a lower temperature T_2. Such a system is the *Carnot engine* illustrated in Figure 2.6. From the conservation of energy, $w = q_1 - q_2$. The total entropy change is

$$\Delta \eta_{tot} = \Delta \eta_1 + \Delta \eta_2 = -\frac{q_1}{T_1} + \frac{q_2}{T_2} \geq 0$$

This equation may be written as a condition on waste heat q_2:

$$q_2 \geq \frac{T_2}{T_1} q_1$$

The efficiency $\mathscr{E}$ of the heat engine is defined as the ratio between the useful work of the engine compared to the heat input. That is,

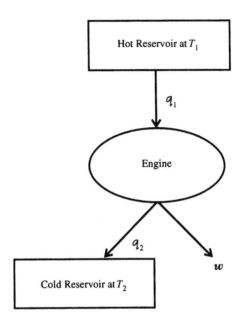

Figure 2.6 Carnot heat engine. Heat q_1 is brought from the hot reservoir to the engine. The engine does work w and rejects heat q_2 into the cold reservoir.

$$\mathscr{E} = \frac{w}{q_1} = 1 - \frac{q_2}{q_1} \qquad (2.30a)$$

The engine is at its highest efficiency when q_2 is as small as possible, which is whenever the cycle is reversible. For a reversible Carnot engine, we have

$$\frac{q_2}{q_1} = \frac{T_2}{T_1}, \qquad \Delta \eta_{tot} = 0, \qquad \text{and} \qquad \mathscr{E} = 1 - \frac{T_2}{T_1} \qquad (2.30b)$$

Thus, the efficiency of a reversible Carnot heat engine depends only on the source and sink temperatures.

2.7 Equilibrium and the Combined First and Second Laws

By using the first and second laws of thermodynamics in combination, we can derive some important results that apply to energy and entropy in the atmosphere and ocean. For any reversible process with expansion work only, we can write the first law as

$$du = dq_{rev} - p\,dv$$

Since $dq_{rev} = Td\eta$ from (2.25a), this becomes

$$du = Td\eta - p\,dv \qquad (2.31)$$

The natural independent variables for internal energy are entropy and volume. If the enthalpy form of the first law is used, (2.14), we have

$$dh = Td\eta + v\,dp \qquad (2.32)$$

The natural independent variables for enthalpy are entropy and pressure.

For many applications in the atmosphere and ocean, it is useful to define a new state function whose natural independent variables are temperature and pressure. The *Gibbs energy, g*, is defined as

$$g = u - T\eta + pv = h - T\eta \qquad (2.33)$$

or in extensive form

$$G = H - T\eta$$

where $\eta = m\eta$ is used to denote extensive entropy and $G = mg$ is the extensive Gibbs energy. In differential form we have

$$dg = -\eta\,dT + v\,dp \qquad (2.34)$$

The natural independent variables of the Gibbs energy are temperature and pressure.

The final basic thermodynamic relationship we consider here is the *Helmholtz energy, a*, defined as

$$a = u - T\eta \qquad (2.35)$$

and in differential form

$$da = -\eta\,dT - p\,dv \qquad (2.36)$$

The extensive form of the Helmholtz energy is $A = ma$. The natural independent variables of the Helmholtz energy are temperature and volume.

Equations (2.31), (2.32), (2.34) and (2.36) are all equivalent forms of the combined first and second laws. The particular form one uses is guided by the specific application.

Consider the following statement of the combined first and second laws (2.31):

$$du = T\, d\eta - p\, dv$$

Equilibrium is a state of balance between a system and its environment, in which small variations in the system will not lead to a general change in its properties, and the system remains constant with time. In a process that occurs at constant entropy and constant volume, the change in internal energy will be zero. In such a process, the equilibrium state is thus specified for that state for which $du = 0$. It can be shown that under conditions of constant η and v that $d^2u > 0$, which says that internal energy is a minimum at equilibrium. Under conditions of constant internal energy and volume, the same version of the first and second laws combined shows that equilibrium is reached when $d\eta = 0$. It can also be shown that under conditions of constant u and v that $d^2\eta < 0$, which states that entropy is a maximum at equilibrium. The drive of thermodynamic systems toward equilibrium is thus a result of two factors. One is the tendency toward minimum energy. The other is the tendency towards maximum entropy. Only if u is held constant can η achieve its maximum; only if η is held constant can u achieve its minimum.

Since processes are rarely studied under conditions of constant entropy or constant energy, it is desirable to obtain criteria for thermodynamic equilibrium under practical conditions such as constant pressure. The four alternative statements of the combined first and second laws: (2.31), (2.32), (2.34), and (2.36), can be used to establish equilibrium criteria under different conditions. Under conditions of constant h and p, equilibrium is reached for $dh = 0$. Under conditions of constant T and p, equilibrium is specified for the condition $dg = 0$. The thermodynamic equilibrium conditions are thus summarized as

At constant η, v:	du =	0,	$d^2u > 0$
At constant η, p:	dh =	0,	$d^2h > 0$
At constant T, v:	da =	0,	$d^2a > 0$
At constant T, p:	dg =	0,	$d^2g > 0$

2.8 Calculation of Thermodynamic Relations

By manipulating the basic thermodynamic equations, we can derive relationships among the thermodynamic variables and thus avoid many difficult laboratory experiments by reducing the body of thermodynamic data to relations in terms of readily measurable functions. The convenience of these relationships will also become apparent through the simplicity introduced into many derivations.

Consider the basic thermodynamic relations (2.31), (2.32), (2.34), and (2.36):

$$
\begin{aligned}
du &= T d\eta - p dv \\
dh &= T d\eta + v dp \\
da &= -\eta dT - p dv \\
dg &= -\eta dT + v dp
\end{aligned}
$$

If we set the left-hand sides of these equations equal to zero, we obtain

$$\left(\frac{\partial \eta}{\partial v}\right)_u = \frac{p}{T} \tag{2.37}$$

$$\left(\frac{\partial \eta}{\partial p}\right)_h = -\frac{v}{T} \tag{2.38}$$

$$\left(\frac{\partial v}{\partial T}\right)_a = -\frac{\eta}{p} \tag{2.39}$$

$$\left(\frac{\partial p}{\partial T}\right)_g = -\frac{\eta}{v} \tag{2.40}$$

We can write expressions for the four functions in functional form as

$$
\begin{aligned}
u &= u\,(\eta, v) \\
h &= h\,(\eta, p) \\
a &= a\,(T, v) \\
g &= g\,(T, p)
\end{aligned}
$$

In differential form the functions can also be written as

$$du = \left(\frac{\partial u}{\partial v}\right)_\eta dv + \left(\frac{\partial u}{\partial \eta}\right)_v d\eta \tag{2.41}$$

$$dh = \left(\frac{\partial h}{\partial p}\right)_\eta dp + \left(\frac{\partial h}{\partial \eta}\right)_p d\eta \tag{2.42}$$

$$da = \left(\frac{\partial a}{\partial v}\right)_T dv + \left(\frac{\partial a}{\partial T}\right)_v dT \tag{2.43}$$

$$dg = \left(\frac{\partial g}{\partial p}\right)_T dp + \left(\frac{\partial g}{\partial T}\right)_p dT \tag{2.44}$$

If we compare (2.41)–(2.44) with (2.31), (2.32), (2.34), and (2.36) and equate coefficients, we obtain

$$\left(\frac{\partial u}{\partial v}\right)_{\eta} = -p \qquad \left(\frac{\partial u}{\partial \eta}\right)_{v} = T \tag{2.45}$$

$$\left(\frac{\partial h}{\partial p}\right)_{\eta} = v \qquad \left(\frac{\partial h}{\partial \eta}\right)_{p} = T \tag{2.46}$$

$$\left(\frac{\partial a}{\partial v}\right)_{T} = -p \qquad \left(\frac{\partial a}{\partial T}\right)_{v} = -\eta \tag{2.47}$$

$$\left(\frac{\partial g}{\partial p}\right)_{T} = v \qquad \left(\frac{\partial g}{\partial T}\right)_{p} = -\eta \tag{2.48}$$

Since du, dh, da, and dg are exact differentials, they obey the Euler condition (2.9). Therefore from (2.31), (2.32), (2.34) and (2.36) we obtain the following set of useful relations called *Maxwell's equations*:

$$\left(\frac{\partial T}{\partial v}\right)_{\eta} = -\left(\frac{\partial p}{\partial \eta}\right)_{v} \tag{2.49}$$

$$\left(\frac{\partial T}{\partial p}\right)_{\eta} = \left(\frac{\partial v}{\partial \eta}\right)_{p} \tag{2.50}$$

$$\left(\frac{\partial p}{\partial T}\right)_{v} = \left(\frac{\partial \eta}{\partial v}\right)_{T} \tag{2.51}$$

$$\left(\frac{\partial v}{\partial T}\right)_{p} = -\left(\frac{\partial \eta}{\partial p}\right)_{T} \tag{2.52}$$

2.9 Heat Capacity

In this section we determine values of the specific heats for air and seawater. The heat capacities of ideal gases and crystalline solids can be determined theoretically by applications of statistical thermodynamics; however, there is not a generally accepted theory for the specific heat of liquids. Here we investigate theoretically the specific heat of ideal gases and describe empirically the specific heat of seawater.

Values of c_p and c_v can be determined for an ideal gas by considering the mechanical degrees of freedom and the equipartition of energy. A *mechanical degree of freedom* refers to an independent mode of motion (a translation, rotation, or vibration) of the molecule in one of three mutually independent directions in space. The total number of degrees of freedom of a mechanical system is equal to the number of variables required to specify the motion of the system. For example, a mass point (e.g., a monatomic molecule) has three degrees of freedom, for motion in each of the x, y, and z directions. For a mechanical system with more than one mass point (e.g., a diatomic or triatomic molecule), additional degrees of freedom arise from rotational and vibrational motions (Figure 2.7). An N-atomic molecule has 3N degrees of freedom:

	Nonlinear molecule	Linear molecule
Translation	3	3
Rotation	3	2
Vibration	3N − 6	3N − 5

Recall from elementary kinetic theory (Section 1.6) that the average molecular kinetic energy of an ideal gas is given by

$$\mathscr{E}_k = \frac{3}{2} n R^* T$$

This suggests that for one mole of a monatomic gas, we can associate $(1/2)R^*T$ thermal energy per mole with each translational degree of freedom. In the case of a more complex molecule, the energy is shared by rotational and vibrational degrees of freedom, rotational modes associated with $(1/2)R^*T$ per mole, and vibrational modes associated with R^*T per mole. Thus the total energy is equally divided among the translational, rotational, and vibrational degrees of freedom. This is called the *equipartition of energy*. The heat capacity of an ideal gas can in principle be determined by summing the contributions to the thermal energy for each of the mechanical degrees of freedom.

The specific heat capacity at constant volume for ideal gases can be determined from the equipartition of energy law to be

$$c_v = (3/2)R \ for \ a \ monatomic \ gas$$
$$c_v = (7/2)R \ for \ a \ diatomic \ gas$$
$$c_v = 6R \ for \ a \ nonlinear \ triatomic \ gas$$

where R is the specific gas constant. The equipartition of energy predicts a heat capacity that is independent of temperature. Real diatomic and polyatomic molecules

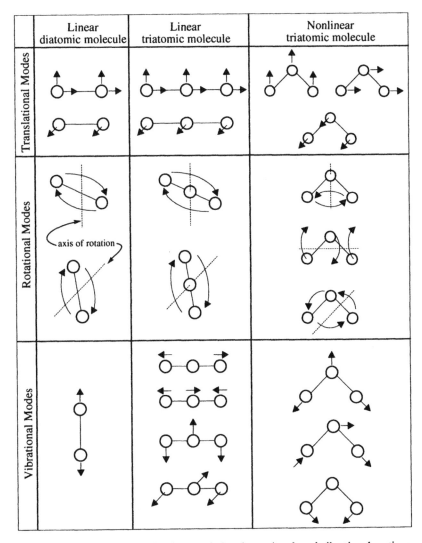

Figure 2.7 Illustration of molecular translational, rotational, and vibrational motions.

have temperature-dependent heat capacities; further, at low temperatures all heat capacities (except for helium) are much lower than the value predicted above. This discrepancy was resolved by the development of quantum mechanics. The contribution of both the rotational and vibrational degrees of freedom to the heat capacity depends on the extent to which the excited vibrational and rotational states are populated for a particular gas, which depends on temperature. The rotationally excited states of the gases in the Earth's atmosphere are fully populated at Earth temperatures, while the Earth is too cold for the vibrationally excited states to be significantly populated. Thus the heat capacity of the major gases in the Earth's atmosphere do not have a contribution from the vibrational modes and are essentially invariant with temperature. The heat capacity of water vapor shows a weak temperature dependence, associated with weak population of excited vibrational states.

For the major atmospheric gases at typical Earth temperatures, the specific heat capacities at constant volume have been determined to be

$$c_v = (3/2)R \text{ for a monatomic gas}$$
$$c_v = (5/2)R \text{ for a diatomic gas}$$
$$c_v = 3R \text{ for a nonlinear triatomic gas}$$

Since $c_p = c_v + R$, we also have

$$c_p = (5/2)R \text{ for a monatomic gas}$$
$$c_p = (7/2)R \text{ for a diatomic gas}$$
$$c_p = 4R \text{ for nonlinear triatomic gas}$$

Air is composed of 98.6% diatomic gases, and thus the values of c_v and c_p for air can be estimated to be 717.76 J K^{-1} kg^{-1} and 1004.86 J K^{-1} kg^{-1}, respectively.

Specific heat capacities of liquids and solids depend on temperature, and are frequently expressed by a polynomial expression with empirically determined coefficients. Heat capacities of liquids are generally greater than those of solids and gases. The specific heat of pure water at surface pressure has been determined empirically to be (Millero *et al.*, 1973)

$$c_p(0,T,0) = 4217.4 - 3.72083\,T + 0.1412855\,T^2$$
$$- 2.654387 \times 10^{-3}\,T^3 + 2.093236 \times 10^{-5}\,T^4 \tag{2.53}$$

where c_p is in J kg^{-1} K^{-1}, T is in °C, and $p = 0$. The influence of salinity is accounted for by

$$c_p(s,T,0) = c_p(0,T,0) + s\,(-7.644 + 0.107276\,T - 1.3839 \times 10^{-3}\,T^2)$$
$$+ s^{3/2}\,(0.17709 - 4.0772 \times 10^{-3}\,T + 5.3539 \times 10^{-5}\,T^2) \tag{2.54}$$

where s is in psu and $p = 0$. Applications of the formula can be checked against

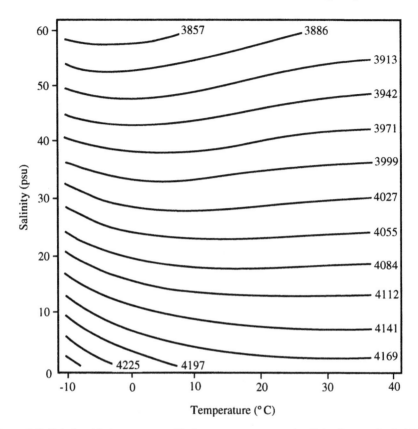

Figure 2.8 Relationship between specific heat, temperature, and salinity for $p = 0$. At high salinities, the specific heat increases with increasing temperature.

$c_p(40, 40, 0) = 3981.050$ J kg^{-1} K^{-1}. For pure water, the specific heat decreases with increasing temperature. The same effect is observed in seawater with low salinities and low temperatures (Figure 2.8). If the salinity exceeds 25 psu, the temperature effect is reversed and c_p increases with increasing temperature. This reversal in sign occurs at lower temperatures for increases in salinity. The specific heat decreases with increasing salinity.

The variation of specific heat with pressure can be derived as follows. We begin with the definition of specific heat (2.15b):

$$c_p = \frac{\partial h}{\partial T} = T\left(\frac{\partial \eta}{\partial T}\right) \tag{2.55}$$

Substituting (2.31) into (2.55), we have

$$c_p = \frac{\partial u}{\partial T} + p\left(\frac{\partial v}{\partial T}\right) \tag{2.56}$$

Taking the derivative of (2.31) with respect to pressure gives

$$T\left(\frac{\partial \eta}{\partial p}\right) = \frac{\partial u}{\partial p} + p\left(\frac{\partial v}{\partial p}\right) \tag{2.57}$$

Subtracting the pressure derivative of (2.56) from the temperature derivative of (2.57) and using Maxwell's relation (2.52) gives the desired result

$$\frac{\partial c_p}{\partial p} = -T\left(\frac{\partial^2 v}{\partial T^2}\right) \tag{2.58}$$

which is determined easily from observations of temperature and specific volume.

The difference $c_p - c_v$ for seawater can be evaluated in the following way. Since entropy is an exact differential, we may write

$$T d\eta = T\left(\frac{\partial \eta}{\partial T}\right)_v dT + T\left(\frac{\partial \eta}{\partial v}\right)_T dv$$

Dividing by dT while holding p constant, we find that

$$T\left(\frac{\partial \eta}{\partial T}\right)_p = T\left(\frac{\partial \eta}{\partial T}\right)_v + T\left(\frac{\partial \eta}{\partial v}\right)_T\left(\frac{\partial v}{\partial T}\right)_p \tag{2.59}$$

From (2.55), we see that $c_p = T(\partial \eta/\partial T)_p$. Since $c_v = T(\partial \eta/\partial T)_v$, we have from (2.59)

$$c_p = c_v + T\left(\frac{\partial \eta}{\partial v}\right)\left(\frac{\partial v}{\partial T}\right)$$

Using (2.51) we can write

$$c_p = c_v + T\left(\frac{\partial p}{\partial T}\right)\left(\frac{\partial v}{\partial T}\right)$$

Using the chain rule for differentiation, we can write $(\partial p/\partial T) = -(\partial p/\partial v) \times (\partial v/\partial T)$ and thus

$$c_p - c_v = -T\left(\frac{\partial p}{\partial v}\right)\left(\frac{\partial v}{\partial T}\right)^2 \qquad (2.60)$$

Hence the value of $c_p - c_v$ is easily obtained from (2.60) by measuring the compressibility (1.31b) and the thermal expansion coefficient (1.31a). Because of the near incompressibility of water, there is very little difference in the values of c_p and c_v. It can be shown from (2.60) that the ratio c_p/c_v for seawater at a salinity of 34.85 psu varies between 1.004 at 0°C and 1.0207 at 30°C. Thus, a distinction is commonly not made between the specific heats at constant pressure and volume for seawater.

2.10 Dry Adiabatic Processes in the Atmosphere

In Section 2.4, the following relationship between pressure and temperature was derived for a reversible adiabatic process for an ideal gas:

$$\frac{T_0}{T} = \left(\frac{p_0}{p}\right)^{R/c_p} \qquad (2.61)$$

The lifting of air parcels by processes such as orographic lifting, frontal lifting, low-level convergence, and vertical mixing causes pressure to decrease, with a corresponding temperature decrease that is specified by (2.61). The lifting of air parcels can be considered a dry adiabatic process as long as condensation does not occur.

If we choose $p_0 = 1000$ mb to correspond to a temperature θ, (2.61) becomes

$$\theta = T\left(\frac{p_0}{p}\right)^{R/c_p} \qquad (2.62)$$

where R/c_p for dry air is evaluated to be

$$\frac{R}{c_p} = \frac{R}{c_v + R} = \frac{R}{\frac{5}{2}R + R} = \frac{2}{7} = 0.286$$

The temperature θ is called the *potential temperature*. It is the temperature a sample of gas would have if it were compressed (or expanded) in an adiabatic reversible process from a given state, p and T, to a pressure of 1000 mb. Since θ is a function of two variables of state (p and T), it is itself a variable of state. θ is thus a characteristic of the gas sample and is invariant during a reversible adiabatic process. Such a quantity is called a *conservative quantity*. Because it is conserved for reversible adiabatic processes in the atmosphere, θ is a useful parameter in atmospheric thermodynamics. Potential temperature and other conserved variables will be used throughout the text to simplify the thermodynamic equations and in the context describing air and water mass characteristics.

Consider an atmospheric temperature profile with a lapse rate $\Gamma = 6°C$ km^{-1}. For atmospheric pressures less than 1000 mb, the potential temperature of a sample of air is greater than the physical temperature since adiabatic compression must be done to lower the parcel to 1000 mb. Conversely, the potential temperature of a sample of air with pressure greater than 1000 mb will be less than the physical temperature. At a pressure level of 1000 mb, $\theta = T$.

A relationship between entropy and potential temperature for the atmosphere is derived by logarithmically differentiating (2.62):

$$d(\ln\theta) = d(\ln T) - \frac{R}{c_p} d(\ln p)$$ (2.63)

Comparing (2.63) with (2.23) shows that

$$d\eta = c_p \, d(\ln\theta)$$ (2.64)

This means that for reversible processes in an ideal gas, potential temperature may be considered an alternative variable for entropy.

Equation (2.62) does not account for water vapor. The specific heat of moist air is

$$c_p = (1 - q_v)c_{pd} + q_v c_{pv} \approx c_{pd}(1 + 0.87q_v)$$ (2.65)

where the subscripts d and v refer to dry air and water vapor, respectively. The ratio R/c_p for moist air can then be determined using (1.23) to be

$$\frac{R}{c_p} = \frac{R_d}{c_{pd}}\left(\frac{1 + 0.608q_v}{1 + 0.87q_v}\right) \approx \frac{R_d}{c_{pd}}(1 - 0.26q_v)$$ (2.66)

The potential temperature of moist air then becomes

$$\theta = T\left(\frac{p_0}{p}\right)^{R_d(1 - 0.26q_v)/c_{pd}}$$ (2.67a)

The difference between the dry-air and moist-air values of θ is generally less than 0.1°C, so that adiabatic expansion or compression of moist air can be treated as if it were dry air. Note that θ is not conserved if a phase change of water occurs (see Section 6.7). We can also define a *virtual potential temperature*, θ_v, by neglecting the water vapor dependence of the exponent of (2.67a) and replacing the temperature by the virtual temperature

$$\theta_v = T_v \left(\frac{p_0}{p}\right)^{R_d/c_{pd}} \tag{2.67b}$$

If we consider the adiabatic ascent of a parcel of air in the atmosphere, the temperature of the parcel will decrease and the potential temperature will remain the same. The rate of decrease of temperature with height in an adiabatic ascent can be determined by considering the first law in enthalpy form for an adiabatic process (2.19b):

$$c_p \, dT = v \, dp$$

If we assume that the ascent of the parcel does not involve any large vertical accelerations and the hydrostatic relation applies, we can substitute the hydrostatic relation into (2.19b) to give

$$c_p \, dT = -g \, dz$$

Recalling that the definition of lapse rate is $\Gamma = -dT/dz$, we can write an expression for the *dry adiabatic lapse rate*, Γ_d, as

$$\Gamma_d = \frac{g}{c_{pd}} \tag{2.68}$$

which has a value of approximately $9.8°C \ km^{-1}$. Both (2.62) and (2.68) describe the temperature evolution of a parcel of air in dry adiabatic ascent, but (2.68) is slightly more restrictive than (2.62) in that it applies only to a hydrostatic process. The adiabatic lapse rate for moist air differs only slightly from (2.68) and can be expressed as

$$\Gamma = \frac{g}{c_{pd}\left(1 + 0.87 q_v\right)}$$

Outside of clouds, diabatic processes such as radiative heating operate on much longer timescales than the characteristic time scale of vertical displacement of the air parcel. Therefore, the lifting of air parcels by processes such as orographic lifting, frontal lifting, low-level convergence, and vertical mixing can be considered dry adiabatic processes as long as condensation does not occur.

2.11 Adiabatic Processes in the Ocean

The adiabatic form of the first law for seawater is written as

$$c_p \, dT = v \, dp \tag{2.69}$$

where the enthalpy of seawater is approximated here by $dh = c_p dT$. To derive the adiabatic lapse rate for seawater, we would like to rewrite the first law in terms of the coefficient of thermal expansion, which is easily evaluated for seawater (Section 1.9). Since entropy is an exact differential, we may write

$$T d\eta = T\left(\frac{\partial \eta}{\partial T}\right) dT + T\left(\frac{\partial \eta}{\partial p}\right) dp \tag{2.70}$$

Substituting (2.55) and (2.52) into (2.70) we have

$$T d\eta = c_p \, dT - T\left(\frac{\partial v}{\partial T}\right) dp \tag{2.71}$$

By comparing (2.71) with (2.32), we obtain

$$v = T\left(\frac{\partial v}{\partial T}\right)$$

For a reversible adiabatic process ($d\eta = 0$), we can express (2.71) as

$$c_p \, dT = T\left(\frac{\partial v}{\partial T}\right) dp = -\frac{T}{\rho^2}\left(\frac{\partial \rho}{\partial T}\right) dp = \frac{\alpha T}{\rho} \, dp$$

or

$$\frac{dT}{dp} = \frac{\alpha T}{\rho c_p} \tag{2.72}$$

where α is the coefficient of thermal expansion, (1.31a). Substitution of the hydro static equation (1.33) for dp yields

$$\Gamma_{ad} = -\frac{\partial T}{\partial z} = \frac{\alpha T g}{c_p} \tag{2.73}$$

Table 2.2 Adiabatic lapse rate in the ocean for selected values of temperature and pressure. Values are given in 10^{-2} K km^{-1}.

p (db) \ T (°C)	-2	0	5	10	15	20	25	30
0	1.7	3.5	7.8	11.6	15.1	18.5	21.7	24.8
1000	3.7	5.4	9.3	12.9	16.2	19.4	—	—
2000	5.7	7.2	10.8	—	—	—	—	—
3000	7.5	8.9	12.3	—	—	—	—	—
4000	9.0	10.6	13.7	—	—	—	—	—
5000	10.9	12.2	15.2	—	—	—	—	—
6000	12.6	13.7	16.2	—	—	—	—	—

Note that for an ideal gas, $\alpha = T^{-1}$ (Section 1.9), and $\Gamma_{ad} = g/c_p$, which is the relationship (2.68).

Because $\alpha = \alpha(s, T, p)$, the adiabatic lapse rate in the ocean is not constant, in contrast to the value for the atmosphere. Table 2.2 shows values of Γ_{ad} in the ocean for selected values of temperature, pressure, and salinity. It is seen that the adiabatic lapse rate increases with increasing temperature and pressure, and varies by over an order of magnitude for the T, p range found in the ocean.

When a water mass is raised adiabatically from the deeper layers to the surface, the pressure decreases and the water mass cools. Conversely, when a mass of water is carried adiabatically from the surface into deeper layers, the temperature increases. Because water is nearly incompressible and has a high heat capacity, the adiabatic temperature change is much smaller than a corresponding change in the atmosphere.

For the atmosphere, the potential temperature is defined in the context of Poisson's equations. In the ocean, potential temperature is defined with respect to a reference temperature, salinity, and pressure (usually taken to be the surface value):

$$\theta(s, T, p; p_0) = T_0 + \int_p^{p_0} \Gamma_{ad}(s, \theta(s, T, p; p_0), p)\, dp \qquad (2.74)$$

where p_0 and T_0 are the reference pressure and temperature. An expression was derived by Bryden (1973) using experimental compressibility data to give θ (°C) as a function of T (°C), s (psu), and p (bars):

2 The First and Second Laws of Thermodynamics

$$\theta(s, T, p) = T - p(3.604 \times 10^{-4} + 8.3198 \times 10^{-5} T - 5.4065 \times 10^{-7} T^2$$
$$+ 4.0274 \times 10^{-9} T^3) - p(s - 35)(1.7439 \times 10^{-5} - 2.9778 \times 10^{-7} T)$$
$$- p^2(8.9309 \times 10^{-7} - 3.1628 \times 10^{-8} T + 2.1987 \times 10^{-10} T^2) \qquad (2.75)$$
$$+ 4.1057 \times 10^{-9}(s - 35)p^2 - p^3(-1.6056 \times 10^{-10} + 5.0484 \times 10^{-12} T)$$

where θ (25, 10, 1000) = 8.4678516°C can be used as a check value.

Potential density, ρ_θ, is defined analogously as the density that a sample of seawater initially at some depth would have if it were lifted adiabatically to a reference level. In the deep ocean, temperature and density profiles are nearly adiabatic, as reflected by the constant values of θ and ρ_θ with depth shown in Figure 2.9. Since θ and ρ_θ are conserved quantities, they can be used in conjunction with the salinity as tracers for water masses.

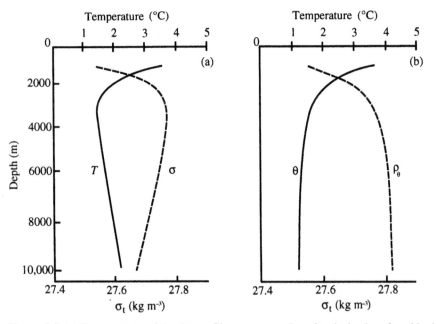

Figure 2.9 (a) Temperature and density profiles, representative of typical values found in the ocean. The nearly adiabatic profiles in the deep ocean produce constant values of potential temperature and potential density, as shown in (b).

Notes

There are many excellent reference text books on thermodynamics. We mention here a few texts that have been useful to us: *Physical Chemistry* (1972) by Moore; *Physical Chemistry* (1984) by Bromberg; and *The Theory of Thermodynamics* (1985) by Waldram.

Problems

1. Calculate the changes in specific internal energy, specific enthalpy, specific entropy, and potential temperature for the following reversible processes in dry air:
a) isothermal expansion from $v = 900$ to 960 cm^3 g^{-1} at $T = 300$ K;
b) isobaric heating from -10 to $+10°C$ at $p = 1000$ hPa;
c) adiabatic compression from $p = 900$ to 950 hPa; at 900 hPa $T = 280$ K.

2. A unit mass of dry air undergoes a Carnot cycle consisting of the following steps:

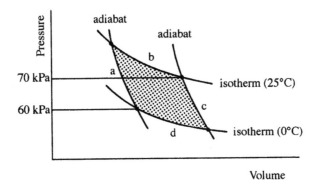

a) adiabatic compression from 60 kPa and $0°C$ to a temperature of $25°C$;
b) isothermal expansion to a pressure of 70 kPa;
c) adiabatic expansion to a temperature of $0°C$;
d) isothermal compression to the original pressure of 60 kPa.
Calculate the work done by the air in this cycle.

3. Consider the system pictured below:

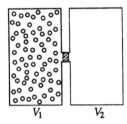

V_1 V_2

Gas is confined to a subvolume V_1 in an insulated rigid container. The container has an adjoining subvolume V_2, initially evacuated, which can be connected to V_1 by opening a valve ($V_1 = V_2$). Suppose the valve is opened and the gas flows out of V_1, filling the entire volume, $V_1 + V_2$.
a) Calculate the work done by the gas in this expansion.
b) Calculate the change in the internal energy of the gas.
c) Calculate the change in entropy of the gas.
d) Is this process adiabatic?
e) Is this process reversible?

4. A hot meteorite falls (velocity 200 km hr^{-1}) into the Atlantic Ocean. The meteorite was originally at a temperature of 1000°C, weighs 1 kg, and has a heat capacity of 0.82 J g^{-1}K^{-1}. If the ocean temperature is 15°C, calculate the change in entropy of the universe as a result of this event.

5. Derive the following identity relating the vertical gradients of temperature and potential temperature:

$$\frac{1}{\theta}\frac{\partial \theta}{\partial z} = \frac{1}{T}\left(\frac{\partial T}{\partial z} + \Gamma_d\right)$$

6. Derive an expression for potential density, ρ_θ, for dry air, which is defined as the density which dry air would attain if it were transformed reversibly and adiabatically from its existing conditions to a standard pressure, p_0. The expression should be a function only of the density, pressure (and standard pressure), the specific heat at constant pressure, and the specific heat at constant volume.

7. Derive the following expressions:
a)

$$\frac{\partial}{\partial v}\left(\frac{\partial \eta}{\partial T}\right) = \frac{1}{T}\left(\frac{\partial c_v}{\partial V}\right)$$

b)

$$\frac{\partial}{\partial T}\left(\frac{\partial \eta}{\partial v}\right) = -\frac{p}{T^2} - \frac{1}{T^2}\left(\frac{\partial u}{\partial v}\right) + \frac{1}{T}\left(\frac{\partial p}{\partial T}\right) + \frac{1}{T}\left(\frac{\partial^2 u}{\partial T \partial v}\right)$$

c)

$$\frac{\partial^2 u}{\partial T \partial v} = \frac{\partial c_v}{\partial v}$$

d) Using the above relationships, show that

$$\frac{\partial u}{\partial v} = T\left(\frac{\partial p}{\partial T}\right) - p$$

This implies that we can predict the change in internal energy associated with an isothermal change for any substance if we know the equation of state for the substance.

e) Evaluate the expression in d) for an ideal gas.

8. The specific volume of seawater can be approximated by a form of the equation of state written in terms of the coefficients of thermal expansion and compressibility:

$$v = v_0\left[1 + \alpha\left(T - T_0\right) + \gamma\left(p - p_0\right)\right]$$

a) Using the results from **7d**, find an expression for $(\partial u/\partial v)$ for seawater.
b) Estimate the error in approximating the internal energy of seawater by $du = c_v\, dT$.

Chapter 3 | Transfer Processes

Knowledge of the transfer of heat and substances such as water vapor and salinity is essential for understanding the evolution of thermodynamic systems. Such transfer processes are a direct consequence of the second law of thermodynamics: transfer occurs as a result of the tendency of thermodynamic systems towards states of maximum entropy and minimum energy. The study of transport processes is part of the general study of how systems approach equilibrium.

Energy can be transferred from one place to another by several different processes. *Conduction* involves the transfer of the kinetic energy of molecules (or heat) through collisions with other molecules. *Advection* transfers energy in a fluid by the physical displacement of matter. *Radiation* transfers energy from one body to another by means of electromagnetic waves, with or without the presence of an intervening physical medium.

Scalar properties such as water vapor and salinity can be transported through *diffusion*, a process analogous to conduction, which involves the random motion of molecules. Advective processes can also transport scalar properties.

3.1 Time-dependent Thermodynamics

The time variation of temperature can be written from (2.18b) as

$$c_p \frac{dT}{dt} = \frac{dq}{dt} + v \frac{dp}{dt} \tag{3.1}$$

Using the definition of potential temperature (2.63) for the atmosphere or (2.73) and (2.74) for the ocean, (3.1) becomes

$$c_p \frac{T}{\theta} \frac{d\theta}{dt} = \frac{dq}{dt} \tag{3.2}$$

Note that use of the variable θ rather than T eliminates the explicit pressure change term in the thermodynamic equation.

74

The total derivative $d\theta/dt$ may be expanded to give

$$d\theta = \left(\frac{\partial\theta}{\partial t}\right)dt + \left(\frac{\partial\theta}{\partial x_j}\right)dx_j \qquad (3.3)$$

Summation notation is used,[1] whereby a variable with no free indices is a scalar and a variable with one free index is a vector. For example

$$\frac{\partial\theta}{\partial x_j} = \frac{\partial\theta}{\partial x_1} + \frac{\partial\theta}{\partial x_2} + \frac{\partial\theta}{\partial x_3} = \frac{\partial\theta}{\partial x} + \frac{\partial\theta}{\partial y} + \frac{\partial\theta}{\partial z} \qquad (3.4)$$

Differentiating with respect to time, we obtain

$$\frac{d\theta}{dt} = \frac{\partial\theta}{\partial t} + u_j\frac{\partial\theta}{\partial x_j} \qquad (3.5)$$

where $u_j = dx_j/dt$ is the velocity vector. The first term on the right-hand side of (3.5) is called the *local derivative* (which is merely the partial derivative with respect to time). The second term is the advection term. We can now incorporate (3.5) into (3.2) and write

$$\frac{\partial\theta}{\partial t} + u_j\frac{\partial\theta}{\partial x_j} = \frac{1}{c_p}\frac{\theta}{T}\frac{dq}{dt} \qquad (3.6)$$

This equation states that the potential temperature at a given point can change by temperature advection due to air motion and by diabatic processes (heating). Heating can arise in the atmosphere and ocean from molecular conduction, radiative transfer, and latent heat associated with phase changes.

The conservation of mass is expressed by the *continuity equation*

$$\frac{\partial\rho}{\partial t} + \frac{\partial}{\partial x_j}\left(\rho u_j\right) = 0 \qquad (3.7)$$

where ρu_j is the momentum per unit volume of the fluid, which is the mass flux per unit area. Equation (3.7) states that the rate at which the mass changes locally is equal

---

[1] Summation notation is used rather than vector notation, since most applications considered will involve derivatives in only one dimension.

to the divergence of ρu_j. Equation (3.7) can be written in the alternate form

$$\frac{1}{\rho}\frac{d\rho}{dt} + \frac{\partial u_j}{\partial x_j} = 0 \tag{3.8}$$

If the fluid or flow is incompressible, then $d\rho/dt = 0$ and

$$\frac{\partial u_j}{\partial x_j} = 0 \tag{3.9}$$

The ocean is nearly incompressible, and except for applications that include sound waves, (3.9) is the form of the continuity equation most often used in the ocean. The incompressible form of the continuity equation is also used frequently in applications to the lower atmosphere.

The mass balance of a scalar quantity, such as water vapor or salinity, can be determined in a manner analogous to the mass continuity equation. Let C be the concentration of a scalar quantity. Conservation of the scalar quantity requires that

$$\frac{\partial(\rho C)}{\partial t} + \frac{\partial(\rho C u_j)}{\partial x_j} = S_C \tag{3.10}$$

where S_C is the body source term for any processes such as change of phase, chemical reaction, molecular diffusion, and boundary fluxes. Using (3.7), (3.10) may be written as

$$\frac{\partial C}{\partial t} + u_j \frac{\partial C}{\partial x_j} = \frac{1}{\rho} S_C \tag{3.11}$$

Processes involved in the heating in (3.6) and the source term in (3.11) are examined for the atmosphere and ocean in the remainder of the chapter.

3.2 Radiant Energy

The Earth receives virtually all of its energy from the sun in the form of electromagnetic radiation. This radiation is absorbed and scattered by the Earth's surface, ocean, and atmosphere. The Earth and its atmosphere emit radiation, some of which is returned to space.

Radiant energy is transmitted from one body to another by means of electromagnetic waves traveling at the speed of light. Radiation is characterized by its *wavelength*, λ, which is the distance from one crest of the wave to the next. Atmospheric radiation spans a broad spectrum of wavelengths. Most of the radiant energy emitted by the sun is in the wavelength range 0.3–4.0 μm, and is referred to as *solar* (or *shortwave*) *radiation*. Most of the radiant energy emitted by the Earth and its atmosphere is in the range 4.0–200 μm, and is referred to as *terrestrial* (or *longwave*) *radiation*.

A number of definitions are needed for the quantitative description of radiant energy. The radiant energy, Q, per unit time coming from a specific direction and passing through a unit area perpendicular to that direction is called the *radiance*, I. The amount of radiant energy, Q, per unit time and area coming from all directions is called the *irradiance* (or radiant flux density), F, which has units watts per square meter (W m^{-2}), while radiance has units watts per square meter per steradian (W m^{-2} sr^{-1}). A *steradian* is a unit from solid geometry that denotes a unit solid angle.

The radiance and irradiance are related as follows:

$$F = \int_0^{2\pi} I \cos Z \, d\omega \tag{3.12}$$

where Z is the angle between the beam of radiation and the direction normal to the surface and $d\omega$ represents the differential of solid angle. The limits 0 and 2π of the integral reflect the hemisphere of directions above the unit area. If the radiant energy comes solely from a single direction, it is called *parallel beam radiation* (or *direct radiation*). If the radiant energy is uniform in all directions, it is called *diffuse* or *isotropic radiation*. In the case of isotropic radiation, we have the following relationship between radiance and irradiance:

$$F = I \int_0^{2\pi} \cos Z \, d\omega = \pi I \tag{3.13}$$

Since the radiant energy is distributed over a spectrum of wavelengths, we define *monochromatic* radiance, I_λ, and irradiance, F_λ, as

$$I = \int_0^\infty I_\lambda \, d\lambda \quad \text{and} \quad F = \int_0^\infty F_\lambda \, d\lambda \tag{3.14}$$

Interactions between radiation and matter can occur via extinction or emission. If the intensity of the radiation decreases then we have *extinction*; if the intensity increases we have *emission*. When considering extinction, there are two possible fates for a photon: it may be absorbed or scattered (reflected). If there is no extinction of the photon by the matter, then the photon has been completely *transmitted* through the matter. The fraction of the incident radiation that is absorbed (*absorptivity*, $\mathcal{A}_\lambda$), transmitted (*transmissivity*, $\mathcal{T}_\lambda$), and reflected (*reflectivity*, $\mathcal{R}_\lambda$) must add up to unity, so that

$$\mathcal{A}_\lambda + \mathcal{T}_\lambda + \mathcal{R}_\lambda = 1 \tag{3.15}$$

where the absorptivity, transmissivity, and reflectivity are usually a function of wavelength.

When matter exists as a dilute gas, it absorbs radiation at discrete wavelengths. These spectral lines are characteristic of the gas and correspond to jumps in the quantum energy levels (electronic, vibrational, roatational) of the gas molecule as photons are either emitted or absorbed. For matter in the liquid or solid state, molecules are so close to each other that liquids and solids tend to emit and absorb in extended continuous regions of the spectrum rather than in discrete spectral lines and bands.

A molecule that absorbs radiation of a particular wavelength can also emit radiation at the same wavelength. The rate at which emission takes place depends only on the temperature of the matter and the wavelength of the radiation. *Kirchoff's law* states that

$$\frac{F_\lambda}{\mathcal{A}_\lambda} = f(\lambda, T) \tag{3.16}$$

That is, for all bodies, the ratio of the emitted radiation to the absorptivity is a function only of the wavelength and temperature. The *emissivity*, ϵ, can be defined as

$$\epsilon_\lambda = \frac{F_\lambda}{f(\lambda, T)} \tag{3.17}$$

which is the ratio of the emitted radiation to the maximum possible radiation that can be emitted at that temperature and wavelength. Combining (3.16) and (3.17), we can write Kirchoff's law as

$$\epsilon_\lambda = \mathcal{A}_\lambda \tag{3.18}$$

which states that the emissivity is equal to the absorptivity. This equation also states that emission can only occur at wavelengths where absorption occurs. If the absorption varies with wavelength, so will the emission. Kirchoff's law is applicable only under conditions of *local thermodynamic equilibrium*, which occurs when a sufficient number of collisions take place between molecules and the translational, rotational, and vibrational energy states are in equilibrium. In the atmosphere, conditions of local thermodynamic equilibrium are not met at heights above about 50 km.

If a body emits the maximum amount of radiation at a particular temperature and wavelength, or equivalently absorbs all of the incident radiation, it is called a *black body*. For a black body, $A_\lambda = 1$ and $R_\lambda = T_\lambda = 0$ for all wavelengths. *Black-body radiation* is characterized by the following properties:

1. The radiant energy is determined uniquely by the temperature of the emitting body.
2. The radiant energy emitted is the maximum possible at all wavelengths for a given temperature.
3. The radiant energy emitted is isotropic.

The theory of black-body radiation was developed by Planck in 1900. Planck determined a semi-empirical relationship that included the concept that energy is quantized. Planck showed from quantum theory that the black-body irradiance, F_λ^*, is given by

$$F_\lambda^* = \frac{2\pi hc^2}{\lambda^5 \left[\exp\left(\frac{hc}{k\lambda T}\right) - 1\right]} \tag{3.19}$$

where h is *Planck's constant* and k is Boltzmann's constant. Equation (3.19) is known as *Planck's radiation law*.

Figure 3.1 plots the black-body irradiance curves for temperatures typical of the Earth and atmosphere determined from (3.19). Note that for each temperature the emission approaches zero for very small and very large wavelengths. For each temperature, there is a maximum of emission at some intermediate wavelength, and this wavelength of maximum emission increases with decreasing temperature. The curve for a warm black body lies above the curve for a cooler black body at each wavelength.

Integration of (3.19) over all wavelengths gives

$$F^* = \int_0^\infty F_\lambda^* \, d\lambda = \sigma T^4 \tag{3.20}$$

where $\sigma = 5.67 \times 10^{-8}$ W m^{-2} K^{-4} is called the *Stefan–Boltzmann constant*. Equation

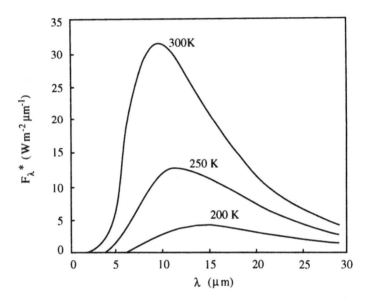

Figure 3.1 Black-body irradiance curves for terrestrial temperatures.

(3.20) is referred to as the *Stefan–Boltzmann law*, whereby the irradiance emitted by a black body varies as the fourth power of the absolute temperature. Evaluation of the Stefan–Boltzmann law at $T = 6000$ K (the approximate emission temperature of the sun) and $T = 300$ K (the approximate emission temperature of the Earth's surface) shows that $F*(6000) = 7.35 \times 10^7$ W m^{-2} and $F*(300) = 4.59 \times 10^2$ W m^{-2}, a difference of five orders of magnitude.

The wavelength of maximum emission for a black body is found by differentiating Planck's law (3.19) with respect to the wavelength, equating to zero, and solving for the wavelength. This yields *Wien's displacement law*:

$$\lambda_{max} = \frac{2897.8}{T} \tag{3.21}$$

where T is in K and λ_{max} is in μm. Evaluation of Wien's displacement law at $T = 6000$ K and $T = 300$ K shows that $\lambda_{max}(6000) = 0.48$ μm and $\lambda_{max}(300) = 9.66$ μm. Thus the wavelength of peak emission from the sun lies in the visible portion of the electromagnetic spectrum, while that from the Earth lies in the infrared.

3.3 Radiative Transfer

Transfer of radiation through the atmosphere and ocean results in extinction of the radiation. Consider first the modification of monochromatic radiation of intensity I_λ as it passes through a thin layer of matter with thickness dx (Figure 3.2). Assume that the matter is cool so that its emission at the given wavelength is negligibly small. From the definition of absorptivity, we can write

$$dI_\lambda = -\mathcal{A}_\lambda I_\lambda \tag{3.22}$$

The absorptivity $\mathcal{A}_\lambda$, of the matter can be shown proportional to the density of the matter and the thickness of the layer, where

$$\mathcal{A}_\lambda = k_\lambda^{abs} \rho\, dx \tag{3.23}$$

and k_λ^{abs} is the constant of proportionality called the *mass absorption coefficient*. As seen from (3.23), the absorption coefficient has units m² kg⁻¹. A *volume absorption coefficient* is commonly defined as $k_\lambda^{v,abs} = \rho\, k_\lambda^{abs}$, with units m⁻¹. Equation (3.22) may now be written as

$$\frac{dI_\lambda}{I_\lambda} = -k_\lambda^{abs} \rho\, dx \tag{3.24}$$

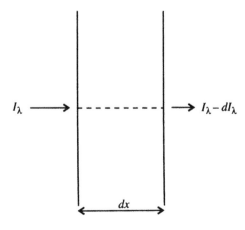

Figure 3.2 Absorption of radiation in a layer of thickness dx.

This equation states that the fractional decrease of the radiance owing to absorption is proportional to the mass per unit area of the absorbing medium. The term on the right-hand side of (3.24) can be written in terms of the nondimensional absorption *optical thickness*, τ_λ

$$d\tau_\lambda = k_\lambda^{abs} \rho \, dx \qquad (3.25)$$

Integration of (3.25) yields

$$I_\lambda(x) = I_\lambda(0) \exp(-\tau_\lambda) \qquad (3.26)$$

where $I_\lambda(0)$ is the incident radiance and $I_\lambda(x)$ is the radiance after penetration to distance x. Equation (3.26) is known as *Beer's law for absorption*.[2] The same equation holds for scattering, by replacing k_λ^{abs} with a *scattering coefficient*, k_λ^{sca} and since all processes are linear, an extinction coefficient can be determined as $k_\lambda^{ext} = k_\lambda^{abs} + k_\lambda^{sca}$.

The transmissivity, $\mathcal{T}_\lambda$, is therefore defined as

$$\mathcal{T}_\lambda = \frac{I_\lambda(x)}{I_\lambda(0)} = \exp(-\tau_\lambda) \qquad (3.27)$$

When $\tau = 1$, the transmissivity has been reduced by the factor $1/e$.

For the transfer of radiation through the atmosphere and ocean, it is convenient to use the vertical distance, z, as a coordinate instead of the distance along the radiation beam, x. For example, energy from the sun is typically not vertical but enters at a *zenith angle*, Z (Figure 3.3). Using a vertical coordinate, we can write

$$dx = -dz \sec Z \qquad (3.28)$$

where the negative sign arises since z is positive upwards, while x is positive in the direction of the incoming radiance. Beer's law may then be written as

$$I_\lambda(x) = I_\lambda(0) \exp(-\tau_\lambda(z) \sec Z) \qquad (3.29)$$

where τ_λ is now measured along the vertical and is called the *optical depth*.

In deriving Beer's law, it was assumed that emission is negligible compared to the

---

[2] This law is also associated with the names Lambert and Bouguer.

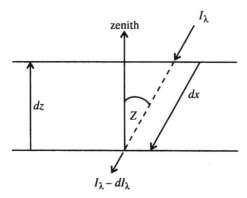

Figure 3.3 Solar radiation enters a layer of the atmosphere or ocean at an angle, Z, called the solar zenith angle. As it travels along a path dx, part of the radiation is absorbed.

incident radiation. This assumption is valid for visible radiation passing through the atmosphere or ocean, since terrestrial temperatures are too cool to emit any significant shortwave radiation (Figure 3.1). In order to consider the transfer of infrared radiation, we must also consider emission. The radiation transfer equation applicable to situations where there is both absorption and emission at the same wavelength is written as

$$dI_\lambda = -\rho k_\lambda^{abs} \left[I_\lambda - I_\lambda^* \right] dx \qquad (3.30)$$

where I_λ^* is the black-body radiance. This equation is known as *Schwarzchild's equation*. More complex radiative transfer equations than (3.30) can be written to include a scattering source term as well, but they are not considered here.

The heating rate due to radiative transfer can be derived from an extensive form of (3.2)

$$mc_p \frac{T d\theta}{\theta \, dt} = \frac{dQ}{dt} \qquad (3.31)$$

From the definition of irradiance, we can write $AdF = dQ/dt$, and then

$$mc_p \frac{dT}{dt} = A \, dF \qquad (3.32)$$

Since $m/A = \rho\,dz$, we have

$$\frac{T}{\theta}\frac{d\theta}{dt} = -\frac{1}{\rho c_p}\frac{\partial F}{\partial z} \tag{3.33}$$

where $\partial F/\partial z$ is the *radiative flux divergence*, and the negative sign arises from the sign convention for F so that $\partial F/\partial z > 0$ implies cooling. Note that $dT/dt = d\theta/dt$ for isobaric heating. Incorporating the hydrostatic equation (1.33) allows (3.33) to be written as

$$\frac{T}{\theta}\frac{\partial\theta}{\partial t} = \frac{g}{c_p}\frac{\partial F}{\partial p} \tag{3.34}$$

The vertical radiative flux divergence may be understood by examining Figure 3.4. Consider two atmospheric levels, a and b, with upwelling and downwelling radiant flux densities as indicated. The net radiant flux density at level a is

$$F_{net,a} = F_a\!\uparrow + F_a\!\downarrow$$

and at level b

$$F_{net,b} = F_b\!\uparrow + F_b\!\downarrow$$

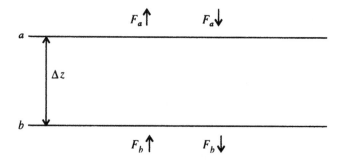

Figure 3.4 Upwelling and downwelling radiant flux densities. The relative magnitudes of F_a and F_b determine whether the layer will warm, cool, or remain the same.

The radiative flux divergence for this situation is

$$\frac{\partial F}{\partial z} \approx \frac{\Delta F}{\Delta z} = \frac{F_{net,\,a} - F_{net,\,b}}{z_a - z_b} \tag{3.35}$$

A positive radiative flux divergence means that more radiation is leaving the layer a,b at a than is coming in at b; therefore, there is a loss of radiation by the layer and the layer cools. A negative radiative flux divergence means that less radiation is leaving the layer a, b at a than is coming in at b; therefore, there is a gain of radiation by the layer and the layer warms.

3.4 Diffusive Transfer Processes

Diffusive transfer processes can transport heat and also concentration of a scalar. Thus there are diffusive components to both dq/dt in (3.6) and S_C in (3.11). In this section we derive equations for diffusive transfer from the second law of thermodynamics.

We know that heat flows from a warmer object to a cooler object until an equilibrium condition, characterized by uniform temperature and maximum entropy, is reached. The entropy increase in such a process is illustrated with an example in which we consider two blocks of copper, a and b, each of unit mass and having specific heat, c. The temperature of block a is $T_a = 400$ K and that of block b is $T_b = 200$ K. The blocks are placed adjacent to each other in a rigid, adiabatic enclosure. After sufficient time, the temperature of both blocks will be $T = 300$ K (Section 2.2). The entropy change (intensive) associated with this process is (Section 2.5)

$$\Delta \eta_{tot} = \Delta \eta_A + \Delta \eta_B = c \ln\left(\frac{300}{400}\right) + c \ln\left(\frac{300}{200}\right) = c \ln\left(\frac{9}{8}\right)$$

Since $9/8 > 1$ and c is positive, the total entropy of the system has increased as a result of the temperature equalization. Such a heat interaction always results in an entropy increase for the system as a whole.

In an analogous manner, we examine the equalization of two concentrations of gases. Consider two volumes, V_1 and V_2, separated by a barrier (Figure 3.5). The volume on the left-hand side of the barrier is V_1 and that on the right side of the barrier is V_2. Initially n_1 moles of gas A are in the left side and n_2 moles of gas B are in the right side. The pressure and temperature are the same in both volumes of gas. The barrier is removed and the gases mix via random molecular motions until each gas is uniformly distributed throughout both volumes. According to Dalton's law of partial pressures (Section 1.7), each gas can be treated independently. Gas A expands from an initial volume V_1 to a final volume $V_1 + V_2$. The entropy change is

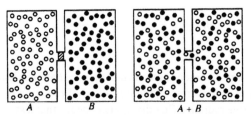

Figure 3.5 Increase in entropy from mixing. Two gases A and B are initially separated by a barrier. When the barrier is removed, the gases mix via random molecular processes, and the entropy of the system increases. This type of transfer process always results in an entropy increase.

$$\Delta \eta_A = R \ln\left[\frac{(V_1 + V_2)}{V_1}\right]$$

and similarly

$$\Delta \eta_B = R \ln\left[\frac{(V_1 + V_2)}{V_2}\right]$$

From Avodagro's law (Section 1.6), we have

$$\frac{V_1}{V_1 + V_2} = \frac{n_1}{n_1 + n_2} = X_1$$

where X_1 is the *mole fraction* of component 1. We determine the total entropy change to be

$$\Delta \eta_{mix} = \Delta \eta_A + \Delta \eta_B = -R\left(\ln X_1 + \ln X_2\right)$$

Since the mole fraction X is by definition less than unity, the entropy of mixing is always a positive quantity.

Equalization of temperature in the transfer of heat from a warmer object to a cooler object and the equalization of the concentration in a mixture of gases both increase the entropy of the system and can be shown to be equilibrium conditions. The second law predicts that processes move in a direction towards increasing entropy. Because of the association of entropy with randomness, the second law is a statement of probability which has meaning only if applied to a statistically significant number of molecules. The certainty and precision of the second law increase with the number of molecules contained in the system and with the time interval to which the law is applied.

The second law dictates the direction in which a process takes place, but not the rate at which the changes occur or the time elapsed before equilibrium is established. Changes in the state of a gas result from transfer by random molecular motions. The general theory of the transfer process is complex, but application of elementary kinetic theory to ideal gases provides a useful conceptual framework for understanding this process.

Consider an ideal gas and assume that some characteristic property γ, which is carried by each molecule varies linearly in the x direction (Figure 3.6). The rate of transfer of γ by a single molecule is expressed by γu_x where u_x is the velocity of the molecule. At any given time, half of the molecules have a velocity component in the positive x direction. The "average" molecule has traveled a distance l, the mean free path, since its last collision with another molecule. The difference in γ over the distance l is $l(\partial \gamma/\partial x)$. As a molecule moves from $x = 0$ to $x = -l$, it transports γ towards $x = -l$. For an ideal gas, one-third of the molecules per unit volume, N, move in the x direction with an average speed $\overline{u_x}$. The flux density, F_γ, is the net rate of transport of γ per unit area:

$$F_\gamma = -\frac{1}{3}N\overline{u_x}\,l\,\frac{\partial \gamma}{\partial x} \tag{3.36}$$

The term $(1/3)N\overline{u_x}\,l$ is the *transfer coefficient*. Note that the transfer coefficient is the ratio of the net rate of transport of γ to the gradient of γ. If l is small in comparison to the dimension of the system, then the molecular transfer coefficient is independent

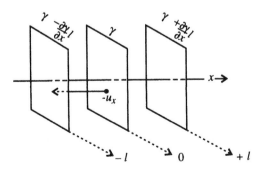

Figure 3.6 Distribution of the property γ and the contribution to the transport of γ by a molecule moving in the $-x$-direction.

of N.

Applying the expression (3.36) to the transfer of heat, we obtain the flux density of heat, F_Q:

$$F_Q = -\frac{1}{3}\overline{u_x}l\rho c_v \frac{\partial T}{\partial x} = -\kappa \frac{\partial T}{\partial x} \tag{3.37}$$

where $\kappa = (1/3)\overline{u_x}l\rho c_v$ is the *thermal conductivity*. Since the temperature of a gas is proportional to the average kinetic energy of the gas molecules, the thermal conductivity may be interpreted as a measure of the transfer of molecular kinetic energy across a kinetic energy gradient.

The flux density equation (3.36) may also be applied to the diffusion of water vapor in air, a transfer process that is important for the growth of cloud drops (Section 5.4). The diffusion of water vapor in air is given by

$$F_{\rho v} = -\frac{1}{3}\overline{u_x}l \frac{\partial \rho_v}{\partial x} = -D_v \frac{\partial \rho_v}{\partial x} \tag{3.38}$$

where ρ_v is the density of water vapor and $D_v = (1/3)\overline{u_x}l$ is the *diffusivity of water vapor in air*.

A differential equation that governs the space and time variations of any conservative quantity is derived from the scalar continuity equation (3.10). The continuity for γ is written as

$$\frac{\partial(\rho\gamma)}{\partial t} + \frac{\partial(\rho\gamma u_i)}{\partial x_i} = S_\gamma \tag{3.39}$$

where the source term includes any nonadvective flux contributions such as molecular diffusion. Incorporating (3.36) into the source term (3.39) and neglecting the bulk fluid advection, we obtain

$$\frac{\partial \gamma}{\partial t} = \frac{\partial}{\partial x_i}\left(D_\gamma \frac{\partial \gamma}{\partial x_i}\right) \approx D_\gamma \frac{\partial^2 \gamma}{\partial x_i^2} \tag{3.40}$$

where we have used D_γ to denote the diffusion coefficient of γ. The three-dimensional heat conduction and water vapor diffusion equations become

$$\rho c_p \frac{\partial T}{\partial t} = \frac{\partial}{\partial x_i}\left(\kappa\frac{\partial T}{\partial x_i}\right) \approx \kappa\frac{\partial^2 T}{\partial x_i^2} \tag{3.41}$$

$$\frac{\partial \rho_v}{\partial t} = \frac{\partial}{\partial x_i}\left(D_v\frac{\partial \rho_v}{\partial x_i}\right) \approx D_v\frac{\partial^2 \rho_v}{\partial x_i^2} \tag{3.42}$$

Equations (3.41) and (3.42) imply that when there is no longer a gradient of temperature or scalar concentration, then the diffusion ceases, and the entropy production stops, as it has reached its maximum.

The general form of the diffusion equation (3.40) is applicable to liquids and solids as well as to gases. The mean free path method is useful in predicting the general pattern of transport behavior of gases in an understandable way, but it is not appropriate as an exact treatment. The diffusion equations are generally applied using empirically determined transfer coefficients. However, the transfer coefficients for real substances are not constant (for example they may depend on temperature), and gradients in the transfer coefficients can thus contribute to the diffusion.

Diffusive transport in the atmosphere and ocean is far smaller than advective transport. However, diffusion is important because it systematically acts to reduce gradients. Diffusive transport has important applications to the atmosphere in the vicinity of a growing cloud drop, where water vapor is diffused towards the cloud drop, and heat is diffused away from the cloud drop (Section 5.4). In the ocean, differential diffusion of heat and salt can result in double-diffusive instability (Section 11.4).

3.5 Turbulence and Turbulent Transport

Turbulence is the small-scale irregular flow superimposed on the mean motion. Turbulence is characterized by irregular swirls of motion called *eddies*. Turbulent motion in the atmosphere and ocean is composed of a spectrum of eddy sizes. A wide spectrum of three-dimensional eddies typically occurs within the atmospheric and oceanic boundary layers, which are the regions of the atmosphere and ocean that are influenced directly by the Earth's surface. The size of turbulent eddies in the boundary layers can range from millimeters to kilometers. Regions of strong vertical gradients in winds or currents are also characterized by the presence of turbulent eddies, even if such regions are distant from the Earth's surface.

It is not possible to predict the behavior of the wide range of eddies using analytical or numerical methods. Therefore, the total motion is typically separated into a mean and turbulent component. The advantage of separating the mean motion from the turbulent eddy motion is that a model can consider the mean motion in a deterministic sense, while determining the turbulent motion using statistical approximations. Consider the following versions of the temperature and scalar conservation equations from (3.6) and (3.11), in the absence of heating and body source terms

$$\frac{\partial \theta}{\partial t} + u_j \frac{\partial \theta}{\partial x_j} = 0 \tag{3.43}$$

$$\frac{\partial C}{\partial t} + u_j \frac{\partial C}{\partial x_j} = 0 \tag{3.44}$$

We assume that the variables can be separated into a slowly varying mean value and a rapidly varying turbulent component. We can thus write

$$u_j = \bar{u}_j + u'_j \tag{3.45}$$

$$\theta = \bar{\theta} + \theta' \tag{3.46}$$

$$C = \bar{C} + C' \tag{3.47}$$

where the overbar denotes the mean value and the prime denotes the fluctuating component associated with the turbulence. The averaging operation that is applied is the *ensemble average*, whereby an ensemble is the set of all turbulent time series that would be obtained under identical exterior conditions, such as wind speed, lapse rate, and surface conditions. Substituting (3.45)–(3.47) into (3.43) and (3.44) yields

$$\frac{\partial \bar{\theta}}{\partial t} + \bar{u}_j \frac{\partial \bar{\theta}}{\partial x_j} + \frac{\partial \overline{u'_j \theta'}}{\partial x_j} = 0 \tag{3.48}$$

$$\frac{\partial \bar{C}}{\partial t} + \bar{u}_j \frac{\partial \bar{C}}{\partial x_j} + \frac{\partial \overline{u'_j C'}}{\partial x_j} = 0 \tag{3.49}$$

where by definition $\overline{u'_j} = \overline{\theta'} = \overline{C'} = 0$. The terms $\overline{u'_j \theta'}$ and $\overline{u'_j C'}$ represent the covariance of the velocity fluctuations with the potential temperature and concentration fluctuations. The covariance terms in (3.48) and (3.49) represent the turbulent fluxes of potential temperature and concentration.

The ability of the turbulent flux terms to transfer heat and scalar properties is illustrated by the following example from the atmospheric boundary layer. Figure 3.7 illustrates the turbulent transfer of heat under conditions when the potential temperature decreases with height (Figure 3.7a) and when the potential temperature increases with height (Figure 3.7b). Consider a turbulent eddy between levels A and B in Figure 3.7a, whereby a negative (downward) fluctuation of u_z warms adiabatically

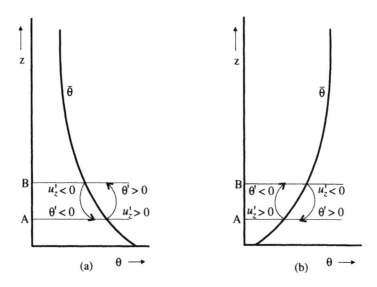

Figure 3.7 The transfer of heat by turbulence. In (a) there is a net transfer of heat upward because the potential temperature profile is such that a descending parcel will be cooler than its surroundings, and an ascending parcel will be warmer than its surroundings. In (b), there is a net transfer of heat downward, since a descending parcel will be warmer than its surroundings, and an ascending parcel will be cooler than its surroundings. (Following Stull, 1988.)

during its descent and θ' is cooler than its surroundings, resulting in an instantaneous product $\overline{u_z' \theta'} > 0$. The upward fluctuation ($u_z' > 0$) will cool adiabatically during its ascent and θ' will be warmer than its surroundings, also resulting in an instantaneous product $\overline{u_i' \theta'} > 0$. The situation is reversed in Figure 3.7b, where the upward fluctuation u_z' is associated with a negative (cool) fluctuation θ' and the downward fluctuation u_z' is associated with a positive (warm) fluctuation θ' and the covariance $\overline{u_i' \theta'} < 0$. This result shows that turbulence can cause a net transport of a quantity such as heat ($\overline{u_z' \theta'} \neq 0$), even though there is no net mass transport ($\overline{u_z'} = 0$). As long as the process is adiabatic (θ is constant), then a decrease of mean potential temperature with height is associated with an upward turbulent transfer of heat, while an increase of mean potential temperature with height is associated with a downward transfer of heat. Note that if the mean potential temperature is constant with height, then there will be no net turbulent transfer of heat. Turbulent transfer of heat is not sufficient to cause a change in mean potential temperature; from (3.48) it is clear that there must be a gradient in the turbulent heat flux to change the temperature.

Because of the nonlinear characteristics of turbulence, a statistical description of turbulence requires an infinite set of equations; this is referred to as the *turbulence closure* problem. The turbulence closure problem has remained one of the great unsolved problems of classical physics. In the atmosphere and ocean, the vertical derivatives of the turbulent flux are typically much larger than the horizontal derivatives, and thus only the vertical derivatives will be considered here. To evaluate turbulent fluxes such as $\overline{u_z'\theta'}$ and $\overline{u_z'C'}$, closure assumptions must be made. The simplest assumption, which is typically referred to as first-order turbulence closure, is to parameterize the turbulent fluxes using an analogy to molecular diffusion, whereby

$$\overline{u_z'\theta'} = -K_\theta \frac{\partial\overline{\theta}}{\partial z} \tag{3.50}$$

$$\overline{u_z'C'} = -K_C \frac{\partial\overline{C}}{\partial z} \tag{3.51}$$

The parameter K is the *eddy diffusion coefficient*. For positive values of K, the turbulent flux has the opposite sign of the mean gradient. The eddy diffusion coefficients have the same dimension as the molecular diffusion coefficients, but their values are typically orders of magnitude larger than the molecular values. The eddy diffusion coefficients vary with the location, state of the fluid, gradient of potential temperature, and the averaging period. The advantage of using the first-order closure (often called the gradient diffusion model) is that the turbulent fluxes can be determined using only mean values. The disadvantage of first-order closure is that a satisfactory method to determine values of K over the necessary range of conditions has not been found. If the mean gradient is zero, then from (3.50) and (3.51) it follows that K is infinitely large, which is not useful. In addition, simple gradient diffusion models are unable to account for counter-gradient diffusion, which is often observed.

3.6 Time-dependent Equations for the Ocean and Atmosphere

In Section 3.1, we developed the following equations ((3.6) and (3.11)) for the time evolution of heat and the concentration of scalar quantities in a thermodynamic system:

$$\frac{\partial\theta}{\partial t} + u_j\frac{\partial\theta}{\partial x_j} = \frac{1}{c_p}\frac{\theta}{T}\frac{dq}{dt}$$

$$\frac{\partial C}{\partial t} + u_j\frac{\partial C}{\partial x_j} = \frac{1}{\rho}S_C$$

By separating the variables into a slowly varying mean value and a rapidly varying turbulent component following Section 3.6, and by using (3.48) and (3.49), we can write (3.6) and (3.11) as

$$\frac{\partial \theta}{\partial t} + u_j \frac{\partial \theta}{\partial x_j} + \frac{\partial \overline{u_z'\theta'}}{\partial z} = \frac{1}{c_p} \frac{\theta}{T} \frac{dq}{dt} \tag{3.52}$$

$$\frac{\partial C}{\partial t} + u_j \frac{\partial C}{\partial x_j} + \frac{\partial \overline{u_z'C'}}{\partial z} = \frac{1}{\rho} S_C \tag{3.53}$$

where we have included only the vertical derivative of the turbulent fluxes and have dropped the overbar over mean values for all terms except the covariances. The ratio θ/T should be interpreted here for a reference state.

The heating term in (3.52) and the body source term in (3.53) can be evaluated using results from Sections 3.3 and 3.4. Explicit inclusion of the molecular diffusion terms from (3.40) and (3.41) yields

$$\frac{\partial \theta}{\partial t} + u_j \frac{\partial \theta}{\partial x_j} + \frac{\partial \overline{u_z'\theta'}}{\partial z} = \frac{\theta}{T}\left[\frac{\partial}{\partial z}\left(\frac{\kappa}{\rho c_p}\frac{\partial T}{\partial z}\right) + \frac{1}{c_p}\frac{dq}{dt}\right]$$

$$\approx \frac{\partial}{\partial z}\left(\frac{\kappa}{\rho c_p}\frac{\partial \theta}{\partial z}\right) + \frac{\theta}{T}\frac{1}{c_p}\frac{dq}{dt} \tag{3.54}$$

$$\frac{\partial C}{\partial t} + u_j \frac{\partial C}{\partial x_j} + \frac{\partial \overline{u_z'C'}}{\partial z} = \frac{\partial}{\partial z}\left(D_C \frac{\partial C}{\partial z}\right) + \frac{1}{\rho} S_C \tag{3.55}$$

where the ratio $\kappa/\rho c_p$ is the *thermal diffusivity* and we include only the z-derivative in the molecular diffusion term. There are two additional heat source terms that must be included in (3.54), which are due to radiative heating and the latent heating associated with a change of phase. We can incorporate radiative heating from (3.34)

$$\frac{\partial \theta}{\partial t} + u_j \frac{\partial \theta}{\partial x_j} + \frac{\partial \overline{u_z'\theta'}}{\partial z} = \frac{\partial}{\partial z}\left(\frac{\kappa}{\rho c_p}\frac{\partial \theta}{\partial z}\right) - \frac{\theta}{T}\left(\frac{1}{\rho c_p}\frac{dF}{dz} + \frac{L}{c_p}\dot{E}\right) \tag{3.56}$$

where $\dot{E}$ is a source term associated with changes of phase of water. Equation (3.56) is generally applicable to both the atmosphere and ocean.

Two scalar equations that are of primary interest in the atmosphere and ocean are atmospheric water substance and salinity. Application of (3.53) to atmospheric water vapor yields

$$\frac{\partial q_v}{\partial t} + u_j \frac{\partial q_v}{\partial x_j} + \frac{\partial \overline{u_z' q_v'}}{\partial z} = \frac{\partial}{\partial z}\left(D_v \frac{\partial q_v}{\partial z}\right) - \dot{E} \qquad (3.57)$$

Application of (3.53) to salinity yields

$$\frac{\partial s}{\partial t} + u_i \frac{\partial s}{\partial x_i} + \frac{\partial \overline{u_z' s'}}{\partial z} = \frac{\partial}{\partial z}\left(D_s \frac{\partial s}{\partial z}\right) + \frac{1}{\rho}S_s \qquad (3.58)$$

where D_s is a diffusion coefficient for salt in seawater. Additional source terms for salinity, S_s, may arise from fluxes of fresh water such as from precipitation or river runoff, or from melting and freezing of sea ice.

Notes

Development of the time-dependent equations applicable to both the atmosphere and ocean is given by Gill (1982) *Atmosphere-Ocean Dynamics.*

Extensive treatments of atmospheric radiative transfer are given by Goody and Yung (1989) *Atmospheric Radiation* and Liou (1980) *An Introduction to Atmospheric Radiation.*

Turbulent transfer in the atmosphere is described in a very readable text by Stull (1988) *An Introduction to Boundary Layer Meteorology.* A unified treatment of turbulence in the atmosphere and ocean is given by Kantha and Clayson (1999) *Small-scale Processes in Geophysical Flows.*

Problems

1. An infrared scanning radiometer aboard a meteorological satellite measures the outgoing radiation emitted from the Earth's surface at a wavelength of 10 μm. Assuming a transparent atmosphere at this wavelength, what is the temperature of the Earth's surface if the observed radiance is 9.8 J m^{-2} s^{-1} μm^{-1} sr^{-1}?

2. Consider an atmospheric constituent whose concentration is constant with height and whose absorption coefficient depends upon pressure according to

$$k_\lambda^{abs}(p) = k_\lambda^{abs}(p_0)\frac{p}{p_0}$$

where p_0 is the sea-level pressure. The mixing ratio, w, of the constituent is defined as the ratio of the density of the constituent, ρ, to the density of air, ρ_a, so that $w = \rho/\rho_a$.
a) Determine an expression for the sunlight intensity, I_λ, that arrives at sea level which enters the atmosphere with intensity, $I_{\lambda 0}$, at zenith angle Z.
b) Determine an expression for the transmissivity of the atmospheric constituent over the depth of the atmosphere.

3. Consider an isothermal atmosphere in hydrostatic balance. In such an atmosphere, the density, ρ_a, of an absorber with constant mass concentration is given by

$$\rho_a = \rho_{a0}\, \exp(-z/H)$$

where H is the scale height of the homogeneous atmosphere and ρ_{a0} is the density of the absorber at the surface.
a) Determine an expression for the optical depth of the absorber.
b) Derive an expression for the radiative heating rate of the absorber, assuming that the radiation is isotropic.

4. Consider a source of salt at the surface of a body of water $(z = 0)$, that is introduced at $t = 0$. The solution of the diffusion equation

$$\frac{\partial s}{\partial t} = D_s \frac{\partial^2 s}{\partial z^2}$$

under these conditions can be written as

$$s(z,t) = \frac{N}{2(\pi D_s t)^{1/2}} \exp\left(\frac{-z^2}{4D_s t}\right)$$

where N is the number of salt molecules introduced at $t = 0$.
a) Confirm that this equation is a solution to (3.40).
b) Derive an expression for the root mean square distance the salt moves from the origin in time t.

Chapter 4 | Thermodynamics of Water

Of all of the planets in the solar system, Earth is unique in possessing abundant water in all three phases (vapor, liquid, and solid). Water vapor may evaporate from the Earth's surface in one location and then return to the surface as precipitation in another. Water plays a dominant role in the radiation balance of the Earth, since all three phases emit and absorb longwave radiation. Some shortwave (solar) radiation is absorbed by all phases of water, although water's principal role in the shortwave radiation balance arises through the scattering of solar radiation by clouds and reflection by surface ice and snow. Water is involved in numerous chemical reactions in both the atmosphere and ocean. The large heat capacity of the ocean allows it to transport large amounts of heat. Hence, water in the atmosphere and ocean is an important modulator of the Earth's climate.

Water is unique; in particular, its liquid phase has the highest specific heat of all liquids (except for NH_3). The high specific heat prevents extreme ranges in temperature. The *molecular viscosity* of water, which is a measure of resistance to the flow of a fluid, is less than most liquids at comparable temperatures. Thus, bulk water flows readily to equalize pressure differences. Water has the highest value of the latent heat of evaporation of all substances and the highest value of the latent heat of fusion (again, except for NH_3). The large values of the latent heat are very important for the transfer of heat and water within the atmosphere. The surface tension of water is the highest of all liquids, controlling the formation and behavior of cloud drops. Water dissolves more substances and in greater quantities than any other liquid. Water also has the highest *dielectric constant* of all liquids except H_2O_2 and HCN. The dielectric constant is a measure of the ability to keep oppositely charged ions in solution separated from one another, which is of importance to the behavior of inorganic dissolved substances such as NaCl. The presence of dissolved salts modifies the thermodynamic properties of seawater relative to pure water.

4.1 Molecular Structure and Properties of Water

To understand the unique properties of water, it is instructive to examine the structure of the individual water molecules and how they interact with one another.

96

A water molecule consists of an oxygen atom bonded to two hydrogen atoms. The structure of the water molecule is determined by the electron configuration around the oxygen atom. Two of the oxygen's eight electrons are near its nucleus, two are involved in the bonding of the hydrogens, and the two pairs of unshared electrons in lone pair orbitals form arms directed towards the corners of a tetrahedron. This simple picture of the bonding suggests that the two H–O bonds should be equivalent and that the H–O–H bond angle should be 90° (a nonlinear molecule). Because of electron–electron repulsion, proton–proton repulsion, and the hybrid nature of the bonds, the bond angles are deformed to give an H–O–H bond angle of 104.5° (Figure 4.1). This bond angle gives rise to a nearly tetrahedral organization of water molecules in the solid phase.

The nonlinear structure of the water molecule has important consequences for the physical characteristics of water. A permanent electric *dipole moment* is produced from the charge separation of the protons at the H⁺ positions and from the unshared electrons at the other end of the molecule (Figure 4.1). Water is thus a *polar molecule.* This permanent dipole moment gives rise to many electromagnetic absorption lines, making water a very important molecule in atmospheric radiative transfer. The nonlinearity of the water molecule also influences the heat capacity of water (Section 2.9).

Because of water's polar structure, water vapor molecules have an attraction to one another and tend to arrange themselves into partly ordered groups, linked by weak intermolecular bonds. As water vapor density increases, the molecules begin to occupy a finite volume and are pressed close enough together for cohesive forces to develop from intermolecular attractions. At high concentrations that can be found

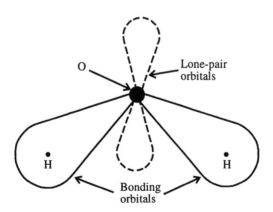

Figure 4.1 Atomic configuration of the water molecule. Bonding with unshared electrons in lone-pair orbitals forms a roughly tetrahedral configuration, with an H–O–H bond angle of 104.5°.

particularly in the lower atmosphere of the tropics, water vapor may form clusters of two or more water vapor molecules called *dimers* or *polymers*, respectively. Intermolecular forces also become important when water vapor approaches condensation. Water vapor at high concentrations or near condensation thus does not behave like an ideal gas.

Recall that there are two properties inherent in the definition of an ideal gas (Section 1.6):

1) The gas molecules occupy no volume.
2) There are no interactive forces between the molecules.

If the gas molecules occupy a volume because of their finite size, then this volume will be approximately independent of pressure. This effective volume is known as the *excluded volume*. The error in neglecting the excluded volume increases substantially as the pressure of the gas increases. The effects of intermolecular attractions in a real gas can be incorporated by reducing the total pressure to account for the attraction of gas molecules to other adjacent gas molecules by weak forces called *van der Waals forces*. The *van der Waals equation of state* is a semi-empirical relation that accounts for the effects of both the excluded volume and the intermolecular forces and is written as

$$\left(p + \frac{an^2}{V}\right)(V - nb) = nR*T \tag{4.1}$$

where a and b are constants. For water, $a = 0.553$ m^3 Pa mole^{-2} and $b = 3.05 \times 10^{-5}$ m^3 mole^{-1}. The term nb represents the excluded volume, and the term an^2/V is the effective pressure reduction associated with the van der Waals forces. The ideal gas approximation for water vapor under atmospheric conditions results in an error of less than 1%. Therefore, we can usually approximate water vapor in the atmosphere as an ideal gas. However, the van der Waals equation of state provides a much better representation than does the ideal gas law of the state of water vapor as it approaches condensation.

As vapor is compressed, intermolecular forces become increasingly important, and water molecules become bonded by hydrogen bonds to form liquid water. A hydrogen bond is an intermolecular bond that forms between a hydrogen atom of one molecule and a highly electronegative atom of another molecule (in this case oxygen). Because of the polar nature of the hydrogen bond, the bond between water molecules is strong, giving liquid water its unusual thermodynamic properties: high boiling and freezing points; large heat capacity; and high latent heat of vaporization. The geometrical configuration of the molecules in the liquid state remains uncertain. Most theories adopt the structure of ice as a starting point. The main differences between liquid water and ice have been hypothesized to include bending or breaking of the hydrogen bonds and the existence of unbonded or "free" liquid molecules within the liquid structure. Theories of water structure continue to proliferate, but there is no single theory that is capable of explaining all of the physical characteristics of water. For example, a satisfactory explanation of the 4°C density maximum of liquid water has remained particularly elusive. One explanation is that this maximum density arises from the increase in volume due to the expansion of the lattice structure and a bending of the hydrogen bonds.

At atmospheric pressures, and temperatures between about −80 and 0°C, the hydrogen-bonded water molecules form tetrahedral structures, becoming a hexagonal solid called "ice-Ih." Water molecules are held together in the ice lattice structure by hydrogen bonding. The hydrogen bond is highly directional, with the hydrogen of one molecule associated with a lone-pair orbital of another molecule. This arrangement leads to a three-dimensional open lattice in which intermolecular cohesion is large (Figure 4.2). Each oxygen atom is located at the center of a tetrahedron, with four other oxygen atoms located at each apex. Each water molecule is bonded to its four nearest neighbors. This tetrahedral coordination results in a crystal structure possessing hexagonal symmetry. An important feature of the structure of ice-Ih is that the oxygen atoms are concentrated in a series of nearly parallel planes called the *basal planes.* The *a*-axis is parallel to the basal plane, and the *c*-axis is perpendicular to the basal plane (Figure 4.2). The crystalline structure of ice is hexagonal, with 24 molecules in the crystal lattice occupying approximately the same volume as 27 free molecules in the liquid state. Thus ice is less dense than liquid water.

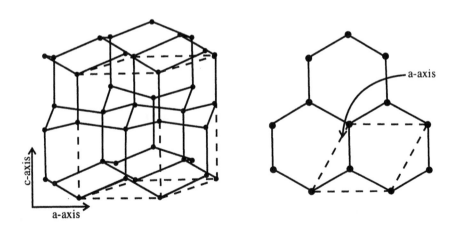

Figure 4.2 Three-dimensional lattice structure of ice. Water molecules are held together in the ice structure by hydrogen bonding. Each water molecule is bonded to its four nearest neighbors. Dashed lines show tetrahedral configuration. View in (b) is perpendicular to view in (a). (After Hobbs, 1974.)

4.2 Thermodynamic Degrees of Freedom

Central to understanding the thermodynamics of water is the equilibrium between the different phases of water, and transitions among the gaseous, liquid, and solid states. In Chapter 2, we considered the thermodynamics of homogeneous systems. A *homogeneous* system is defined as one with uniform chemical composition whose intensive properties are uniform throughout. A *heterogeneous* system is made up of two or more homogeneous parts with abrupt changes in properties at the boundaries of these parts.

A homogeneous system consists of only one *phase*. Each physically or chemically different, homogeneous, and mechanically separable part of a system constitutes a distinct phase. A solution of liquid water and ice is a two-phase system. A system consisting entirely of gases has only one phase, since all gases mix readily in all proportions. With liquids, one, two, or more phases can arise, depending on whether the liquids mix readily. For example, a system of oil and water constitutes two phases, while water and alcohol constitutes one phase. A solid dissolved in a liquid is also one phase. Many different solid phases can coexist. However, differences in shape or degree of subdivision do not constitute a distinct phase.

The number of *components* in a system is the minimum number of distinct chemical species necessary to specify completely the chemical composition of all the phases in the system. More specifically, it is the number of constituents whose concentrations may be independently varied. In a solution of NaCl in water, the chemical species include H_2O, NaCl, Na^+, Cl^-, H^+, and OH^-, yet the number of components is two. Once the amounts of H_2O and NaCl are specified, the concentrations of each of the other species can be determined. It is noted here that for purposes of atmospheric thermodynamics (but not for atmospheric chemistry), dry air gases are regarded as a single component. Table 4.1 gives some examples of thermodynamic systems and the associated numbers of phases and components.

Table 4.1 Examples of some thermodynamic systems and their associated numbers of components and phases.

Examples	Components	Phases
Liquid water with ice	1	2
Mixture of two gases	2	1
Oil and vinegar	2	2
Water and alcohol	2	1
Sugar in water	2	1
Sand in water	2	2
Two blocks of copper	1	1

In summary, a component is the minimum number of distinct chemical species needed to completely specify all phases of the system. A phase is each physically or chemically different, homogeneous, and mechanically separable part of a system. A homogeneous system consists of one phase and one component. In a heterogeneous system, there is more than one phase and/or more than one component. We adopt the notation χ for the number of components and φ for the number of phases.

In our studies of the atmosphere and ocean, we will consider the following systems:

1) moist air (dry air + water vapor): $\chi = 2$; $\varphi = 1$;
2) liquid cloud (dry air + water vapor + liquid water drops): $\chi = 2$; $\varphi = 2$;
3) cloud drops (liquid water + a soluble aerosol particle): $\chi = 2$; $\varphi = 1$;
4) mixed-phase cloud (dry air + water vapor + liquid water drops + ice particles): $\chi = 2$; $\varphi = 3$;
5) ice cloud (dry air + water vapor + ice particles): $\chi = 2$, $\varphi = 2$;
6) ocean (water + salt, with or without sea ice): $\chi = 2$; $\varphi = 1, 2$.

How is our knowledge of the state of the substance influenced by the number of phases and components? For a gas, we know that two intensive variables of the system can be varied independently. The number of the intensive state variables that can be independently varied without changing the number of phases is the number of *thermodynamic degrees of freedom* of the system.

The number of degrees of freedom equals the total number of intensive variables required to specify the complete system minus the number of these variables that cannot be independently varied. The total number of intensive variables which can define each phase is $\chi - 1$ plus pressure and temperature. For φ coexisting phases, the total number of intensive variables defining the system is $\varphi(\chi - 1)$ plus temperature and pressure. The number of variables that cannot be independently varied is given by $\chi(\varphi - 1)$. The *Gibbs phase rule* relates the number of degrees of freedom, f, the number of phases, φ, and the number of components, χ, in the following way:

$$f = 2 + \varphi(\chi - 1) - \chi(\varphi - 1) = \chi - \varphi + 2 \qquad (4.2)$$

where the number "2" refers to the degrees of freedom associated with temperature and pressure of all phases. The Gibbs phase rule states that the total number of degrees of freedom equals the number of components minus the number of phases plus two. This rule enables us to determine the number of intensive variables which may be freely specified in determining the state, without changing the number of components and/or phases.

If we apply the Gibbs phase rule to water, a one-component system, we have

$$f = 3 - \varphi \qquad (4.3)$$

There are three possibilities:

1) $\varphi = 1, f = 2$: *bivariant system*
     Two state variables completely specify the state (e.g., water vapor whose state is specified by T, p);
2) $\varphi = 2, f = 1$: *univariant system*
     One free state variable (e.g., liquid and vapor in equilibrium whose state is specified only by T); or
3) $\varphi = 3, f = 0$: *invariant system*
     Occurs at only one point (T, p), called the *triple point*.

Since the maximum number of degrees of freedom of a one-component system is two, any one-component system can be represented by a two-dimensional diagram. Figure 4.3 gives a schematic representation of the p, T phase diagram for water. The diagram is divided into three areas: solid ice, liquid, and vapor. Within these single-phase areas, the system is bivariant ($f = 2$), and pressure and temperature can be independently varied. Separating the $f = 2$ areas are lines connecting the points at which two phases can coexist at equilibrium, representing a univariant ($f = 1$) system.

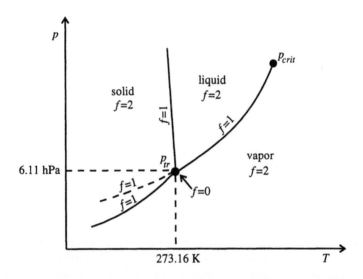

Figure 4.3 p, T phase diagram for water. The three curves indicate those points for which two phases coexist at equilibrium. The dashed curve is the extension of the vapor-pressure curve for liquid water to temperatures below 273.16 K. The solid curve below 273.16 K connects the points at which ice and vapor coexist at equilibrium. p_{crit} indicates the pressure and temperature values beyond which liquid water and water vapor are no longer distinguishable from one another. p_{tr} indicates the triple point, the unique p, T point at which all three phases coexist.

The line dividing the liquid from the vapor is the vapor pressure curve of liquid water. At any given temperature, there is one and only one pressure at which water vapor is in equilibrium with liquid water. The vapor pressure curve has a natural upper limit at the *critical point* (p_{crit} in Figure 4.3), beyond which the liquid phase is no longer distinguishable from the vapor phase (this is because the surface tension of water goes to zero at this point; Section 5.1). The critical point temperature for water is $T_{crit} = 647$ K and $p_{crit} = 218.8$ atm.

The extension of the vapor pressure curve to temperatures less than 0°C (dashed line in Figure 4.3) indicates that liquid water may be cooled below its freezing point without solidifying. At these temperatures, the liquid is referred to as *supercooled water*. The line dividing the vapor from the ice phase is the *sublimation*-pressure curve of ice (solid line below 0°C in Figure 4.3), giving the pressure of water vapor in equilibrium with solid ice. Note that at any given temperature below 0°C, the equilibrium vapor pressure over liquid is greater than the equilibrium vapor pressure over ice. The difference between the saturation vapor pressures over liquid and ice has important consequences for the thermodynamics of mixed-phase clouds and the formation of precipitation (Sections 5.3, 8.2).

The line dividing the ice region from the liquid region (*fusion* curve) shows how the melting temperature of ice (or conversely the freezing temperature of water) varies with pressure. The slope of the melting-point curve shows that the melting point of ice decreases with increasing pressure; thus water expands upon freezing. The three $f = 1$ lines intersect at the *triple point* of water (p_{tr} in Figure 4.3), corresponding to a temperature of $T_{tr} = 273.16$ K and pressure of $p_{tr} = 6.11$ hPa. Since three phases coexist at the triple point, the system is invariant. There are no degrees of freedom and neither pressure nor temperature can be altered even slightly from the triple point values without causing the disappearance of one of the phases.

An alternative perspective can be gained by examining the p,V phase diagram for water (Figure 4.4). The curves on the diagram are isotherms. At high temperatures, the isotherms correspond to ideal gas behavior in the water vapor. Below the critical temperature and within the region circumscribed by the dashed lines, the vapor and liquid regions ($f = 2$) are separated by a zone of discontinuity, where liquid water and vapor coexist ($f = 1$). Thus if vapor at point A is isothermally compressed, it follows the isotherm until reaching B. At that point, condensation begins and liquid forms; the volume decreases substantially and latent heat is released. As condensation proceeds, the specific volume becomes smaller while pressure and temperature remain constant (line between B and C). Point C is reached when all vapor has condensed into liquid. Further isothermal compression follows the compression curve of the liquid, which shows a much larger slope (smaller compressibility) than for the vapor. The same type of process can be described for the sublimation region which occurs below the triple point line, T_{tr}.

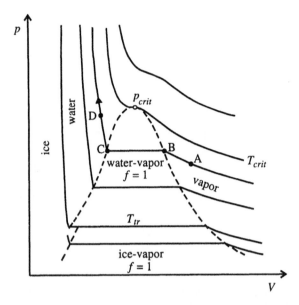

Figure 4.4 p,V phase diagram for water. Solid lines are isotherms. Vapor initially at A is compressed isothermally to B, where liquid water begins to form. Further compression leads to increased formation of liquid water until at C, all remaining water vapor has been condensed. Any further compression causes a sharp rise in the pressure. The curve between points C and D (liquid) is steeper than that between points A and B (vapor), indicating that liquid water is less easily compressed than water vapor.

4.3 Phase Equilibria

The Gibbs phase rule allows us to construct qualitatively a phase diagram for water. It now remains to determine the slopes of the $f = 1$ lines in Figure 4.3 that relate pressure and temperature at equilibrium between two phases. To determine the slopes, we consider the combined first and second laws for a heterogeneous system.

Consider the equation for the Gibbs function for a reversible process (2.34) in extensive form:

$$dG = -\eta dT + Vdp \qquad (4.4)$$

where η is here an extensive entropy. This equation applies to a homogeneous system. To extend the equation for the Gibbs function to an open system where a new phase may form in the system or a new component may be added, we write

$$dG = -\eta dT + V dp + \sum_i \sum_j \frac{\partial G}{\partial n_{ij}} dn_{ij} \tag{4.5}$$

where i is the number of components and j is the number of phases.

The derivative $(\partial G/\partial n)$ is the *chemical potential*, μ:

$$\mu = \frac{\partial G}{\partial n} \tag{4.6}$$

The chemical potential is the change in the Gibbs function of the system with a change in the number of moles of a given component or phase. Note that $\mu = g$. Equation (4.5) applies to an *open system*. We can change the amount of any component i or phase j in an open system by adding or removing dn of the component. Using the definition of chemical potential, we may write (4.5) as

$$dG = -\eta dT + V dp + \sum_i \sum_j \mu_{ij} dn_{ij} \tag{4.7}$$

Under conditions of constant temperature and pressure for a system consisting of one component, we have

$$dG = \sum_j \mu_j dn_j \tag{4.8}$$

For a closed system of one component with three phases (e.g., the water example considered in Section 4.2), we allow the number of moles of a given phase to vary under possible phase transitions, but require the total number of moles to remain constant, so that

$$n = n_1 + n_2 + n_3 = \text{constant}$$

and $dn = 0$. For a closed system at constant temperature and pressure, we therefore have

$$dG = \sum_j \mu_j dn_j = 0 \tag{4.9}$$

In a system containing several phases, certain thermodynamic requirements for the existence of equilibrium may be derived. The conditions for equilibrium between the two phases, 1 and 2, are:

1) Thermal equilibrium: $T_1 = T_2$. If $T_1 \neq T_2$, then heat would flow from one phase to the other and there would be no equilibrium.
2) Mechanical equilibrium: $p_1 = p_2$. If $p_1 \neq p_2$, then one phase would be expanding at the expense of the other and there would be no equilibrium.
3) Chemical equilibrium: $\mu_1 = \mu_2$. If $\mu_1 \neq \mu_2$, then a transfer of n moles of phase 1 to phase 2 would change the Gibbs function.

As seen from Figure 4.3, if we fix the pressure and then heat a condensed phase, the temperature will increase until the equilibrium value is reached where two phases may coexist. The temperature and pressure remain constant until one of the phases disappears. During a phase change, heat is added to (or removed from) the system without changing the temperature or pressure of the system. During this phase transition, entropy and the specific volume will increase. The enthalpy change during the phase transition is

$$\Delta h = L \qquad (4.10)$$

where L is the *latent heat* of the phase transition (sometimes called the *molar heat*). The *latent heat of fusion*, $L_{il} = h_l - h_i$ is the latent heat associated with the solid–liquid phase transition, where the subscript l refers to liquid and the subscript i to ice. The *latent heat of vaporization*, $L_{lv} = h_v - h_l$ is associated with the liquid–vapor phase transition; and the *latent heat of sublimation*, L_{iv} is associated with the solid–vapor phase transition. Note that $L_{il} = L_{iv} - L_{lv}$. In a phase change process at constant pressure, the entropy change can easily be shown from (2.32) and (4.10) to be

$$\Delta \eta = \frac{\Delta h}{T} = \frac{L}{T} \qquad (4.11)$$

A molecular interpretation of the latent heat provides additional insight. Consider first the vaporization of a liquid. By virtue of the differences in density between the liquid and vapor and thus the average distance between molecules, the molecular interactions are strong in a liquid and weak in a gas. The latent heat of vaporization is a very rough measure of the average intermolecular potential energy in the liquid. During vaporization, the majority of the latent heat is used to overcome the cohesive forces holding the molecules together in the liquid form. The latent heat of fusion is much less than the heat of vaporization, since the density difference between a solid and a liquid is relatively small, and the cohesive forces holding the solid together do not differ greatly from those holding the liquid together.

We are now able to derive the slopes of the $f = 1$ lines in Figure 4.3. Chemical equilibrium ($\mu_1 = \mu_2$) implies that $G_1 = G_2$ at equilibrium and $dg_1 = dg_2$, so that

$$dg_1 = -\eta_1 \, dT + v_1 \, dp$$

$$dg_2 = -\eta_2 \, dT + v_2 \, dp \tag{4.12}$$

Since $dg_1 = dg_2$ at equilibrium, we may write

$$-\eta_1 \, dT + v_1 \, dp = -\eta_2 \, dT + v_2 \, dp \tag{4.13}$$

Collecting terms we have

$$\frac{dp}{dT} = \frac{\eta_2 - \eta_1}{v_2 - v_1} = \frac{\Delta\eta}{\Delta v} = \frac{\Delta h}{T\Delta v} = \frac{L}{T\Delta v} \tag{4.14}$$

which is known as the *Clapeyron equation* or the *first latent heat equation*. This equation can be used to evaluate the slope of each of the $f = 1$ lines on the p, T phase diagram (Figure 4.3).

First, consider the solid–liquid equilibrium line. Equation (4.14) may then be written:

$$\frac{dp}{dT} = \frac{L_{il}}{T(v_l - v_i)} \tag{4.15}$$

Inverting this equation gives the variation of the melting point with pressure:

$$\frac{dT}{dp} = \frac{T(v_l - v_i)}{L_{il}} \tag{4.16}$$

Because the specific volume of liquid water is less than the specific volume of ice, the melting point decreases with increasing pressure.

For the liquid–vapor equilibrium,

$$\frac{dp}{dT} = \frac{L_{lv}}{T(v_v - v_l)} \tag{4.17}$$

At the triple point $v_v = 206 \text{ m}^3 \text{ kg}^{-1}$ and $v_l = 10^{-3} \text{ m}^3 \text{ kg}^{-1}$, so that $v_v \gg v_l$ and v_l can be neglected relative to v_v. We may then write the Clapeyron equation as

$$\frac{dp}{dT} \approx \frac{L_{lv}}{Tv_v} \tag{4.18}$$

If we substitute the ideal gas law for v_v, we obtain

$$\frac{dp}{dT} = \frac{L_{lv}\,p}{R_v\,T^2}$$

(4.19)

Equation (4.19) is the *Clausius–Clapeyron equation.*

The *boiling point temperature* is defined to be the temperature at which the vapor pressure is equal to the atmospheric pressure.[1] Equation (4.19) can be inverted to determine the variation of the boiling point temperature with atmospheric pressure:

$$\frac{dT}{dp} = \frac{R_v\,T^2}{L_{lv}\,p}$$

(4.20)

This equation clearly shows the well-known decrease of boiling point temperature with decreasing pressure.

Integration of (4.19) requires that some assumption be made about $L_{lv}(T)$. The variation of L_{lv} with temperature is slow, so we incur little error if we assume that L_{lv} is constant over a small range of temperature variation. Using the notation that e denotes the water vapor pressure, and assuming that L_{lv} is constant, (4.19) is easily integrated:

$$\int_{e_1}^{e_2} d\left(\ln e\right) = \int_{T_1}^{T_2} \frac{L_{lv}}{R_v\,T^2}\,dT$$

(4.21)

to yield

$$\ln\frac{e_2}{e_1} = -\frac{L_{lv}}{R_v}\left(\frac{1}{T_2} - \frac{1}{T_1}\right)$$

(4.22)

or

$$e_2 = e_1\exp\left[-\frac{L_{lv}}{R_v}\left(\frac{1}{T_2} - \frac{1}{T_1}\right)\right]$$

(4.23)

---

[1] If the boiling point temperature is defined to be the temperature at which bursting vapor bubbles appear on the surface of the water, then the boiling temperature is slightly higher than 100°C at sea–level pressure, because of the internal pressure and surface tension associated with the bubbles.

The values of e_1 and e_2 are the *saturation vapor pressure* at T_1 and T_2, respectively. Recall that in arriving at (4.23), we have assumed that the vapor phase obeys the ideal gas law and that L_{lv} is constant.

Analogously, the sublimation–pressure curve, which defines equilibrium between the vapor and ice phases, may be determined from

$$\frac{de}{dT} = \frac{L_{iv}}{T(v_v - v_i)} \tag{4.24}$$

Since $v_i = 1.091 \times 10^{-3}$ m^3 kg^{-1} at the triple point, $v_v \gg v_i$, and we have

$$\frac{de}{dT} \approx \frac{L_{iv}}{Tv_v} \tag{4.25}$$

Again, using the ideal gas law and assuming that L_{iv} is constant, we can integrate to obtain

$$e_2 = e_1 \exp\left[-\frac{L_{iv}}{R_v}\left(\frac{1}{T_2} - \frac{1}{T_1}\right)\right] \tag{4.26}$$

where e_1 and e_2 are the *saturation vapor pressures with respect to ice* at T_1 and T_2, respectively. As can be seen from Figure 4.3, the saturation vapor pressure with respect to ice is less than the saturation vapor pressure with respect to water. This difference is a consequence of the latent heat of sublimation being larger than the latent heat of vaporization.

To integrate the Clausius–Clapeyron equation more precisely, we must include the variation of L_{lv} and L_{iv} with T. The latent heat of vaporization is defined by (4.10) to be the difference in enthalpy, Δh, between the two phases. For any process, the change of Δh with temperature and pressure can be represented by the general equation

$$d\Delta h = \left(\frac{\partial \Delta h}{\partial T}\right)dT + \left(\frac{\partial \Delta h}{\partial p}\right)dp \tag{4.27}$$

and hence

$$\frac{d\Delta h}{dT} = \frac{\partial \Delta h}{\partial T} + \left(\frac{\partial \Delta h}{\partial p}\right)\frac{dp}{dT} = \Delta c_p + \left(\frac{\partial \Delta h}{\partial p}\right)\frac{dp}{dT} \tag{4.28}$$

The last term on the right-hand side of the equation is small, and we have

$$\frac{dL_{lv}}{dT} = c_{pv} - c_{pl} \tag{4.29}$$

Thus the rate of change of the latent heat of vaporization with absolute temperature is equal to the difference between the specific heat at constant pressure of the vapor and the specific heat of the liquid. Equation (4.29) is *Kirchoff's law*, also called the *second latent heat equation*. The variation of the latent heat of fusion with temperature can be shown analogously to be

$$\frac{dL_{il}}{dT} = c_{pl} - c_{pi} \tag{4.30}$$

Equations (4.29) and (4.30) can be integrated by assuming that the specific heats are constant. Although the specific heat capacities depend weakly on both temperature and pressure, c_{pv} and c_{pl} vary by only 1% over the temperature range 0°C to 30°C (Table 4.2). The latent heat of vaporization is seen to decrease with increasing temperature. At the critical temperature, T_{crit}, the latent of vaporization becomes zero as the specific heats of water vapor and liquid become equal.

The Clausius–Clapeyron equation describes the equilibrium between gaseous and condensed phases. Because the Clausius–Clapeyron equation was derived from the combined first and second laws, the equilibrium that is described is a statistical equilibrium. Consider a system consisting of a layer of liquid water overlain by a layer of water vapor (Figure 4.5). If the vapor pressure is equal to the saturation vapor pressure of the liquid, then there is equilibrium between the two phases. This does not mean that an individual water molecule cannot undergo a phase transition.

Table 4.2 Latent heats of condensation and sublimation for water, and specific heat capacities of ice (c_{pi}), liquid (c_{pl}), and vapor (c_{pv}).

T (°C)	L_{lv} (10^6 J kg^{-1})	L_{iv} (10^6 J kg^{-1})	c_{pi} (J kg^{-1} K^{-1})	c_{pl} (J kg^{-1} K^{-1})	c_{pv} (J kg^{-1} K^{-1})
−40	2.603	2.839	1814	4773	1856
−30	2.575	2.839	1885	4522	1858
−20	2.549	2.838	1960	4355	1861
−10	2.525	2.837	2032	4271	1865
0	2.501	2.834	2107	4218	1870
10	2.477			4193	1878
20	2.453			4182	1886
30	2.430			4179	1898
40	2.406			4179	1907

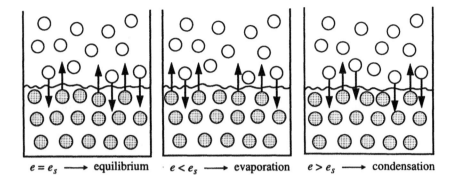

$e = e_s$ ⟶ equilibrium $e < e_s$ ⟶ evaporation $e > e_s$ ⟶ condensation

Figure 4.5 Schematic representation illustrating how the relative magnitudes of the vapor pressure of the water vapor layer, e and the saturation vapor pressure of the liquid below it, e_s determine the phase transitions between water and water vapor. When $e = e_s$, the vapor and liquid are in equilibrium and there is no net exchange between water and water vapor. When $e < e_s$, there is a net transfer of water into water vapor (*i.e.*, net evaporation). When $e > e_s$, there is a net transfer of water vapor into water (*i.e.*, net condensation).

Due to random molecular motions, some liquid molecules will leave the surface (becoming vapor) and some vapor molecules will return. While individual water molecules can undergo phase transitions, the net migration of vapor molecules to the liquid phase and liquid molecules to the vapor phase is zero under equilibrium conditions. If the overlying vapor pressure is less than the saturation vapor pressure of the liquid, there will be a net migration of molecules from the liquid to vapor phase. This is called *evaporation*. Conversely, *condensation* is defined as a net migration of water molecules from the vapor to the liquid, when the vapor pressure exceeds the saturation vapor pressure of the liquid. During condensation the entropy decreases, since liquid is a "less random" state, while during evaporation the entropy increases.

More generally, we have the following terms for the phase transitions:

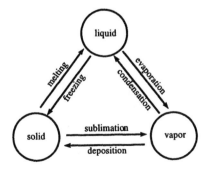

4.4 Atmospheric Humidity Variables

In the previous section, the gaseous phase under consideration was taken to be pure water vapor. In the atmosphere we have a mixture of dry air gases and water vapor. In the following, we refer to the partial pressure of the water vapor by e, the partial pressure of dry air by p_d, and the total atmospheric pressure by p. The saturation vapor pressure with respect to liquid water is denoted by e_s and the saturation vapor pressure with respect to ice is e_{si}. If we write the integrated forms of the Clausius–Clapeyron equation for atmospheric water vapor (assuming that the latent heat does not vary with temperature), we have

$$e_s = e_{s,tr}\, \exp\left[\frac{L_{lv}}{R_v}\left(\frac{1}{T_{tr}} - \frac{1}{T}\right)\right] \tag{4.31}$$

$$e_{si} = e_{s,tr}\, \exp\left[\frac{L_{iv}}{R_v}\left(\frac{1}{T_{tr}} - \frac{1}{T}\right)\right] \tag{4.32}$$

where the reference pressure and temperature referring to the triple point are commonly used ($e_{s,tr} = 6.11$ hPa; $T_{tr} = 273.16$ K).

Application of the Clausius–Clapeyron equation to determining the saturation vapor pressure in the atmosphere is not strictly valid because:

1) the total pressure is not the sum of the partial pressures of two ideal gases (i.e., Dalton's law of partial pressures is not strictly valid);
2) the condensed phase is under a total pressure that is augmented by the presence of dry air; and
3) the condensed phase is not purely liquid water, but contains dissolved air.

The departure from the ideal case can be shown to be less than 1%.

As a result of the departures from an ideal gas, and because of the variations in the latent heat with temperature, values of the saturation vapor pressure calculated from the Clausius–Clapeyron equation, especially in its simplest integrated form (Equation 4.31), are not exact. Empirical values (Appendix D) of the saturation vapor pressure are used when high accuracy is needed.

A sixth-order polynomial can be shown to fit to observations to within their accuracy:

$$e_s = a_1 + \sum_{n=2}^{7} a_n (T - T_{tr})^{n-1} \tag{4.33}$$

where the coefficients for the saturation vapor pressure over water and over ice are given in Table 4.3 and where $T_{tr} = 273.15$ K. This expression provides the high accuracy needed for numerical cloud models.

Table 4.3 Coefficients of the sixth-order polynomial fits to saturation vapor pressure for the temperature range –50° to 50°C for both liquid water and ice. (After Flatau *et al.*, 1992.)

Coefficient	Liquid water	Ice
a_1	6.11176750	6.10952665
a_2	0.443986062	0.501948366
a_3	0.143053301E-01	0.186288989E-01
a_4	0.265027242E-03	0.403488906E-03
a_5	0.302246994E-05	0.539797852E-05
a_6	0.203886313E-07	0.420713632E-07
a_7	0.638780966E-10	0.147271071E-09

Values of the saturation vapor pressure are used in the determination of some of the commonly used atmospheric humidity variables. The *relative humidity*, $\mathcal{H}$, is defined as

$$\mathcal{H} = \frac{e}{e_s} \tag{4.34a}$$

and $\mathcal{H}_i$, the *relative humidity with respect to ice saturation*, is defined as

$$\mathcal{H}_i = \frac{e}{e_{si}} \tag{4.34b}$$

The relative humidity is the ratio of the actual partial pressure of water vapor in the air to the saturation vapor pressure, and is a function only of e and T. It is commonly multiplied by 100 and expressed as a percentage. At temperatures below 0°C, it is necessary to specify whether the relative humidity is being evaluated relative to the saturation vapor pressure over liquid water or over ice.

Comparing (4.32a) and (4.32b) for water and ice shows that, at a given subfreezing temperature,

$$\frac{e_s(T)}{e_{si}(T)} = \exp\left[\frac{L_{il}}{R_v T_{tr}}\left(\frac{T_{tr}}{T} - 1\right)\right] \tag{4.35}$$

This relation indicates that $e_s(T)/e_{si}(T) > 1$ for all subfreezing temperatures and that the ratio increases as the temperature decreases. Table 4.4 shows that an atmosphere

Table 4.4 Variation of $\mathcal{H}_i$ with T for constant $\mathcal{H} = 1$.

$T\,(°C)$	$\mathcal{H}_i$
0	1.0
−10	1.10
−20	1.22
−30	1.34
−40	1.47

saturated with respect to liquid water is supersaturated with respect to ice, and that the degree of supersaturation increases with the supercooling.

The *water vapor mixing ratio*, w_v, is the ratio of the mass of water vapor present to the mass of dry air. It is thus defined, after substituting from the ideal gas law, as

$$w_v = \frac{m_v}{m_d} = \frac{\rho_v}{\rho_d} = \varepsilon \frac{e}{p-e} \tag{4.36}$$

where $\varepsilon = M_v/M_d = 0.622$ (Section 1.7). A value of the *saturation mixing ratio*, w_s, is given by

$$w_s = \varepsilon \frac{e_s}{p-e_s} \tag{4.37}$$

Since $p \gg e$ and $p \gg e_s$,

$$\mathcal{H} \approx \frac{w_v}{w_s} \tag{4.38}$$

is an approximate definition of the relative humidity.

The water vapor mixing ratio can be related to the specific humidity, q_v, which was originally defined in Section 1.7, as

$$q_v = \frac{m_v}{m_d + m_v} = \varepsilon \frac{e}{p - (1-\varepsilon)e} = \frac{w_v}{1+w_v} \tag{4.39}$$

Since both w_v and q_v are always smaller than 0.04, $q_v \approx w_v$.

In summary, given T, p, and one of the humidity variables (for example, e), all of the other humidity variables ($\mathcal{H}$, w_v, q_v, etc.) can easily be determined.

The total mass of water vapor in a column of unit cross-sectional area extending from the surface to the top of the atmosphere is called the *precipitable water, W_v* (or water vapor path):

$$W_v = \int_0^\infty \rho_v \, dz \tag{4.40}$$

The term precipitable water is used because if all the vapor in the column were to be condensed into a pool of liquid at the base of the column, the depth of the pool would be W_v/ρ_l. To obtain a relationship between precipitable water and specific humidity, we can write (4.40) in terms of pressure by incorporating the hydrostatic equation, (1.33):

$$W_v = \frac{1}{g} \int_p^{p_0} \frac{\rho_v}{\rho_a} \, dp = \frac{1}{g} \int_p^{p_0} q_v \, dp \tag{4.41}$$

where p_0 is the surface pressure, corresponding to $z = 0$.

4.5 Colligative Properties of Water Solutions

A *solution* is a homogeneous system, or a single-phase system, that contains more than one component. Two substances that are mutually soluble are said to be *miscible*. Solutions may be gaseous, liquid, or solid. All non-reacting gases are miscible in all proportions. Liquids and solids can dissolve a wide range of gases, or other liquids and solids as a consequence of the second law of thermodynamics. Our discussion of water solutions is motivated by the fact that both seawater and cloud drops are solutions.

The composition of solutions is described by the mole fraction. A two-component solution containing n_A moles of component A and n_B moles of component B has a mole fraction of component A, X_A:

$$X_A = \frac{n_A}{n_A + n_B} \tag{4.42}$$

The component with the largest mole fraction is commonly referred to as the *solvent*, and the other component as the *solute*. If water is the solvent, the solution is said to be *aqueous*.

Colligative properties of a solution depend only on the mole fraction of the solute and not on the particular identity of the solute. In this section, we discuss two colligative properties of aqueous solutions: the lowering of the saturation vapor pressure and the freezing point depression.

4.5.1 Vapor Pressure Depression

The effect of the mole fraction of the solute on the vapor pressure of the solvent is given by *Raoult's law*. This law states that the vapor pressure (p_A) of solvent A above the solution is given by

$$p_A = X_A p_A^\circ \tag{4.43a}$$

where p° is the vapor pressure of the pure phase. In an aqueous solution, $p^\circ = e_s$. If the solute is *volatile* (i.e., it has a vapor pressure), we can also write

$$p_B = X_B p_B^\circ \tag{4.43b}$$

A solution that follows Raoult's law is known as an *ideal solution*. It is easily seen that $p = p_A + p_B$ for an ideal solution. An ideal solution is characterized by complete uniformity of intermolecular forces; that is, a molecule in a solution cannot differentiate between an A and a B molecule. An ideal solution is a hypothetical solution whose properties are approached but seldom encountered in real solutions. The ideal solution concept is useful because it enables us to establish a reference state for consideration of more complex solutions.

We would like to find the ratio of the vapor pressure over an aqueous solution to the vapor pressure of pure water. If we use the subscript *soln* to denote solution and *solt* to denote solute, we have

$$\frac{p_{soln}}{e_s} = \frac{X_{H_2O} e_s + X_{solt} p_{solt}^\circ}{e_s} \tag{4.44}$$

A dilute aqueous solution is defined as $X_{solt} \ll X_{H_2O}$. For a dilute solution, (4.44) can be written as

$$\frac{p_{soln}}{e_s} \approx X_{H_2O} = \frac{n_{H_2O}}{n_{H_2O} + n_{solt}} = 1 - \frac{n_{solt}}{n_{H_2O}} \tag{4.45}$$

It is thus seen that $p_{soln} < e_s$ in a dilute aqueous solution and that p_{soln} decreases as n_{solt} increases.

To understand the lowering of the solution vapor pressure relative to that of pure water, it is useful to imagine a substance having essentially zero vapor pressure to be dissolved in water. The molecules of the solute are distributed uniformly through the water, and some of the solute molecules will therefore occupy positions in the surface layer (Figure 4.6b). With the addition of the solute, the proportion of the surface area occupied by water molecules is

$$\frac{n_{H_2O}}{n_{H_2O} + n_{solt}} < 1 \tag{4.46}$$

It follows that the number of water molecules escaping from the surface, and therefore the equilibrium vapor pressure of the solution, should be reduced relative to that of pure water.

Solids can dissolve in water in two different ways. The molecules of the solid can remain intact, or the molecules can break up into positively and negatively charged ions. When sugar dissolves in water, the sugar molecule does not break up. When common salt (NaCl) dissolves in water, the salt molecule breaks up into Na^+ and Cl^- ions. Aqueous solutions containing charged ions are electrically conducting and are called *electrolytic solutions.*

Seawater is an electrolytic solution whose chemistry is dominated by the presence of six ions (Na^+, K^+, Mg^{2+}, Ca^{2+}, Cl^-, and SO_4^{2-}). To understand the properties of seawater as an electrolytic solution, we consider a dilute solution with a single

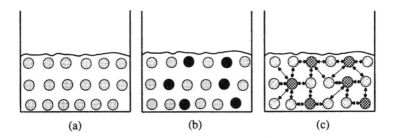

(a)   (b)   (c)

Figure 4.6 Schematic diagram illustrating the reduction in saturation vapor pressure in a solution. In (a) and (b), pure water molecules are represented by dotted circles, and nonelectrolytic molecules are represented by solid circles. The reduction in saturation vapor pressure in the solution (b) arises from the reduction in surface area occupied by the water molecules, since fewer water molecules are able to escape from the surface. In an electrolytic solution (c), both the solute (cross-hatched circles) and solution (hatched circles) become ionized, further reducing the saturation vapor pressure because of the attractive forces exerted by the solute ions on the water molecules.

electrolyte (NaCl). Raoult's law for dilute solutions was modified for dilute electrolytic solutions by van't Hoff, who found that an electrolytic solution effectively contains an increase in the number of moles of solute, so that

$$n_{solt}^{eff} = i n_{solt} \qquad (4.47)$$

where i is the *van't Hoff dissociation factor*. For strong electrolytic solutions, i is equal to the number of ions formed in solution; for weak electrolytic solutions it is less than this number but still greater than one. The ratio of the vapor pressure of a dilute electrolytic solution to that of pure water is thus

$$\frac{p_{soln}}{e_s} = 1 - \frac{i n_{solt}}{n_{H_2O}} \qquad (4.48)$$

Since $i > 1$, it is seen that the partial pressure over an electrolytic solution is less than that of an equivalent nonelectrolytic solution. The reduction of vapor pressure for an electrolytic solution arises from the attraction between the highly polar water molecules and the ions (Figure 4.6c), which reduces the escaping tendency for the water molecules below that for a nonelectrolytic solution. For seawater with salinity of 35 psu (assuming $i = 2$), the saturated vapor pressure is $p_{soln} \approx 0.98 e_s$.

4.5.2 Equilibrium Concentration of Gases

Seawater contains all of the atmospheric gases in solution. Pressure causes gas molecules to enter a liquid; intermolecular forces keep them from exiting. *Henry's law* states that at a given temperature, the equilibrium concentration of a gas in a liquid is proportional to the partial pressure of that gas above it. Each gaseous component of air separately satisfies Henry's law. Henry's law is written as

$$X_{gas} = \frac{p_{gas}}{k} \qquad (4.49)$$

where p_{gas} is the partial pressure of the gas and k is the Henry's law constant, which represents the fraction of the dissolved gas that occurs in non-dissociated form. The parameter k depends on the particular gas, the salinity or ionic strength of the solution, temperature, and total pressure. Henry's law differs from Raoult's law (4.43a) since the proportionality constants are different, $k \neq p_A^o$. The equilibrium concentrations of gases by volume are therefore controlled only by its atmospheric partial pressures.

Equilibrium concentrations of atmospheric gases in seawater generally decrease with increasing salinity, increase with increasing pressure, and decrease with increasing temperature. The effect of temperature on gas solubility is well known to consumers of soda pop: the amount of CO_2 gas that escapes when the pressure is released upon opening the bottle is greater if the bottle is warm. Carbon dioxide has a much lower saturation vapor pressure than nitrogen or oxygen, and also interacts chemically with seawater. This causes CO_2 to be about 60 times more abundant in the ocean than in the atmosphere. Oceanic motions, biological activity, and chemical reactions act as local sources and sinks of dissolved gases, in addition to the gaseous exchanges with the atmosphere.

4.5.3 Freezing Point Depression

The presence of solute in water also modifies the freezing point of the solution relative to pure water. When the solution is cooled, a solid ice phase forms at some temperature below the normal freezing point of water. If the solute is soluble in the liquid water but completely insoluble in the solid water phase, the ice is pure solid water and the liquid phase is a solution. At the freezing point, the solid and liquid are in equilibrium and must therefore have the same vapor pressure (Figure 4.7). For pure liquid water and ice, this equilibrium is at the triple point, which occurs at the intersection of the vapor pressure curve for liquid and ice. The addition of solute lowers the vapor pressure of the liquid water, in accordance with Raoult's law. As seen in Figure 4.7, the vapor pressure curve for the solution intersects the vapor pressure curve for pure ice at a lower temperature. This freezing point depression increases as the amount of solute increases and the vapor pressure over the solution decreases.

Equilibrium between the ice and aqueous solution is characterized by $\mu_i = \mu_l$ and $T_i = T_l$ (Section 4.3), so we can write

$$d\left(\frac{\mu_i}{T}\right) = d\left(\frac{\mu_l}{T}\right) \tag{4.50}$$

We can expand each of these differentials as

$$d\left(\frac{\mu_i}{T}\right) = \frac{\partial(\mu_i/T)}{\partial T}\, dT + \frac{\partial(\mu_i/T)}{dX_i}\, dX_i \tag{4.51}$$

$$d\left(\frac{\mu_l}{T}\right) = \frac{\partial(\mu_l/T)}{\partial T}\, dT + \frac{\partial(\mu_l/T)}{dX_l}\, dX_l \tag{4.52}$$

At equilibrium, we can equate (4.51) and (4.52):

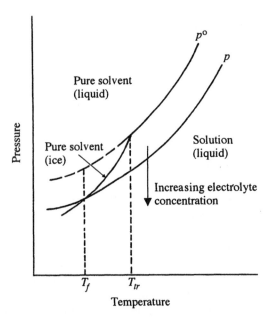

Figure 4.7 Freezing point depression. The presence of solute in water lowers the freezing temperature below that of pure water. The vapor pressure curve for ice intersects the vapor pressure curve for the pure solvent (p^o) at the freezing point of the pure liquid, T_{tr}. Note that the vapor pressure curve for the solution intersects the ice curve at a lower temperature, T_f, than the intersection of the pure solvent curve and the ice curve. This freezing point depression increases as the amount of solute increases.

$$\frac{\partial\left(\mu_i/T\right)}{\partial T}\, dT = \frac{\partial\left(\mu_l/T\right)}{\partial T}\, dT + \frac{\partial\left(\mu_l/T\right)}{dX_l}\, dX_l \qquad (4.53)$$

where $dX_i = 0$ for the pure solid phase.

The $\partial\left(\mu/T\right)/\partial T$ terms in (4.53) can be evaluated from (2.33) and (2.48):

$$\mu = g = h - T\eta$$

$$\frac{\partial g}{\partial T} = -\eta$$

to be

$$\frac{\partial(\mu/T)}{\partial T} = \frac{\partial g/T}{\partial T} = \frac{1}{T}\frac{\partial g}{\partial T} - \frac{\mu}{T^2} = -\frac{\eta}{T} - \frac{\mu}{T^2} = -\frac{h}{T^2}$$

so we have

$$\frac{\partial(\mu_i/T)}{\partial T} = -\frac{h_i}{T^2}; \qquad \frac{\partial(\mu_l/T)}{\partial T} = -\frac{h_l}{T^2} \qquad (4.54)$$

The term $\partial(\mu_l/T)/\partial X_l$ in (4.53) is evaluated as follows. First assume that the liquid and its vapor are in equilibrium, so that the chemical potential is the same in both phases, $\mu_l = \mu_v$. We can therefore write

$$\frac{\partial(\mu_l/T)}{\partial X_l} = \frac{1}{T}\frac{\partial \mu_l}{\partial X_l} = \frac{1}{T}\frac{\partial \mu_v}{\partial X_l} = \frac{1}{T}\frac{\partial \mu_v}{\partial p}\frac{\partial p}{\partial X_l}$$

From (2.48), we can write

$$\frac{\partial \mu}{\partial p} = \frac{\partial g}{\partial p} = v$$

and thus obtain

$$\frac{\partial(\mu_l/T)}{\partial X_l} = \frac{v}{T}\frac{\partial p}{\partial X_l} = \frac{R_v}{p_l}\frac{\partial p}{\partial X_l} \qquad (4.55)$$

For an ideal solution, we can show from Raoult's law (4.43a) that

$$\frac{\partial p}{\partial X_l} = p_l^\circ = \frac{p_l}{X_l}$$

and (4.55) becomes

$$\frac{\partial(\mu_l/T)}{\partial X_l} = \frac{R_v}{X_l} \qquad (4.56)$$

By incorporating (4.54) and (4.56) into (4.53), we can write

$$-\frac{h_i}{R_v T^2} dT = -\frac{h_l}{R_v T^2} dT + \frac{dX_l}{X_l}$$

Collecting terms we have

$$\frac{dT}{T^2} = \frac{R_v}{L_{il}} d(\ln X_l)$$

where we have used the definition of the latent heat of fusion (4.10), $L_{il} = h_l - h_i$. If the latent heat of fusion is taken to be independent of temperature, then we can integrate to yield

$$\frac{T - T^\circ}{TT^\circ} = \frac{R_v}{L_{il}} \ln X_l$$

where T° is the freezing point of the pure solvent. The quantity $\Delta T_f = (T - T^\circ)$ is called the *freezing point depression*. The value of ΔT_f is usually small, in which case $TT^\circ \approx T^{\circ 2}$. We can expand the term $(\ln X_l)$ in a power series

$$\ln X_l = \ln(1 - X_{solt}) = -X_{solt} - \frac{1}{2} X_{solt}^2 - \frac{1}{3} X_{solt}^3 - \cdots$$

For dilute solutions, $X_{solt} \ll 1$ and only the first term in the series is needed. With these approximations, we can write

$$\Delta T_f = \frac{R_v T^{\circ 2}}{L_{il}} X_{solt} \tag{4.57}$$

Thus in a dilute solution where the solvent freezes out as a pure solid, the freezing depression is proportional to the mole fraction of the solute. For dilute ionic solutions, we have

$$\Delta T_f = i \frac{R_v T^{\circ 2}}{L_{il}} X_{solt} \tag{4.58}$$

where i is the van't Hoff dissociation factor.

In principle, the colligative properties of seawater could be determined from theory. However, because of the complexity of the seawater solution and the need to derive very accurate values, empirical relationships are used. The freezing point depression for seawater is given by Millero (1978) as

$$T_f(s) = -0.0575\,s + 1.710523 \times 10^{-3} s^{3/2} - 2.154996 \times 10^{-4} s^2 - 7.53 \times 10^{-3} p \quad (4.59)$$

where T is in °C, s is in psu, and p is in bars ($p = 0$ at the surface). This formula fits measurements to an accuracy of $\pm$ 0.0004 K. For many applications, we can approximate $T_f(s) = -0.055s$ at the surface, which illustrates that the relationship is nearly linear with salinity. The freezing point of seawater with a salinity of 35 psu is −1.92°C.

4.6 Simple Eutectics

In Section 4.2, we constructed a phase diagram for pure water, a one-component system. Here we consider the phase diagram for a two-component system consisting of H_2O and NaCl, as a prelude to consideration of seawater. For a two-component system ($\chi = 2$), the Gibbs phase rule (4.2) takes the form

$$f = \chi - \varphi + 2 = 4 - \varphi$$

There are four possibilities:

$$\varphi = 1; f = 3 \quad \text{(trivariant system)}$$
$$\varphi = 2; f = 2 \quad \text{(bivariant system)}$$
$$\varphi = 3; f = 1 \quad \text{(univariant system)}$$
$$\varphi = 4; f = 0 \quad \text{(invariant system)}$$

A complete representation of a two-component system requires three dimensions. For convenience of graphical representation, we typically employ a two-dimensional space at constant temperature or pressure. This accounts for one of the degrees of freedom, reducing the number of available degrees of freedom of the system by one. So we have

$$f = \chi - \varphi + 1 = 3 - \varphi$$

for a two-component system at constant pressure.

An important consideration in constructing a phase diagram for a solution is the amount of solute (NaCl) that enters into the ice. Thermodynamic arguments can be made that suggest a finite amount of solute should exist in the ice structure at equilibrium. However, this amount is so small that for most purposes the ice can be considered as a pure phase. Therefore, as ice forms from a salt solution, essentially all of the solute is rejected back into the liquid melt. As the temperature decreases and more ice forms, the melt that coexists in equilibrium with the ice becomes increasingly saline.

Figure 4.8 shows the water-rich portion of the H_2O–NaCl phase diagram at constant pressure. The phase relations specify the number and composition of the different phases that coexist at the different temperatures. Consider the constant-pressure cooling of a H_2O–NaCl liquid solution with 35 psu NaCl at an initial temperature of +5°C (point A). The H_2O–NaCl liquid solution is referred to as *brine*.[2] Brine is a bivariant system ($f = 2$) since we have fixed one degree of freedom (pressure). If we cool the brine, we see from Figure 4.8 that ice will start to form from a 35 psu NaCl solution at –2°C (point B). The *liquidus curve* is the $f = 1$ line in Figure 4.8 separating the brine

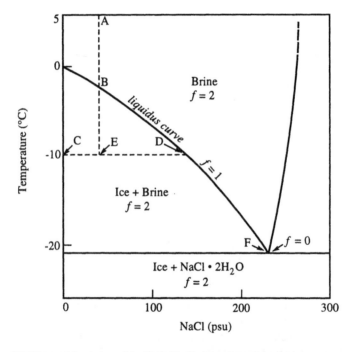

Figure 4.8 Water-rich portion of the H_2O–NaCl phase diagram at constant pressure, showing the phases that coexist at various temperatures. A solution with 35 psu NaCl, initially at A is cooled isobarically until ice begins to form at B, corresponding to a temperature of –2°C. As the solution cools to E, more and more ice forms, leaving an increasingly saline brine, as indicated by the liquidus curve. The ratio DE/CE gives the amount of ice present relative to the amount of liquid. Point F indicates the eutectic point, at which the three phases (brine, pure solid ice, and pure solid salt) coexist in equilibrium.

---

[2] The term brine is sometimes reserved only for a salt-saturated solution; however, the convention in ocean-ography is to use the term brine to refer to seawater with salinity that is higher than normal.

($f = 2$) from the ice plus brine ($f = 2$). Recall that since the NaCl is not incorporated into the ice structure, the ice coexists with increasingly concentrated brine as the cooling proceeds. As the system continues to cool, more ice forms from the brine, causing the remaining brine to decrease in volume and become more saline. The composition of the melt is determined by the liquidus curve. For example, at $-5°C$ ice coexists with brine that contains 80 psu NaCl, while at $-10°C$ the brine composition is 140 psu. At a temperature of $-10°C$, the system consists of the pure solid ice phase (point C) and brine (point D). The ratio of the amount of ice to the amount of liquid is DE/CE. If cooling is continued to $-21.2°C$ a third phase, the solid salt NaCl $\cdot 2H_2O$ (sodium chloride dihydrite), is observed to form at the *eutectic temperature* (point F). At the eutectic point there are three phases in equilibrium: brine, pure solid ice, and the pure solid salt NaCl $\cdot 2H_2O$. At constant pressure, the eutectic point is fixed and $f = 0$. The solid eutectic mass is a very fine-grained mixture of the two components, with all brine having disappeared.

When seawater freezes, a similar but more complex series of events takes place. Seawater is a system that has more than eight components and does not exhibit a eutectic point. As the solution cools, ice forms and the remaining brine becomes increasingly saline. As the cooling continues, different solid salts precipitate from the brine. However, they precipitate over a temperature range, as opposed to a fixed eutectic temperature. At a temperature of $-70°C$, there is still a measurable amount of brine in the ice, with at least five solid salts: $CaCO_3 \cdot 6H_2O$, $Na_2SO_4 \cdot 10H_2O$, KCl, $MgCl_2 \cdot 12H_2O$. The temperature of initial salt formation for each of the solids is given in Table 4.5.

As sea ice forms, small brine pockets are created within the ice. The salinity of a brine pocket is shown in Figure 4.9, which establishes a relationship between the freezing point of brine and its salt content. For an ideal solution, a linear relation between salt content and the freezing temperature results (4.58). In Figure 4.9, a linear relation is seen to about $-8°C$, the temperature at which $Na_2SO_4 \cdot 10H_2O$ precipitates. At temperatures less than $-8°C$ the relation is linear again to about $-23°C$, when NaCl $\cdot 2H_2O$ begins to precipitate.

Table 4.5 Properties of solid salts presumed to occur in sea ice. (Following Weeks and Ackley, 1986.)

Salt composition	Eutectic temperature of salt in water solution (°C)	Temperature of initial salt formation in seawater (°C)
$CaCO_3 \cdot 6H_2O$	—	-2.2
$Na_2SO_4 \cdot 10H_2O$	-3.6	-8.2
$MgCl_2 \cdot 8H_2O$	-33.6	-18.0
NaCl $\cdot 2H_2O$	-21.2	-22.9
KCl	-11.1	-36.8
$MgCl_2 \cdot 12H_2)$	-33.6	-43.2

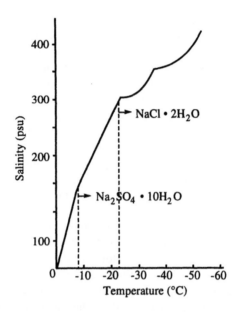

Figure 4.9 Freezing point of brine as a function of salinity. Breaks in the curve occur at –8°C and –23°C and again at very low temperatures due to precipitation of various salts, as indicated on the figure. (After Assur, 1958.)

Notes

Comprehensive treatises on water are given in *Water, A Comprehensive Treatise* (1982) by Franks; *The Structure and Properties of Water* (1969) by Eisenberg and Kauzmann; and *Ice Physics* (1974) by Hobbs.

An extensive treatment of the thermodynamics of solutions is given in *Thermodynamics* (1996) by Gokcen and Reddy.

A discussion of seawater as an ionic solution is given in *Chemical Oceanography* (1992) by Millero and Sohn. Brine in sea ice is discussed by Assur (1958).

Problems

1. Evaluate and compare the values of saturation vapor pressure determined at −20, 10, and 30°C in the following ways:
a) Use the integrated form of the Clausius–Clapeyron equation (4.23).
b) Integrate (4.21) by including the temperature variation of the latent heat of vaporization, following $L_{lv} = 2.501 \times 10^6 - 2400 \, (T - 273)$.
c) Evaluate the expression in (4.33).

2. At what temperature does water boil in Los Alamos, New Mexico, U.S., where the surface pressure is 786 hPa?

3. Using the integrated forms of the Clausius–Clapeyron equation, (4.23) and (4.26), derive an expression to determine the temperature at which the maximum difference occurs between the equilibrium vapor pressure over water and that over ice. Determine this temperature. Describe qualitatively what would happen in a closed, insulated system consisting of water vapor at the saturation vapor pressure with respect to liquid, and a population of equal numbers of cloud drops and ice crystals. Assume that initially the drops, ice crystals, and vapor are at −13°C. Is this an equilibrium situation? If not, what would happen to this system if it were allowed to reach equilibrium?

4. Calculate the changes in U, H, and η at 1 atm for 1 g of ice initially at −10°C, which is converted to steam (complete vaporization; 100% quality steam) at 100°C due to isobaric heating. Evaluate the error in ΔH made by ignoring the temperature dependence of L_{il}, L_{lv}, and c_p.

5. An air mass has a temperature of 30°C and a relative humidity of 50% at a pressure of 1000 mb. Determine the following (note: you may use Appendix D for values of e_s):
a) water vapor pressure;
b) mixing ratio;
c) specific humidity;
d) specific heat at constant pressure;
e) virtual temperature.

6. Derive an expression for the precipitable water between the surface and altitude z, assuming that the water vapor mixing ratio decreases exponentially with height with a scale height H_w and the density of moist air decreases exponentially with height with a scale height H:

$$w(z) = w_0 \exp(-z/H_w)$$
$$\rho(z) = \rho_0 \exp(-z/H)$$

7. Develop a relationship between the precipitable water through the entire vertical extent of the atmosphere and the sea surface temperature, T_0. Assume:
a) the vertical profile of specific humidity, q_v, has the following form:
$q_v = q_0(p/p_0)\mathcal{H}_0$, where $\mathcal{H}_0$ is the surface air relative humidity;
b) the saturated vapor pressure can be approximated by the following expression:
$e_s \sim b \exp [a (T_s - T_0)]$; and
c) the specific humidity can be approximated by the water vapor mixing ratio.

8. A crude estimate of the surface moisture flux from evaporation, $\dot{E}_0$, over a large, homogeneous body of water is given by

$$\dot{E}_0 = \rho_a\, C_q\, u_x\, (q_0 - q_a)$$

where ρ_a is density, $C_q = 1.3 \times 10^{-3}$ (dimensionless), and u_x is wind speed. Subscripts a and 0 represent values at 10 m above the surface and the surface value, respectively. Calculate $\dot{E}_0$ under the following conditions: $u_x = 5$ m s^{-1}, $p_a = 1000$ hPa, $q_a = 20$ g kg^{-1} and $T_0 = 30°C$ (assume that q_0 corresponds to saturation) and compare for the following situations:
a) a "fresh water" lake;
b) the ocean with $s = 35$ psu.

9. Calculate and compare the freezing temperature of seawater for $s = 30$ psu determined using the following:
a) theoretical value assuming an ideal, nonelectrolytic solution (4.57);
b) theoretical value assuming an electrolytic solution (4.58);
c) empirical relationship (4.59).

10. Your city has 10 km of streets, each 8 m wide. A snowfall equivalent to a sheet of ice 2 cm thick has fallen, and the temperature is $-5°C$. How much salt is required to melt all of the ice on the city streets?

Chapter 5 | Nucleation and Diffusional Growth

In Chapter 4, equilibrium between the phases of water was examined. In the case of condensation, it was implied that an excess of vapor pressure over the equilibrium value would cause condensation to occur, with a net migration of water molecules from the vapor to the liquid phase. In the example shown in Figure 4.5, a net migration of molecules occurred between two existing bulk phases.

In this chapter, we consider *nucleation,* a process whereby a stable element of a new phase first appears within the initial or "parent" phase. Phase transition does not occur under conditions of thermodynamic equilibrium, since a strong energy barrier must be surmounted for a phase to be nucleated if the new phase has higher atomic order than the parent phase. The energy barrier arises when a surface must be formed between the two phases.

Homogeneous nucleation refers to nucleation of a pure phase of one component. *Heterogeneous nucleation* refers to nucleation that occurs in the presence of a foreign substance, which can reduce the energy barrier to nucleation. Most of the nucleation processes in the atmosphere and the ocean occur through heterogeneous nucleation. The specific nucleation processes of interest here are the nucleation of water drops from water vapor, the nucleation of ice crystals from water drops or from vapor, and the nucleation of ice crystals in seawater to form the initial sea ice cover.

Once a phase has been nucleated, it can undergo diffusional growth if environmental conditions are favorable. Diffusion of water vapor to a cloud drop or ice crystal results in growth by condensation and deposition, respectively. Diffusional cooling of the initial sea ice cover causes further growth of the sea ice.

5.1 Surface Tension

In contrast to gases, which expand to fill the volume of their container, condensed phases can sustain a free boundary, or *surface*, and occupy a definite volume. The interface between a liquid and its vapor is not a surface in the mathematical sense, but rather a zone that is several molecules thick, over which the concentration of water molecules varies continuously.

A molecule at the surface of a liquid is free to move along the surface or into the interior of the liquid, in which case its place on the surface is taken by another molecule. However, a molecule cannot move freely from the interior of the liquid to the surface. The water molecules that form the free surface of a body of water are subjected to intermolecular attractive forces exerted by the neighboring liquid water molecules just beneath the surface. Therefore, the force field surrounding a molecule at the surface of a liquid is not symmetrical, and the molecule experiences a net force from the other molecules towards the interior of the liquid (Figure 5.1). If the surface area is increased, for example by changing the shape of the container, water molecules must be moved from the interior to the surface. To move a molecule from the interior to the surface of the liquid, work must be done against the intermolecular forces. The energy of a molecule at the surface of a liquid is thus higher than the energy in the interior. We can therefore consider the formation of a liquid surface as representing an increase in potential energy. For bulk water, such as the liquid in a drinking glass, the surface tension is a negligible part of the total potential energy, and the surface is flat rather than curved. The surface potential energy is a significant fraction of the total potential energy for small drops, since they have a large surface-to-volume ratio and curvature. A sphere has the smallest surface area for a given volume, and hence the smallest surface energy. The spherical shape of cloud drops is a consequence of the minimum-energy principle (Section 2.7).

We can extend the thermodynamic equations to include surface effects with the introduction of *surface tension work*, W_{st}. For the work required to extend a liquid surface against its vapor, we can write

$$dW_{st} = \sigma \, dA \tag{5.1}$$

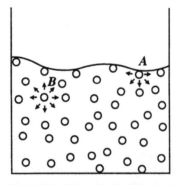

Figure 5.1 Water molecules at the surface of a liquid are subjected to a different attractive force field than those in the interior. A surface molecule, A, experiences a net attractive force towards the interior of the liquid. An interior molecule, B, experiences a symmetric force field exerted by its neighboring molecules and therefore does not freely move to the surface.

where dA is the change in surface area and σ is the *surface tension* between two phases. The surface tension is defined as the surface potential energy per unit surface area. The augmented form of the combined first and second laws for a thermodynamic system for which the surface energy is significant can be written in terms of the Gibbs function as[1]

$$dG = -\eta dT + V dp + \sigma dA \qquad (5.2)$$

5.2 Nucleation of Cloud Drops

A small fraction of collisions between water vapor molecules in the atmosphere are inelastic, leading to the formation of molecular aggregates (Section 4.1). The size of an aggregate can be increased by inelastic collisions between molecular aggregates, or by accretion of individual molecules. Most aggregates have a short lifetime, since they disintegrate under continual molecular bombardment. When an aggregate attains a size sufficient for survival, then nucleation of a water drop has occurred.

To nucleate an embryo drop, energy must be supplied to form the drop surface. This energy comes from the latent heat of condensation. A water drop can be nucleated when the incremental change of surface potential energy associated with condensing new water is less than the latent heat associated with condensation. Surface tension work is proportional to the surface area of the drop ($\propto r^2$) and latent heat release is proportional to the mass of the drop ($\propto r^3$). This implies the existence of a critical radius where surface tension work and latent heat release are in balance; this critical radius thus represents the threshold for nucleation.

Because of surface tension effects, the escaping tendency of the water molecules in a spherical drop is reduced, and equilibrium vapor pressure over a spherical drop can be substantially greater than values derived for bulk water without significant surface tension effects. The higher value of saturation vapor pressure over a curved surface relative to a bulk flat surface at the same temperature is a consequence of the work that must be performed on the system to increase the drop's surface area.

For a drop with surface area $A = 4\pi r^2$, the surface tension work is determined from (5.1) to be

$$dW_{st} = \sigma_{lv} 8\pi r\, dr \qquad (5.3)$$

where σ_{lv} is the surface tension between the liquid and vapor phases. Since the work

---

[1] Strictly speaking, since the surface tension is a change in work potential, it is defined in terms of the Helmholtz function. However, the interpretation of $\sigma = dG/dA$ is generally accepted but it cannot be demonstrated in a thermodynamically rigorous manner.

of expansion against a difference in pressure must equal the work done in changing the surface drop area,

$$\sigma_{lv} dA = \Delta p \, dV \tag{5.4}$$

where Δp is the pressure differential between the external ambient pressure and the internal pressure of the drop. For a spherical drop, we can write

$$\sigma_{lv} 8\pi r \, dr = \Delta p \, 4\pi r^2 \, dr \tag{5.5}$$

or

$$\Delta p = \frac{2\sigma_{lv}}{r} \tag{5.6}$$

Note that the smaller the drop, the larger the pressure differential. Under standard atmospheric conditions, the internal pressure of a drop with 1 μm radius is 1.5 atm. The existence of this excess internal pressure in a small drop is a fundamental consequence of the surface tension.

The surface tension for the vapor–liquid interface of water, σ_{lv}, is a function of temperature and is given by

$$\sigma_{lv} = 0.0761 - 1.55 \times 10^{-4} T \tag{5.7}$$

where σ_{lv} is in N m^{-1} and T is in °C. The expression in (5.7) is accurate for terrestrial temperatures; for higher temperatures (e.g., near the critical temperature), a higher-order polynomial expression is needed to obtain accurate values. Note that $\sigma_{lv} = 0$ at $T = 374$°C; this is defined as the critical temperature of water (Section 4.2).

To determine the conditions for nucleation of a water drop, we use (4.7) and (5.2) to write the Gibbs function

$$dG = -\eta dT + V dp + \sigma_{lv} dA + \mu_l \, dn_l + \mu_v \, dn_v \tag{5.8}$$

If we assume that nucleation occurs at constant temperature and pressure, and that $dn_l = -dn_v$, we have

$$dG = \sigma_{lv} 8\pi r \, dr + \left(\mu_l - \mu_v \right) dn_l \tag{5.9}$$

where we have incorporated (5.3).

Since dG is an exact differential, we can consider nucleation to occur in two stages: first, the bulk condensation of supersaturated water vapor onto a plane surface; and second, the formation of drops from the bulk water. For the first stage, we can write (5.9) as

$$dG = (\mu_l - \mu_v)\, dn_l$$

The term $(\mu_l - \mu_v)$ can be evaluated as follows. For an isothermal process involving one mole of water vapor, we can write

$$d\mu_v = dG = R^* T\, d(\ln e)$$

or

$$\mu_v = \mu_v^0 + R^* T \ln\frac{e}{e_0}$$

where μ_v^0 is a reference chemical potential that varies only with temperature and e_0 is the corresponding reference vapor pressure. At saturation, we can write

$$d\mu_{vs} = dG = R^* T\, d(\ln e_s)$$

and

$$\mu_{vs} = \mu_v^0 + R^* T \ln\frac{e_s}{e_0}$$

Since $\mu_l = \mu_{vs}$ when the two phases are in equilibrium over a plane surface, we can therefore write

$$\mu_l - \mu_v = R^* T \ln\left(\frac{e_s}{e}\right) \tag{5.10}$$

The number of moles of water vapor, dn_v, that condense onto spherical drops can be written

$$dn_v = \frac{1}{M_v}\, dm_v = \frac{\rho_l}{M_v}\, 4\pi r^2\, dr \tag{5.11}$$

Substitution of (5.10) and (5.11) into (5.8) yields

$$dG = \left(-R_v T \ln\frac{e}{e_s}\, \rho_l\, 4\pi r^2 + \sigma_{lv}\, 8\pi r\right) dr \tag{5.12}$$

Integration of (5.12) for the nucleation process gives

$$\Delta G = 4\pi r^2 \sigma_{lv} - \frac{4}{3}\pi r^3 \rho_l R_v T \ln S \qquad (5.13)$$

where $S = e_s(r)/e_s$ is the *saturation ratio* and e_s is the saturated vapor pressure over a plane surface (4.31), (4.33).

Equation (5.13) is illustrated in Figure 5.2 by plotting ΔG versus r for several different values of S at constant temperature. It is seen that each constant S curve has a maximum at radius, r^*, corresponding to the critical saturation ratio, S^*. For a drop to grow when $r < r^*$, ΔG must be added to the drop by increasing S. However, if $r > r^*$, as r increases then ΔG decreases and the drop grows spontaneously without increasing S. Since in the atmosphere S cannot usually continue to increase, cloud drops generally grow only if they have attained a size of r^*. If the drop can reach r^* (the peak of the ΔG curve), a slight addition of molecules allows the drop to grow spontaneously. When $r = r^*$, the incremental change of surface potential energy associated with condensing new water is equal to the latent heat associated with condensation. At this point, nucleation occurs. Note from Figure 5.2 that smaller values of S^* are associated with larger values of r^*. As the saturation ratio increases, the energy peak (or "barrier") is lower and the value of r^* is larger.

Values of the critical radius r^* can be found by differentiating (5.13) with respect to r and setting the derivative equal to zero:

$$\left[\frac{d(\Delta G)}{dr}\right]_{T,S} = \sigma_{lv}8\pi r^* - 4\pi r^{*2}\rho_l R_v T \ln S = 0$$

Solving for r^* yields

$$r^* = \frac{2\sigma_{lv}}{\rho_l R_v T \ln S} \qquad (5.14a)$$

We can write equivalently

$$\ln S = \frac{2\sigma_{lv}}{\rho_l R_v T r^*} \qquad (5.14b)$$

or

$$e_s(r) = e_s \exp\left(\frac{2\sigma_{lv}}{\rho_l R_v T r^*}\right) \qquad (5.14c)$$

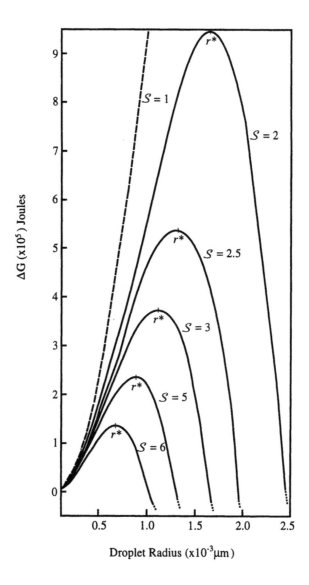

Figure 5.2 ΔG vs. r for several values of S at constant temperature. Each curve represents an energy barrier to embryo growth for a particular supersaturation. For radii less than the critical radius (i.e., $r < r^*$), growth occurs only by increasing S. However, if a drop can reach the critical radius, a slight addition of molecules will push the drop over the barrier into the region where $r > r^*$. In this region, the drop can grow spontaneously, since an increase in r is accompanied by a decrease in the Gibbs energy. Note that higher values of S correspond to lower Gibbs energy peaks, as expected. (From Byers, 1965.)

Equations (5.14) are different forms of *Kelvin's equation*. A plot of S versus r^* is shown in Figure 5.3, the curve representing the critical radius for nucleation as a function of the saturation ratio. If the saturation ratio in the environment remains constant, a drop with a radius above the curve will grow spontaneously since its equilibrium vapor pressure will remain lower than the vapor pressure of its environment. A drop with radius below the curve will evaporate. Figure 5.3 shows that values of saturation ratio of order $S = 3$ (corresponding to $\mathcal{H} = 300\%$) are required for the homogeneous nucleation of water drops. These high values of S imply a significant barrier to homogeneous nucleation of water drops in the atmosphere. Values of *supersaturation* $(S-1)$ are rarely observed to exceed 1% in the atmosphere. Homogeneous nucleation of water drops is not possible under these conditions, and we must look to heterogenous nucleation to understand nucleation of cloud drops in the atmosphere.

In the atmosphere, cloud drops form by heterogeneous nucleation. Some aerosol particles are *hygroscopic*; that is, they "attract" water vapor molecules to their surface through chemical processes or through physical forces such as those caused by the presence of permanent dipoles. A hygroscopic aerosol particle can *deliquesce* into a saturated salt solution at relative humidities signifcantly below 100%. For example, at $T = 25°C$, the deliquescent point of NaCl is about $\mathcal{H} = 75\%$ and that for $(NH_4)_2SO_4$ (ammonium sulfate) is $\mathcal{H} = 80\%$. *Cloud condensation nuclei* (CCN) are a subset of hygroscopic aerosol particles that nucleate water drops at supersaturations less than 1%. Soluble particles such as NaCl and $(NH_4)_2SO_4$ lower the equilibrium vapor pressure of a water solution relative to pure water (Section 4.5) and thus partially counteract the effects of surface tension. There are both natural and anthropogenic sources of CCN. Sulfate particles are produced anthropogenically by the burning of sulfur-containing fuels. Volcanic eruptions are also a source of sulfate particles.

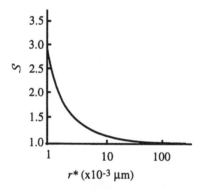

Figure 5.3 Equilibrium saturation ratio for pure water drops as a function of radius. Values are calculated from the Kelvin equation (5.14). The curve represents an unstable equilibrium. A drop above the equilibrium curve will grow, while a drop below the equilibrium curve will evaporate.

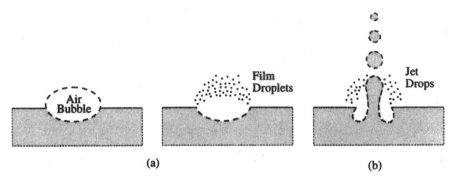

Figure 5.4 Production of sea-salt aerosols from (a) bursting air bubbles on the surface of the ocean and (b) mechanically breaking surface waves.

The ocean is a source of two types of aerosol important for the condensation of water in the atmosphere. Air bubbles at the surface of the ocean burst, ejecting small drops into the atmosphere (Figure 5.4). After evaporation these drops leave behind sea-salt particles with diameters smaller than about 0.3 μm. When air bubbles in breaking waves burst, larger drops are ejected in jets; upon evaporation of the drops, giant sea-salt particles (>2 μm) may be left in the atmosphere. It has been estimated that the rate of production of sea-salt aerosol particles over the oceans by this mechanism is 100 cm^{-2} s^{-1}.

An additional source mechanism of aerosols from the oceans is associated with organosulfides produced by micro-organisms in the ocean. In particular, the compound dimethylsulfide (DMS; $(CH_3)_2S$) is produced by marine phytoplankton in the upper layers of the ocean and represents the major flux of reduced sulfur to the marine atmosphere. Since DMS is rather volatile and insoluble, it passes rapidly from the seawater into the atmosphere. Once in the atmosphere, DMS is oxidized and forms sulfate particles.

The number of CCN per unit volume of air that have critical supersaturation values less than $(S-1)$ is approximated by

$$N_{CCN} = c_1 (S-1)^k \tag{5.15}$$

where c_1 and k are parameters that depend on the particular air mass. Maritime conditions have values that are typically $c_1 = 50$ cm^{-3} and $k = 0.4$, whereas continental conditions have typical values around $c_1 = 4000$ cm^{-3} and $k = 0.9$ (for $S-1$ in %). Urban regions have exceptionally large numbers of CCN. The fraction of the total atmospheric aerosol population that serve as CCN in clouds is about 1% of the total aerosol population in continental regions and up to 20% of the total aerosol concentration over the ocean.

Recall Raoult's law from Section 4.5 for an electrolytic solution (4.48):

$$\frac{e_s(n_{solt})}{e_s} = 1 - \frac{i n_{solt}}{n_{H_2O}}$$

Since $n = m/M$, we can write for a solution drop

$$\frac{e_s(n_{solt})}{e_s} = 1 - \frac{3 i m_{solt} M_v}{4 \pi M_{solt} \rho_l r^3} = 1 - \frac{b}{r^3} \tag{5.16a}$$

where

$$b = 3 i M_v \frac{m_{solt}}{4 \pi M_{solt} \rho_l} \tag{5.16b}$$

For a given mass of solute, the vapor pressure required for equilibrium decreases as the cube of the drop radius. As r increases through condensation, the mole fraction of the solute decreases. Thus the depression of the equilibrium vapor pressure for a given mass of solute decreases as r increases.

Combination of the curvature (5.14c) and the surface tension (5.16) effects on saturation vapor pressure gives the ratio of the saturation vapor pressure of a solution drop to the saturation vapor pressure of pure water over a flat surface:

$$\frac{e_s(r, m_{solt})}{e_s} = \left(1 - \frac{b}{r^3}\right) \exp(a/r) \tag{5.17}$$

where $a = 2\sigma_{lv}/(\rho_l R_v T)$. If r is not too small, (5.17) can be written as

$$\frac{e_s(r, n_{solt})}{e_s} = 1 + \frac{a}{r} - \frac{b}{r^3} \tag{5.18}$$

For given values of T, M_{solt}, and m_{solt}, (5.18) describes the dependence of saturation ratio on the size of the solution drop.

Equilibrium curves for drops containing a given nucleus mass (referred to as *Kohler curves*) are shown in Figure 5.5. Depending on whether the solute or curvature effect dominates, the saturation ratio may be greater or less than unity. The peaks in the Kohler curves correspond to critical values of the supersaturation and radius, S^* and r^*. For $r < r^*$, the drops grow only in response to an increase in relative humidity, and are termed *haze particles*. A condensation nucleus is said to be *activated* when the

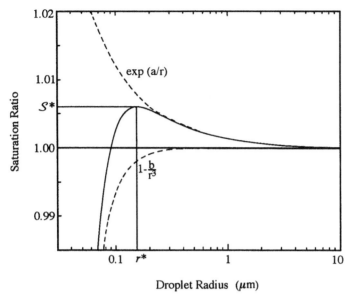

Figure 5.5 Equilibrium saturation ratio for a solution drop as a function of drop radius. The solution effect $(1-b/r^3)$ dominates when the radius is small. For values of $r < r^*$, drop growth occurs only in response to an increase in the relative humidity. If the relative humidity at the critical radius slightly exceeds the critical saturation ratio, then the drop can grow spontaneously, and will continue to grow as long as the ambient saturation ratio remains higher than the equilibrium saturation ratio of the drop.

drop formed on it grows to size r^*. Once the drop grows only slightly beyond r^*, its equilibrium value of S is less than S^*, and the drop grows spontaneously without requiring further increase in S. For typical sizes of condensation nuclei found in the atmosphere, the critical supersaturation for heterogeneous nucleation on a soluble particle is less than 1% (Table 5.1), in agreement with observations.

Table 5.1 Critical values of radius and supersaturation for typical condensation nuclei in the atmosphere (values assume that the nuclei are NaCl and that the temperature is 273 K).

m_{nuclei} (g)	r_{nuclei} (μm)	r^* (μm)	S^*-1 (%)
10^{-16}	0.0223	0.19	0.42
10^{-15}	0.0479	0.61	0.13
10^{-14}	0.103	1.9	0.042
10^{-13}	0.223	6.1	0.013
10^{-12}	0.479	19.0	0.0042

5.3 Nucleation of the Ice Phase

Nucleation of ice in the atmosphere may occur from supersaturated vapor (deposition) or from supercooled water (freezing). While the theory of liquid water nucleation from the vapor phase is fairly well established, there is considerable controversy over the mechanisms of ice nucleation in the atmosphere.

Pure bulk water is not observed to exist in the liquid phase at temperatures significantly below 0°C. However, when water is divided into small drops, its freezing temperature is observed to become much lower. The smaller the drop, the lower the statistical probability that it will freeze at a given temperature since there are fewer water molecules from which to form a stable, ice-like structure. Analogous to (5.14a), an expression for critical radius for homogenous ice nucleation from a liquid water drop can be derived:

$$r^* = \frac{2\sigma_{il}}{\rho_i R_v T \ln\left(\dfrac{e_s}{e_{si}}\right)} \tag{5.19}$$

where σ_{il} is the surface tension at the ice-liquid interface. Using the Clausius–Clapeyron equation (4.31), it can be shown that (following McDonald, 1964)

$$r^* = \frac{2\sigma_{il} T_{tr}}{\rho_i \overline{L_{il}} \left(T_{tr} - T\right)} \tag{5.20}$$

where the subscript i refers to ice, $T_{tr} = 273.15$ K is the nominal freezing temperature of bulk water, and $\overline{L_{il}}$ represents the average latent heat of fusion over the temperature range T to T_{tr}. The numerical value of σ_{il} is known approximately to be about 0.002 N m^{-1}. Table 5.2 shows typical homogeneous freezing temperatures of water drops. Small drops of pure water freeze at a temperature of about −40°C. Equation (5.20) agrees with observations to within 2°C.

The homogeneous nucleation of ice from a pure supercooled water drop in the atmosphere is not possible, since cloud drops contain at least one CCN. In accordance with (4.65), the freezing temperature of a solution is depressed relative to pure water. However, as a drop grows, the solution becomes increasingly dilute. In the upper troposphere, temperatures reach −40°C and colder. Therefore, homogenous freezing nucleation of dilute solution drops is believed to be the primary nucleation mechanism for cirrus clouds.

Most clouds contain ice particles by the time the temperature has reached −20°C, indicating the importance of heterogeneous ice nucleation. The class of aerosols that act as condensation nuclei and those that act as ice nuclei are almost mutually exclusive. CCN particles are soluble in water, promoting nucleation by lowering the saturation vapor pressure. In contrast, ice nuclei promote nucleation by providing a

Table 5.2 Typical homogeneous freezing temperatures of water drops. (Data from Young, 1993.)

Drop diameter (μm)	T_f (°C)
1	−42.3
10	−38.2
100	−34.8
1,000	−32.2
10,000	−30.0

substrate upon which the ice lattice can form. For an aerosol to be an effective ice nucleus, some combination of lattice matching, molecular binding, and low interfacial energy with ice is needed. Soil particles, particularly clay minerals, are effective ice nuclei since their lattice structure is similar to that of ice. Anthropogenic sources of ice nuclei include by-products of combustion and smelting, such as metallic oxides. Biogenic ice nuclei consist of bacterial cells, and sources from leaves and the ocean have been identified. Biogenic ice nuclei possess hydrogen bonding capability, with positions of the hydrogen bonding sites matching those found in ice.

Four different types of heterogeneous ice nucleation have been hypothesized: deposition nucleation, immersion freezing, contact freezing, and condensation freezing. *Deposition nucleation* occurs when a small amount of water is adsorbed on the surface of a nucleus and freezes, then additional water vapor is deposited. *Immersion freezing* occurs when an ice nucleus is present within the drop. As the drop cools, the likelihood of nucleating an ice crystal increases. The amount of supercooling required for immersion nucleation decreases for larger sizes of nuclei. *Contact freezing* occurs when an ice nucleus makes external contact with a supercooled drop and very quickly initiates freezing. A necessary condition for contact freezing is that the aerosol particle must make contact with the drop. This contact can occur via bombardment of the aerosol particle by air molecules or by aerosol transport in a temperature or vapor gradient. *Condensation freezing* occurs when a transient water drop forms before the freezing occurs, and then freezing occurs via contact or immersion nucleation.

Table 5.3 summarizes the ice nucleation thresholds for various substances. It is seen that a particular aerosol particle may nucleate ice in different ways, depending on the history of the aerosol interaction with the cloud and the ambient temperature and humidity conditions. Given the complexity of the processes and the difficulty in making measurements, the dominant modes of heterogeneous ice nucleation in the atmosphere, and even the possible modes, remain controversial. Note the small supercooling required for ice nucleation by AgI (silver iodide) in Table 5.3. Because the crystal lattice of AgI very nearly duplicates that of ice, AgI has been used as a

Table 5.3 Comparison of the threshold temperatures for ice nucleation of various substances for various nucleation modes. Thresholds represent the highest temperatures at which ice nucleation has been observed to occur. (Following Young, 1994.)

Substance	Contact freezing	Condensation freezing	Deposition nucleation	Immersion
Silver iodide	−3	−4	−8	−13
Cupric sulfide	−6	n/a	−13	−16
Lead iodide	−6	−7	−15	n/a
Cadmium iodide	−12	n/a	−21	n/a
Metaldehyde	−3	−2	−10	n/a
1,5-Dihydroxy-naphlene	−6	−6	−12	n/a
Phloroglucinol	n/a	−5	−9	n/a
Kaolinite	−5	−10	−19	−32

cloud-seeding agent, whereby it is injected into supercooled clouds to modify the cloud microphysical processes. Cloud seeding has been attempted for precipitation enhancement and suppression, hail and lightning suppression, and the dispersal of fog. The effectiveness of cloud seeding, however, remains controversial.

Observations in some clouds show substantially larger numbers of ice crystals than expected, relative to the number of ice nuclei present. Freezing drops with radius greater than 12 μm may splinter into several ice crystals, and existing ice crystals may fracture through collisions with other ice crystals. Secondary ice crystal production is seen in clouds with large drops and large updrafts.

5.4 Diffusional Growth of Cloud Drops

Cloud drops grow by diffusion of water vapor to the drop. Water vapor is transferred to the drop by molecular diffusion as long as the vapor pressure surrounding the drop exceeds the saturation vapor pressure of the drop. As water condenses on the drop, latent heat is released, which warms the drop and reduces its growth rate. As a result of the latent heat release, the drop becomes warmer than the environment, and heat is diffused away from the drop. Condensation can thus be considered as a double diffusive process, with water vapor diffused towards the drop and heat diffused away from the drop. Evaporation of a drop occurs in reverse, as water diffuses away from the drop and heat diffuses toward the drop.

First we derive an equation for the diffusional growth of a single drop that exists in a vapor field of infinite extent. The diffusion equation for water vapor (3.42) is

$$\frac{\partial \rho_v}{\partial t} = D_v \frac{\partial^2 \rho_v}{\partial x_i^2}$$

where we assume here that D_v is not a function of x_i. For stationary (steady-state) conditions, the mass flux of water vapor, dm/dt, on a sphere of radius r is equal to the flux of vapor across the drop surface:

$$\frac{dm}{dt} = 4\pi r^2 D_v \frac{d\rho_v}{dr} \tag{5.21}$$

Integrating (5.21) from the surface of the drop to infinity (a distance from the drop sufficiently far so that the vapor pressure is unaffected by diffusion to the drop) and assuming that the growth rate, dm/dt, remains constant, we obtain

$$\frac{dm}{dt} \int_r^\infty \frac{dr}{r^2} = 4\pi D_v \int_{\rho_v(r)}^{\rho_v(\infty)} d\rho_v \tag{5.22}$$

and therefore

$$\frac{dm}{dt} = 4\pi r D_v \left[\rho_v(\infty) - \rho_v(r) \right] \tag{5.23}$$

Latent heat liberated by condensation at the drop surface is diffused away from the drop according to (3.41), and we can write analogously to (5.23)

$$\frac{dQ}{dt} = -L_{lv} \frac{dm}{dt} = 4\pi r \kappa \left[T(r) - T(\infty) \right] \tag{5.24}$$

Writing the term dm/dt in terms of a change in radius, we obtain

$$\frac{dm}{dt} = \rho_l \frac{dV}{dt} = \rho_l 4\pi r^2 \frac{dr}{dt} \tag{5.25}$$

Combination of (5.23) and (5.24) determines the drop growth rate as influenced by both diffusion of water vapor and heat. An approximate expression for the growth rate of a drop by diffusion has been determined from (5.24)–(5.26) to be (Mason, 1971):

$$r\frac{dr}{dt} = \frac{S-1}{\left(\dfrac{L_{lv}^2\rho_l}{\kappa R_v T^2} + \dfrac{\rho_l R_v T}{e_s(T)D_v}\right)} = \frac{S-1}{\mathcal{K}+\mathcal{D}} \tag{5.26}$$

If $S < 1$, then (5.26) describes the evaporation of a cloud drop. The solution (5.26) depends only on the ambient environmental conditions (S, T, p) and does not require determination of the drop temperature. Note that curvature and solute effects have been ignored in this derivation; their effects are small once the drop size has increased beyond a few microns.

The term $\mathcal{K}$ in (5.26) represents the thermodynamic term associated with heat conduction, and $\mathcal{D}$ is associated with the diffusion of water vapor. The coefficients of thermal conductivity and water vapor diffusivity vary with temperature, and selected values are given in Table 5.4.

Since $\mathcal{K}$ and $\mathcal{D}$ depend on the ambient temperature and pressure, (5.26) cannot be integrated analytically. If we assume that ambient conditions in the atmosphere remain constant (i.e., S, $\mathcal{K}$ and $\mathcal{D}$ are constant), we can integrate (5.26) as

$$r(t) = \left[r_o^2 + \frac{2(S-1)}{\mathcal{K}+\mathcal{D}}(t-t_o)\right]^{1/2} \tag{5.27}$$

Because of the square–root dependence, (5.27) is often called the *parabolic growth law*. Table 5.5 shows the growth rate of drops of different sizes, as determined from (5.27).

Table 5.4 Coefficients of atmospheric thermal conductivity and water vapor diffusivity at a pressure of 1000 mb (from Houghton, 1985). Since D_v varies with pressure, a value of D_v for an arbitrary pressure p (hPa) can be obtained by multiplying the tabulated value by $(1000/p)$.

T (°C)	κ (J m^{-1} s^{-1} K^{-1}) × 10^{-2}	D_v (m^2 s^{-1}) × 10^{-5}
−40	2.07	1.62
−30	2.16	1.76
−20	2.24	1.91
−10	2.32	2.06
0	2.40	2.21
10	2.48	2.36
20	2.55	2.52
30	2.63	2.69

Table 5.5 Drop growth rate calculated from (5.26), with $r_0 = 0.75$ μm. The drops are growing on nuclei of NaCl at $(S - 1) = 0.05\%$, $p = 900$ mb, $T = 273$ K. (After Mason, 1971.)

Mass (g)	10^{-14}	10^{-13}	10^{-12}
Radius (μm)	Time (seconds) to grow from initial radius, r_0		
1	2.4	0.15	0.013
2	130	7.0	0.61
4	1,000	320	62
10	2,700	1,800	870
20	8,500	7,400	5,900
30	17,500	16,000	14,500
50	44,500	43,500	41,500

Smaller drops have a faster growth rate (dr/dt) than larger drops. However, large drops have a greater rate of mass buildup (dm/dt). It is clear from Table 5.5 that diffusional growth of drops is not sufficient to produce a rain drop even over a period of a half day. Additional mechanisms are required to explain the observed rapid formation of rain drops (see Section 8.2).

In natural clouds, there is not an infinite source of water vapor, and drops compete for water vapor and otherwise influence each other's growth. The rate of change of supersaturation is therefore determined as a balance between the production of supersaturation (by cooling, for example) and condensation (which decreases the ambient supersaturation). Assuming that supersaturation is produced initially by adiabatic cooling in an updraft u_z, the rate of change of the saturation ratio is given by

$$\frac{dS}{dt} = a_1 u_z - a_2 \frac{dw_l}{dt} \tag{5.28}$$

where w_l is the liquid water mixing ratio (mass of liquid water per mass of dry air), and dw_l/dt is the rate of condensation. The term $a_1 u_z$ is thus a "source" term, representing the increase of the saturation ratio due to cooling in adiabatic ascent, and $a_2(dw_l/dt)$ is the "sink" term, representing the decrease in supersaturation due to the diffusion of water vapor to the growing drops.

The term a_1 can be derived in the following way. Assuming ascent without condensation, (5.28) becomes

$$\frac{dS}{dt} = a_1 u_z \tag{5.29}$$

Using the definition $S = e/e_s$, we can write

$$\frac{dS}{dt} = \frac{\left(e_s \frac{de}{dt} - e \frac{de_s}{dt}\right)}{e_s^2} \tag{5.30}$$

Using Dalton's law of partial pressures (1.13), we can apply the hydrostatic equation (1.33) to water vapor

$$\frac{de}{dz} = -g\rho_v \tag{5.31}$$

Using the chain rule, we can write (5.31) as

$$\frac{de}{dt}\frac{dt}{dz} = -g\rho_v \tag{5.32}$$

or equivalently

$$\frac{de}{dt} = -\frac{eg}{R_v T}u_z \tag{5.33}$$

where we have incorporated the ideal gas law. The Clausius–Clapeyron equation (4.19) can be expanded using the chain rule as

$$\frac{de_s}{dt}\frac{dt}{dz}\frac{dz}{dT} = \frac{L_{lv}e_s}{R_v T^2} \tag{5.34}$$

or equivalently

$$\frac{de_s}{dt} = -\frac{L_{lv}e_s}{R_v T^2}\frac{g}{c_p}u_z \tag{5.35}$$

where $dT/dz = -g/c_p$ since no condensation has occurred. Incorporating (5.33) and (5.35) into (5.30) yields the coefficient a_1. An analogous procedure is used to derive a_2. Values of a_1 and a_2 thus derived are

$$\alpha_1 = \frac{1}{T}\left(\frac{L_{lv}g}{R_v c_p T} - \frac{g}{R_d}\right) \tag{5.37}$$

$$\alpha_2 = \rho_a\left(\frac{R_v T}{\varepsilon e_s(T)} + \frac{\varepsilon L_{lv}^2}{pTc_p}\right) \tag{5.38}$$

If the production of saturation ratio occurs via isobaric cooling rather than by adiabatic cooling, then

$$\frac{dS}{dt} = \alpha_3 \frac{dT}{dt} - \alpha_4 \frac{dw_l}{dt} \tag{5.39}$$

where dT/dt is the isobaric cooling rate resulting from radiative cooling or other isobaric processes. The terms α_3 and α_4 can be shown to be

$$\alpha_3 = -\frac{L_{lv}}{R_v T^2} \tag{5.40a}$$

$$\alpha_4 = \frac{p}{\varepsilon e_s} \tag{5.40b}$$

By using either (5.29) or (5.39) with (5.26), and providing a distribution of CCN and an updraft velocity or an isobaric cooling rate, the evolution of a spectrum of drops can be calculated. The results of such a calculation are shown in Figure 5.6. From an initial spectrum of CCN, drops are nucleated and grow in a steady updraft. The supersaturation increases from zero at the cloud base to a maximum value of 0.5% at about 10 m above the cloud base. Above this height, $(S-1)$ decreases as condensation depletes the water vapor concentration. The peak supersaturation is not sufficient to nucleate the two smallest sizes of nuclei; they remain as haze drops, growing only when the supersaturation is increasing. The larger nuclei become activated as cloud drops, which undergo rapid growth when the supersaturation is a maximum.

The *drop size spectrum, n(r)*, is defined as the number of drops per unit volume of air with radii in the interval $r + dr$. The total number concentration of the drops, N, is

$$N = \int_0^\infty n(r)\,dr \tag{5.41}$$

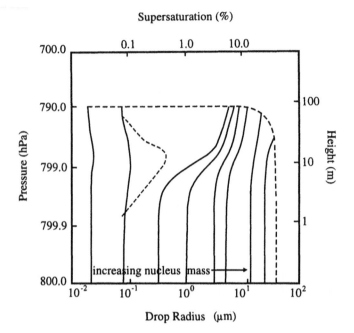

Figure 5.6 Evolution of a cloud drop spectrum from an assumed updraft velocity and initial distribution of CCN. Solid lines show the sizes of drops growing on nuclei of different masses. The dashed line shows how the supersaturation varies with height. The smallest drops grow slightly during the increase in supersaturation, but then evaporate again when the supersaturation decreases. Larger drops become activated and grow rapidly during the increase in supersaturation. (From Rogers and Yau, 1989.)

Variations in drop sizes associated with a spectrum of aerosol sizes give rise to a spectrum of drop sizes. As the activated drops grow, their spread in size becomes smaller, in accord with the parabolic growth law. Figure 5.7a shows the narrowing of the drop size spectrum with time (or with height above cloud base for drops growing in an updraft) according to the simple diffusional growth theory. Figure 5.7b shows observed drop size spectra from a shallow non-precipitating cloud. In general, observed drop size spectra are significantly broader than modeled spectra. Also, the observed drop size spectrum broadens with height above cloud base, while the simple diffusional growth model indicates a narrowing of the drop size spectrum. The discrepancy between the modeled and observed drop size spectra is of concern, because the drop size spectra is very important in formation of precipitation and in the interaction of clouds with radiation (see Sections 8.2 and 8.3).

Explaining the observed broadening of drop size spectra remains a major challenge to cloud physicists. Numerous theories have been proposed to explain the spectral broadening. For example, "giant" particles within a distribution of CCN may act as

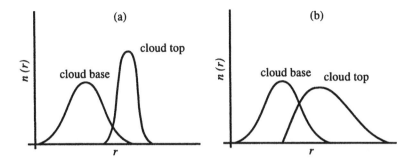

Figure 5.7 (a) Modeled cloud drop size spectra according to the simple diffusional growth model. (b) Observed spectra for a shallow, non-precipitating cloud. At cloud top, the observed spectrum is much broader than the modeled spectrum.

embryos for large drops. Another theory is that small-scale turbulence and the associated fluctuations in supersaturation may cause drop spectral broadening in some circumstances. Variations in the spatial distribution of drops may give rise to local supersaturation fluctuations and thus contribute to broadening of the drop size spectra. Entrainment of dry air into the cloud has also been hypothesized to contribute to drop spectral broadening. Because of the difficulty in making measurements at these small scales, the relative importance of each of these processes in broadening the drop spectrum by condensational growth remains uncertain.

5.5 Ice Crystal Morphology and Growth

Ice crystal growth by diffusion is regulated to a large extent by the surface properties of the ice crystal lattice. Unlike diffusional growth of a liquid water drop, water vapor molecules cannot be incorporated into the crystal lattice at their arrival position. A water vapor molecule can only be incorporated into steps or corners of the lattice, and must migrate across the crystal surface until it either reaches such a site or returns to the vapor. The necessity of incorporating incoming water molecules into the growing ice lattice reduces the growth rate of the ice crystal.

The temperature and saturation ratio are the primary influences on the *habit* assumed by a growing ice crystal, determining whether a crystal grows preferentially along the basal (*c*-axis) or the prism (*a*-axis) faces. Figure 5.8 shows various ice crystal habits. Needles occur singly or in bundles, and columnar forms include columns, pyramids, and bullets. Plates include simple hexagons, sectored hexagons, stars, and highly branched dendrites. Figure 5.9 shows the dependence of ice crystal habit on temperature and vapor density excess. Needles and columnar ice crystals represent growth primarily on the *c*-axis, while plates represent growth that has

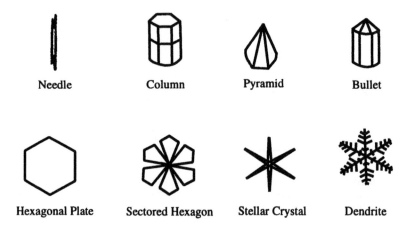

Needle   Column   Pyramid   Bullet

Hexagonal Plate   Sectored Hexagon   Stellar Crystal   Dendrite

Figure 5.8 Examples of various ice crystal habits.

occurred primarily on the a-axis. Combination crystals such as columns with plates at the ends can arise from exposure to successively different ambient temperatures and humidities.

The diffusional growth of ice crystals in a water vapor field (and conversely, ice crystal sublimation) is treated in essentially the same way as the growth and evaporation of liquid water drops. In a manner analogous to that used to derive (5.26), the following equation can be derived for the growth rate of an ice crystal:

$$\frac{dm}{dt} = \frac{4\pi C(S_i - 1)}{\left(\dfrac{L_{iv}^2}{\kappa R_v T^2} + \dfrac{R_v T}{e_{si}(T)D} \right)} \qquad (5.42)$$

In (5.41) the saturation ratio with respect to ice, $S_i = e/e_{si}$, is used in place of S, and the latent heat of sublimation replaces the latent heat of vaporization. Another difference between (5.41) and (5.26) is the appearance in (5.41) of the factor C instead of r. Since ice crystals are nonspherical, a radius cannot be assigned to them. The diffusion of vapor to an ice crystal can be addressed in a manner derived from an analogous situation in electricity, where the capacitance, C, is used in dealing with irregularly shaped objects. For a sphere, $C = r$. For a disk, which can be used to approximate a plate, $C = 2r/\pi$. Because of kinetic effects and the nature of growth of the ice crystal lattice, the growth rate of small ice crystals in some situations may be only half the rate predicted by (5.42).

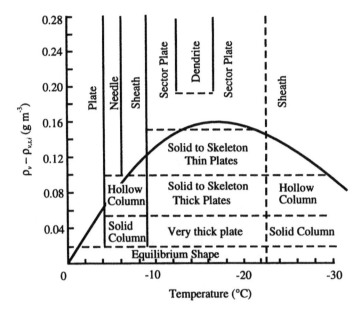

Figure 5.9 Dependence of ice crystal habit on environmental conditions. Also shown is the excess vapor density over ice equilibrium in a saturated atmospere (thick curve). Note that the excess vapor density is a maximum at around −17°C, which corresponds to the temperature at which ice crystal growth is a maximum. (From Pruppacher and Klett, 1978.)

When ice crystals first nucleate in a cloud, they are typically found in the presence of water drops, with the ambient vapor pressure approximately equal to saturation vapor pressure over liquid water. As was shown in Figure 4.4 and Table 4.4, this results in a supersaturation with respect to ice. A water-saturated cloud has a high supersaturation with respect to ice, and hence provides a very favorable environment for diffusional growth of ice crystals.

5.6 Formation of the Initial Sea Ice Cover

Nucleation of the initial ice particles in seawater in response to surface cooling occurs via heterogeneous nucleation. The sources of ice-forming nuclei are solid impurities (both organic and inorganic) that occur in seawater and also snow particles that fall onto the ocean surface. Supercooling required to form the initial ice cover probably does not exceed a few tenths of a degree Celsius.

Central to the formation of a surface ice cover on a body of fresh water is the fact that the density of the ice is less than the density of the liquid. The density of pure liquid water has a maximum value of 1000 kg m^{-3} at $T_\rho = 3.98°C$, while the density of pure ice is 916.4 kg m^{-3} at the freezing point (0°C). The freezing of bulk fresh water (e.g., a lake) in response to surface cooling results in a layer of ice on the surface that is less dense than the water below and thus "floats."

The addition of salt causes the freezing temperature of water, T_f, to decrease according to (4.59) and the temperature of maximum density, T_ρ, to decrease approximately according to (1.30). Values of T_f and T_ρ are plotted as a function of salinity in Figure 5.10. If $s = 24.695$ psu, then $T_f = T_\rho$. If $s > 24.695$ psu, which is usually the case for seawater, then $T_f > T_\rho$. Therefore, surface cooling causes a density increase at the surface, which results in vertical mixing that continues until the water reaches T_f. Water at the surface is denser than the water below and thus will sink. A layer several meters thick must cool to the freezing point in order to support the initial sea ice formation.

Observations show that the first crystals that form near the surface are minute spheres of pure ice. With growth, these spheres evolve into thin disks. Ice crystals growing in seawater must dissipate both heat and solute into the surrounding liquid, whereas freezing cloud drops need only to diffuse heat. As the radius of curvature of the disk increases and the surface area to volume ratio decreases, the ability of the disk to dissipate heat and solute decreases. At a critical radius the disk becomes

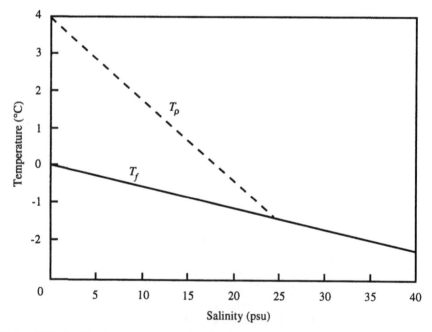

Figure 5.10 The freezing point temperature, T_f, and the temperature of maximum density, T_ρ, as a function of salinity.

unstable and dendritic growth begins, which substantially increases the ratio of the surface area to volume and thus the ability for the growing crystal to dissipate heat and solute. Because of this increased ability to diffuse heat, the disk-to-dendrite transition is also accompanied by an increase in crystal growth rate, resulting in the formation of stellar dendrites. Under calm conditions, the dendritic crystals will grow rapidly until they overlap and form a continuous ice skim, with the basal plane of the crystals floating in the plane of the water surface, giving rise to a generally vertical orientation for the c-axis. The vertical orientation of the c-axis lends itself to the more rapid removal of latent heat.

Wind-induced turbulence in the upper ocean influences the formation of the initial ice cover. Turbulent mixing modifies the freezing process relative to calm conditions by introducing more freezing nuclei to the surface area and stirring the ice crystals over a depth of several meters in the upper ocean. Thus more crystals form per unit volume and abrasion of ice crystals increases. This results in the breaking off of arms of the dendrites, causing secondary ice crystal formation. The small isolated ice crystals thus formed are termed *frazil ice*, with crystal sizes generally less than 2 to 4 mm in diameter.

Once the ice fraction of frazil exceeds 0.3 to 0.4, sufficient bonding has occurred between individual crystals so that their mobility is reduced and transition to a solid cover begins. This transition is usually marked by the formation of *pancake ice*, consisting of rounded masses of semi-consolidated slush 0.3 to 3.0 m in diameter (Figure 5.11). Pancake ice forms as frazil crystals are brought into contact and form

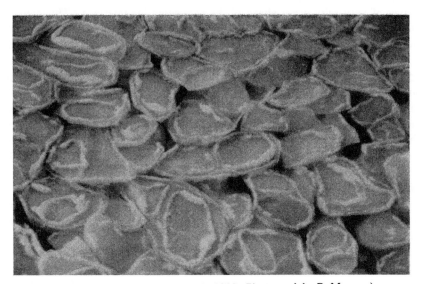

Figure 5.11 Pancake ice. (From Lock, 1990. Photograph by R. Masson.)

bonds. Oscillatory motions and repeated bumping between pancake elements results in the characteristically round shapes to the pancakes and elevated rims. The overall orientation of crystals in pancakes is essentially random. With continued cooling, the pancakes are welded together by the freezing of grease ice. Frazil crystals may be advected downwind where accumulations of up to 1 m in thickness can form at obstructions, such as the edge of an ice floe.

5.7 Formation of Sea Ice Transition and Columnar Zones

Once a continuous cover of ice has formed across the sea ice surface, ice crystal growth by freezing becomes more ordered. Growth parallel to the c-axis is much slower than growth perpendicular to the c-axis because the density of potential bonding sites is greater on a prism face than on a basal plane. Thus crystals whose c-axes are parallel to the sea surface grow downward in the water more quickly than those crystals with a more vertical orientation. The region over which the crystal orientation shifts from being random to having all c-axes horizontal is the *transition zone*, which typically has a depth of 5 to 10 cm.

Sea ice crystals with the favored orientation grow more quickly and dominate the crystal structure below the transition layer, called the *columnar zone*. In the columnar zone, all c-axes are oriented within a few degrees of the horizontal plane, and grow in the direction of the temperature gradient, which is essentially in the vertical direction. Crystal diameter in the upper half–meter of the columnar zone is typically less than 2 cm. Ice in the columnar zone is referred to as *congelation ice*.

When seawater freezes, the salts rejected from the ice lattice result in a thin boundary layer of very salty water ahead of the advancing ice interface. This causes a gradient in salinity, where the salinity of the layer nearest to the ice is greater than the salinity in the underlying water. Therefore, the freezing temperature, T_f, at the ice–water interface will be lower than that in the underlying water. This produces a downward flux of salt and an upward flux of heat through the boundary layer. The thermal diffusivity, $\kappa/\rho c_p$, of seawater is two orders of magnitude larger than the diffusivity of salt in seawater, D_s. Since heat diffuses more rapidly than salt, the flux of heat through the interfacial boundary layer is much more rapid than the flux of salt, and the slope of the actual temperature profile may be less than the slope of the freezing temperature (controlled by salinity) in the layer near the interface (Figure 5.12). This is called *constitutional supercooling*.

Under conditions of supercooling, the ice–water interface initially develops an array of parallel knife-edged cells which deepen until the cells separate into a line of individual platelets (Figure 5.13). The length and spacing of these platelets act to minimize the amount of supercooling ahead of the interface.

Concentrated brine accumulates in the grooves that develop between the rows of cells and is eventually trapped when adjacent cells develop lateral connections (Figure 5.14). The *brine pockets* are typically long and narrow, on the order of 0.05 mm

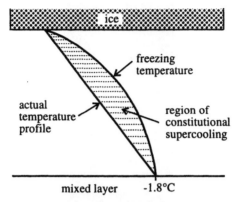

Figure 5.12 As seawater freezes, the layer near the ice–water interface becomes increasingly salty, and the freezing temperature of the layer decreases relative to the underlying water layer. The salt and temperature gradients cause a downward flux of salt and an upward flux of heat. Since the heat diffuses upward more rapidly than the salt diffuses downward, the slope of the actual temperature profile in the layer may be less than T_f. In this region where the temperature is lower than T_f, the water is said to be "constitutionally supercooled." (After Maykut, 1985.)

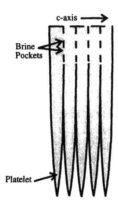

Figure 5.13 Platelets formed during supercooling of the layer near the ice–water interface. (After Maykut, 1985.)

in diameter. The horizontal spacing of brine pockets has been observed to range from 0.5 mm to 1 mm. The spacing of the brine pockets corresponds to the spacing of the grooves between the knife-edged cells (Figure 5.13). Therefore, as seawater freezes, the salt is not completely rejected into the water below, but some salt becomes entrapped in the ice in the form of brine pockets.

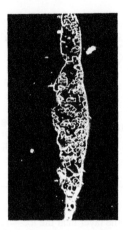

Figure 5.14 As seawater freezes, some of the salt becomes trapped in the ice to form brine pockets. The figure shows the characteristic long, narrow form typical of brine pockets. (From Sinha, 1977.)

Notes

Recent texts that provide more detailed discussion of the nucleation and diffusional growth of cloud particles are *Microphysical Processes in Clouds* (1993) by Young and *Microphysics of Clouds* (1997) by Pruppacher and Klett. Ice nucleation and diffusional growth is described in detail in *Physics of Ice* (1974) by Hobbs. *A Short Course in Cloud Physics* (1989; Chapters 6–9) by Rogers and Yau provides a relatively simple but very readable treatments of these topics.

Storm and Cloud Dynamics by Cotton and Anthes (1989) provides an overview of the theories to explain the observed broadening of cloud drop size spectra.

An extensive treatment of the formation of the initial sea ice cover and sea ice microstructure is given by Weeks and Ackley (1986).

Problems

1. Drops frequently form above the ocean by the mechanical disruption of waves in strong winds. How much work is required to break up 1 kg of seawater ($T = 20°C$, $s = 35$ psu) into tiny drops (called *spray drops*), each in the form a sphere with radius 1 μm? The surface tension between seawater and air can be evaluated from

$$\sigma\,(\text{air–seawater}) = 75.63 - 0.144\,T + 0.221\,s$$

where T is in °C, s is in psu, and σ is 10^{-3} N m^{-1}.

2. An issue of considerable environmental concern is contaminants in the ocean. The spreading of a drop of insoluble oil on a water surface involves the buoyancy of oil on water as well as the surface interaction between the three substances: oil, water, and air.

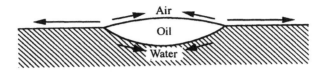

The rim of an oil lens is pulled outward if the air–water surface tension is larger than the sum of the air–oil and oil–water surface tensions. The spreading of an oil slick thus depends on the sign of the *spreading coefficient, Sc*

$$Sc = \sigma\,(\text{air–water}) - \sigma\,(\text{air–oil}) - \sigma\,(\text{oil–water})$$

If *Sc* is positive, the oil will spread until it becomes distributed as a thin film on the water. If *Sc* is negative, then the density difference between oil and water may still push the rim of the oil patch outward, but the surface tension pulls it inward, and the oil will tend to form lens-shaped globules. Compare the spreading of olive oil and paraffin oil on seawater at a temperature of 20°C. The surface tension between sea-water and air can be evaluated using the expression in **1**. The surface tensions of the two types of oil at 20°C are (following Krauss and Businger, 1994):

	Surface tension ($\times$ 10^{-3} N m^{-1})	
	air–oil	oil–water
Olive oil	32.0	20.6
Paraffin oil	26.5	48.4

3. Using the approximate expression

$$\frac{e_s\left(r, n_{solt}\right)}{e_s} = 1 + \frac{a}{r} - \frac{b}{r^3}$$

show that the peak in the Kohler curve occurs at

$$r^* = \left(\frac{3b}{a}\right)^{1/2} \qquad S^* = 1 + \left(\frac{4a^3}{27b}\right)^{1/2}$$

Calculate the critical size r^* and critical saturation ratio S^* for a drop assuming a temperature of 280 K, the nuclei is $(NH_4)_2SO_4$ ($\iota = 2$) with a mass of 10^{-15} g.

4. Determine how large drops must be beyond the critical radius before solute (Raoult) effects are negligibly small relative to the curvature (Kelvin) effect.

5. Consider a parcel of air at $T = -5°C$ and $p = 800$ hPa. Assume that a slight supersaturation exists with $\mathcal{H} = 100.5\%$ (with respect to liquid).
a) Compute how long it would take to grow a cloud drop from an initial radius of 1 μm to a drop radius of 10 μm, 100 μm, and 1000 μm.
b) Compute how long it would take to grow a spherical ice ball from an initial radius of 1 μm to a radius of 10 μm, 100 μm, and 1000 μm.

6. Derive expression for α_2 in the following equation:

$$\frac{dS}{dt} = \alpha_1 \frac{dz}{dt} - \alpha_2 \frac{dq_l}{dt}$$

7. An analytic expression of the following form has been used to describe drop size spectra:

$$n(r) = Ar^2 \exp(-Br)$$

where A and B are parameters. For a drop size spectrum represented by this relationship, determine the following:
a) the total drop concentration per volume of air:

$$N = \int_0^\infty n(r)\, dr$$

b) the mean drop radius:

$$\bar{r} = \frac{1}{N} \int_0^\infty r n(r)\, dr$$

c) the coefficients A and B for $N = 200$ cm^{-3} and $\bar{r} = 10$ μm;
d) the liquid water mixing ratio, w_l:

$$w_l = \frac{\rho_l}{\rho_a} \frac{4}{3} p \int_0^\infty r n(r)\, dr$$

where ρ_l is the density of water and ρ_a is the density of air.

Chapter 6 | Thermodynamic Transformations of Moist Air

In this chapter we consider the thermodynamic processes that result in the formation and dissipation of clouds. Based on microphysical considerations, we found in Chapter 5 that the liquid phase is nucleated at relative humidities only slightly greater than 100%. For simplicity, we assume here that clouds form in the atmosphere when the water vapor reaches its saturation value and $\mathcal{H} = 100\%$.

In a closed system consisting of moist air, the water vapor mixing ratio remains constant through the course of thermodynamic transformations as long as condensation does not occur. However, vapor pressure and relative humidity do not remain the same during such transformations. For example, in an adiabatic expansion the vapor pressure decreases, since it remains proportional to atmospheric pressure.

The relative humidity was defined in Section 4.4 as

$$\mathcal{H} \approx \frac{w_v}{w_s(T)}$$

where w_v is the water vapor mixing ratio and w_s is the saturation mixing ratio. For initially unsaturated air to become saturated, the relative humidity must increase. An increase in relative humidity can be accomplished by increasing the amount of water vapor in the air (i.e., increasing w_v), and/or by cooling the air, which decreases $w_s(T)$. The amount of water vapor in the air can increase by evaporation of water from a surface or via evaporation of rain falling through unsaturated air. The temperature of the atmosphere can decrease by isobaric cooling (e.g., radiative cooling) or by adiabatic cooling of rising air. An additional mechanism that can increase the relative humidity is the mixing of two unsaturated parcels of air.

In this chapter, we begin by writing the combined first and second laws of thermodynamics for a system that consists of moist air plus condensed water. To understand the changes in thermodynamic state associated with the formation and dissipation of clouds, we apply the combined first and second laws to the following idealized thermodynamic reference processes associated with phase changes of water:

- isobaric cooling;
- adiabatic isobaric processes;
- adiabatic expansion;
- adiabatic isobaric freezing.

159

Although real clouds nearly always involve more than one of these reference processes in their formation, consideration of the individual processes provides a convenient framework for understanding mechanisms that cause clouds to form and dissipate.

6.1 Combined First and Second Laws

To understand thermodynamic processes in moist and cloudy air, consider the combined first and second laws for a system that consists of two components (dry air and water) and two phases (gas and liquid). For the present, we ignore surface and solute effects in the condensed phase. Following Section 4.3, the combined first and second laws are written as

$$dU = T\,d\eta - p\,dV + \mu_d\,dn_d + \mu_v\,dn_v + \mu_l\,dn_l$$

$$dH = T\,d\eta + V\,dp + \mu_d\,dn_d + \mu_v\,dn_v + \mu_l\,dn_l$$

$$dG = -\eta\,dT + V\,dp + \mu_d\,dn_d + \mu_v\,dn_v + \mu_l\,dn_l$$

where the subscripts d, v, and l refer to dry air, water vapor, and liquid water, respectively.

The exact differential of the enthalpy, dH, where $H = H(T, p, m_d, m_v, m_l)$, can be expanded as follows:

$$dH = \left(\frac{\partial H}{\partial T}\right)dT + \left(\frac{\partial H}{\partial p}\right)dp + \left(\frac{\partial H}{\partial m_d}\right)dm_d + \left(\frac{\partial H}{\partial m_v}\right)dm_v + \left(\frac{\partial H}{\partial m_l}\right)dm_l$$

If the system is closed, then $dm_d = 0$ and $dm_v = -dm_l$, and therefore

$$dH = \left(\frac{\partial H}{\partial T}\right)dT + \left(\frac{\partial H}{\partial p}\right)dp + \left[\left(\frac{\partial H}{\partial m_v}\right) - \left(\frac{\partial H}{\partial m_l}\right)\right]dm_v \qquad (6.1a)$$

Since $(h_v - h_l) = L_{lv}$ (Section 4.3), we have

$$dH = \left(\frac{\partial H}{\partial T}\right)dT + \left(\frac{\partial H}{\partial p}\right)dp + L_{lv}\,dm_v \qquad (6.1b)$$

To evaluate $\partial H/\partial T$ and $\partial H/\partial p$, consider the total enthalpy as the sum of the individual contributions from the dry air, water vapor, and liquid water, so that $H = m_d h_d + m_v h_v + m_l h_l$. We can then write

$$\frac{\partial H}{\partial T} = m_d c_{pd} + m_v c_{pv} + m_l c_l \tag{6.2a}$$

Recall that in Section 2.9 we established that there is little difference between the specific heats of liquid water at constant pressure and volume, so henceforth we do not distinguish between them. In Section 2.3, we found that $\partial H/\partial p = 0$ for an ideal gas. For liquid water, $\partial H/\partial p \neq 0$, but the value is small and thus neglected here. We can therefore write (6.1) as

$$dH = \left(m_d c_{pd} + m_v c_{pv} + m_l c_l \right) dT + L_{lv} dm_v \tag{6.2b}$$

In the atmosphere, the mass of water vapor is only a few percent of the mass of dry air (Section 1.1), and the mass of condensed water is a small fraction of the mass of water vapor. Thus $m_d \gg m_v \gg m_l$ and we can approximate (6.2b) by

$$dH \approx m_d c_{pd} dT + L_{lv} dm_v \tag{6.3}$$

The enthalpy of a system consisting of moist air and a liquid water cloud is not only a function of temperature (as was the ideal gas), but also a function of the latent heat associated with the phase change. In intensive form, we have

$$dh \approx c_{pd} dT + L_{lv} dw_v \tag{6.4}$$

In a similar manner, we can write an equation for internal energy[1] as

$$du = \left(c_{vd} + w_v c_{vv} + w_l c_l \right) dT + L_{lv} dw_v \tag{6.5}$$

and an approximate form as

$$du \approx c_{vd} dT + L_{lv} dw_v \tag{6.6}$$

where w_l is the liquid water mixing ratio introduced in (5.28).

---

[1] Mixing ratio is used here instead of specific humidity to avoid confusion of the notation q (specific humidity) with q (heat). Note that a liquid water specific humidity, q_l, can be defined analogously to the liquid water mixing ratio, w_l.

Depending on how the thermodynamic system is defined, the term $L_{lv} dw_v$ may be included as part of the enthalpy, or it may constitute an external heat source. For a closed system, we can write

$$dq = c_{pd} \, dT + L_{lv} \, dw_v - v \, dp \tag{6.7a}$$

and for an adiabatic process,

$$0 = c_{pd} \, dT + L_{lv} \, dw_v - v \, dp \tag{6.7b}$$

Now consider a system that consists of moist air, with an external heat source associated with evaporation from a water source (such as moist air over a lake). The first law of thermodynamics can be written as

$$dq = dh - v \, dp$$

where $dh = c_{pd} \, dT$ and $dq = L_{lv} \, dw_l = -L_{lv} \, dw_v$. We can then write

$$-L_{lv} \, dw_v = c_{pd} \, dT - v \, dp \tag{6.8}$$

Note that (6.7) and (6.8) are mathematically equivalent; however, in (6.7b) the term $L_{lv} dw_v$ is part of the enthalpy, while in (6.8) the term $L_{lv} dw_v$ is a heat source. This example illustrates the care that must be taken to interpret correctly the thermodynamic equation in the context in which the system is defined.

The combined first and second law for a system consisting of moist air and a liquid water cloud can be written using (4.7) and (2.33) as

$$T d\eta = dH - V \, dp - \sum_j \mu_j \, dn_j \tag{6.9}$$

Including only the liquid–vapor phase change, we can incorporate (6.2) into (6.9) and write

$$T d\eta = \left(m_d c_{pd} + m_v c_{pv} + m_l c_l \right) dT + L_{lv} \, dm_v - V \, dp - \mu_v \, dm_v - \mu_l \, dm_l \tag{6.10}$$

If the system is closed, then $dm_d = 0$ and $dm_v = -dm_l$, and analogously to (6.1b) we can write (6.10) in intensive form as

$$d\eta = \left(c_{pd} + w_v c_{pv} + w_l c_l \right) d(\ln T) - R_d \, d(\ln p_d) - w_v R_v \, d(\ln e) + \frac{L_{lv} + A_{lv}}{T} \, dw_v \tag{6.11}$$

In (6.11) we have separated the expansion work term into components (neglecting the expansion work of liquid water). The *affinity for vaporization*, A_{lv}, is defined (following Dutton, 1986) as $A_{lv} = \mu_l - \mu_v$, which can be evaluted following (5.10). If the liquid and vapor phases are in equilibrium ($\mu_v = \mu_l$), then $A_{lv} = 0$. In subsaturated or supersaturated conditions, the affinity term can be of the order of several percent of the latent heat of vaporization. Using the first and second latent heat equations (4.19) and (4.29), we can write (6.11) as

$$d\eta = \left(c_{pd} + w_t c_l\right) d(\ln T) - R_d \, d(\ln p_d) + d\left(\frac{L_{lv} w_v}{T}\right) + w_v \, d\left(\frac{A_{lv}}{T}\right) \qquad (6.12)$$

where w_t is the *total water mixing ratio* ($w_t = w_v + w_l$).

Analogous arguments can be used to incorporate the ice phase into the entropy equation. The complete thermodynamic equation for moist air and clouds that includes all three phases of water is written as

$$\begin{aligned} d\eta = \left(c_{pd} + w_t c_l\right) d(\ln T) - R_d \, d(\ln p_d) + w_v \, d\left(\frac{A_{lv}}{T}\right) + d\left(\frac{L_{lv} w_v}{T}\right) \\ - w_i \, d\left(\frac{A_{il}}{T}\right) - d\left(\frac{L_{il} w_i}{T}\right) \end{aligned} \qquad (6.13)$$

where the total water mixing ratio, w_t, in (6.13) includes the *ice water mixing ratio*, w_i. The *affinity for freezing*, A_{il}, is defined analogously to that for vaporization as $A_{il} = \mu_i - \mu_l$. The affinity for freezing can reach 20% of the latent heat of fusion.

6.2 Isobaric Cooling

A thermodynamic process can be approximated as isobaric if vertical motions are small and there is only a small departure from a reference pressure. In the absence of condensation, the first law of thermodynamics for an isobaric process in moist air is written (following 2.16) as

$$dq = dh = c_p \, dT$$

where c_p can be approximated as the dry air value, or alternatively the contribution from water vapor can be incorporated following (2.65). As moist air cools, relative humidity increases: w_v remains the same, but as the temperature decreases then w_s decreases. If the cooling continues, w_s will become equal to w_v and $\mathcal{H}$ will equal unity; at this point, the air has reached saturation. Further cooling beyond saturation results in condensation.

The temperature at which saturation is reached in an isobaric cooling process is the *dew-point temperature*, which is illustrated in Figure 6.1a. The dew-point temperature, denoted by T_D, can be defined by

$$e = e_s\left(T_D\right) \tag{6.14}$$

or equivalently by

$$w_v = w_s\left(T_D\right) \tag{6.15}$$

We can determine the dew-point temperature by inverting either (6.14) or (6.15), which can be done using (4.31) and (4.36).

Analogously to the dew-point temperature, we define the *frost-point temperature* as the temperature at which ice saturation occurs. The frost-point temperature, T_F, is thus defined as

$$e = e_{si}\left(T_F\right) \tag{6.16}$$

or equivalently as

$$w_v = w_{si}\left(T_F\right) \tag{6.17}$$

In Figure 6.1b, it is seen that if the vapor pressure is initially below the triple–point pressure of water (point 1), isobaric cooling results in deposition once the frost point is reached (point 2). As described in Section 5.3, saturation with respect to ice is not sufficient to initiate the ice phase in the atmosphere. Deposition occurs at the frost point only if ice crystals already exist in the atmosphere. Since $T_F > T_D$, the formation of frost on the ground must occur by deposition rather than by freezing of condensed water vapor; grass and other structures provide a good substrate for initiating the ice phase by deposition.

Although the units of the dew-point temperature are kelvins, the dew-point temperature is a measure not of temperature but of atmospheric humidity. By examining Figure 6.1 and the Clausius–Clapeyron relationship (4.19), it is seen that

$$\frac{d(\ln e)}{dT_D} = \frac{L_{lv}}{R_v T_D^2} \tag{6.18}$$

and that e and T_D give equivalent information about the amount of water vapor in the atmosphere. A relationship between T_D and $\mathcal{H}$ can be obtained by integrating (6.18) between T and T_D:

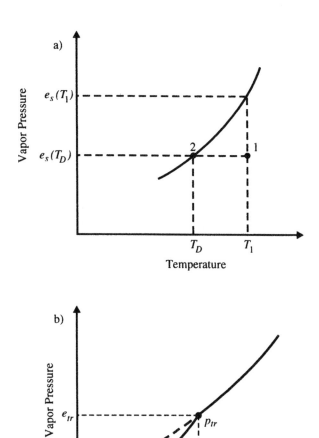

Figure 6.1 a) Relationship between temperature and vapor pressure in an isobaric cooling process. Air initially at temperature T_1 (point 1) is cooled isobarically until it reaches saturation (point 2). The temperature at point 2 defines the dew-point temperature, T_D. b) Air at T_1 (point 1) cools isobarically until it reaches saturation. If the saturation is reached with respect to ice (point 2), the temperature is called the frost point, T_F.

$$\ln \frac{e_s}{e} = -\ln \mathcal{H} = \frac{L_{lv}}{R_v}\left(\frac{1}{T_D} - \frac{1}{T}\right)$$

or equivalently

$$\mathcal{H} = \exp\left[-\frac{L_{lv}}{R_v}\left(\frac{T - T_D}{T T_D}\right)\right] \tag{6.19}$$

The term $T - T_D$ in (6.19) is called the *dew-point depression.* Figure 6.2 illustrates that dew-point depression is inversely proportional to relative humidity and that a relative humidity of 100% corresponds to a dew-point depression of zero.

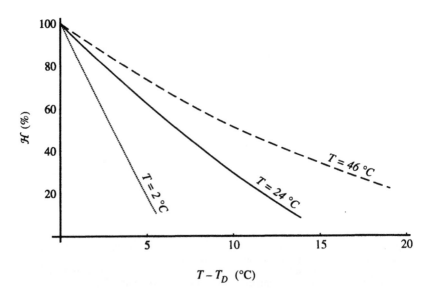

Figure 6.2 Dew–point depression. As the relative humidity increases, the difference between the ambient temperature and the dew-point temperature (i.e., the *dew–point depression*) decreases. As the ambient temperature decreases, the dew–point depression becomes less sensitive to changes in the relative humidity.

Thus, through (6.14), (6.15), and (6.19), the dew-point temperature is shown to be a humidity variable. If temperature, dew-point temperature and pressure are given, then the values of mixing ratio, relative humidity, and vapor pressure can be calculated. Analogously, the frost-point temperature can be related to all of the other humidity variables. In an isobaric process in the absence of condensation, the dew-point and frost-point temperatures are conservative; that is, they do not change during the cooling process until condensation is reached.

Once the air is cooled slightly below the dew-point temperature, condensation begins. After condensation begins, the first law of thermodynamics for an isobaric process is written following (6.4) in the approximate form

$$dq = dh = c_p \, dT + L_{lv} \, dw_v \qquad (6.20)$$

Assuming that condensation occurs at saturation ($\mathcal{H} = 1$) and that the water vapor mixing ratio is equal to the saturation vapor mixing ratio $w_v = w_s$, we can write

$$w_t = w_s + w_l \qquad (6.21)$$

In a closed system, w_t remains constant, so

$$dw_l = -dw_s$$

Using the approximation $w_s \approx \varepsilon \, e_s / p$ from Section 4.4 and the Clausius–Clapeyron relation (4.19), we can write

$$dw_l = -dw_s = -\varepsilon \frac{de_s}{p} = -\frac{\varepsilon L_{lv} e_s}{p R_v T^2} dT \qquad (6.22a)$$

Incorporating (6.22a) into (6.20) and using $R_d = R_v/\varepsilon$, we obtain

$$dw_l = -\left(\frac{L_{lv} e_s}{c_p p R_d T^2 + L_{lv} e_s} \right) dq \qquad (6.22b)$$

Combination of (6.22b) with (6.20) gives a relationship between dq and dT during isobaric condensation:

$$dq = -\left(c_p + \frac{L_{lv} e_s}{p R_d T^2} \right) dT \qquad (6.22c)$$

Integration of (6.22b) (which is most easily done numerically, since e_s is a function of T) allows determination of the amount of isobaric cooling, Δq, required to condense an amount of liquid water, Δw_l. Analogously, integration of (6.22c) allows determination of the temperature change, ΔT, in response to the isobaric cooling, Δq. Before condensation occurs, we have $\Delta q = -c_p \Delta T$. Once condensation begins, it is seen from (6.22c) that the temperature drops much more slowly in response to the isobaric cooling, because the heat loss is partially compensated by the latent heat released during condensation.

Once condensation begins, the dew-point temperature decreases, since the water vapor mixing ratio is decreasing as the water is condensed. Relative humidity remains constant, at $\mathcal{H} = 1$.

Isobaric cooling is a primary formation mechanism for certain types of fog and stratus clouds (see Section 8.4). The equations derived in this section are equally applicable to isobaric heating. In this instance, an existing cloud or fog can be dissipated by evaporation that ensues from isobaric heating (e.g., solar radiation).

6.3 Cooling and Moistening by Evaporation of Water

Consider a system composed of unsaturated moist air plus rain falling through the air. Because the air is unsaturated, the rain will evaporate. If there are no external heat sources ($\Delta q = 0$), and the evaporation occurs isobarically ($dp = 0$), we can write an adiabatic, isobaric (or *isenthalpic*) form of the enthalpy equation (6.20) as

$$0 = dh = c_p \, dT - L_{lv} \, dw_l = c_p \, dT + L_{lv} \, dw_s \qquad (6.23)$$

where c_p can be approximated as the dry–air value, or alternatively the contributions from water vapor and liquid water can be incorporated following (6.2a). Since $dh = 0$, (6.23) can be used to determine a relationship between temperature and humidity variables for isenthalpic processes in the atmosphere that involve a phase change of water.

If we allow just enough liquid water from the rain to evaporate so that the air becomes saturated, we can integrate (6.23)

$$c_p \int_T^{T_W} dT = -L_{lv} \int_{w_l}^0 dw_s$$

where w_l represents the amount of water that must be evaporated to bring the air to saturation. During the evaporation process, latent heat is drawn from the atmosphere, and the final temperature, referred to as the *wet-bulb temperature* (T_W), is cooler than the original temperature. Integration gives

$$c_p\left(T_W - T\right) = -L_{lv}\left[w_s\left(T_W\right) - w_v\right]$$

or alternatively

$$T_W = T - \frac{L_{lv}}{c_p}\left[w_s\left(T_W\right) - w_v\right] \tag{6.24}$$

where the temperature dependence of L_{lv} has been neglected. Given w_v and T, this expression is implicit for T_W and must be solved numerically. However, if T and T_W are given, then w_v is easily determined. T_W can be measured using a *wet-bulb thermometer*, whereby a wetted muslin wick is affixed to the bulb of a thermometer. Concurrent measurement of the "dry-bulb" temperature by a normal thermometer can then provide a means of determining the water vapor mixing ratio and therefore atmospheric humidity. For this reason, (6.24) is often referred to as the *wet-bulb equation*.

The wet-bulb temperature is thus defined as the temperature to which air would cool isobarically as the result of evaporating sufficient liquid water into the air to make it saturated. As such, the wet-bulb temperature in the atmosphere is conservative with respect to evaporation of falling rain. Calculations for given values of T and w show that $T_D < T_W < T$. This can be shown graphically. Since e increases while T decreases during the approach to T_W, the Clapeyron diagram looks like:

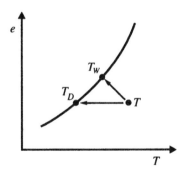

If ice is the evaporating phase, we can determine an analogous *ice-bulb temperature*, T_I:

$$T_I = T - \frac{L_{iv}}{c_p}\left[w_{si}(T_I) - w_v\right] \tag{6.25}$$

It is easily shown that $T_I > T_W$.

6.4 Saturation by Adiabatic, Isobaric Mixing

We have seen in Sections 6.2 and 6.3 how unsaturated air can be brought to saturation by isobaric cooling and by the adiabatic, isobaric evaporation of falling rain. There is an additional isobaric process that can bring unsaturated air to saturation. Under some circumstances, the isobaric mixing of two samples of unsaturated air leads to saturation. One example of this process occurs when your breath produces a puff of cloud on a cold day.

Consider the isobaric mixing of two moist air masses, with different temperatures and humidities but at the same pressure. Condensation is assumed not to occur. For adiabatic, isobaric mixing, we can write the first law of thermodynamics from (2.16) as

$$0 = dH \approx m_1\, c_{pd}\, dT_1 + m_2\, c_{pd}\, dT_2$$

where dT_1 and dT_2 correspond to the temperature change of the air masses upon mixing and we have ignored the heat capacity of the water vapor in accordance with (6.4). Upon integration from an initial state where the air masses are unmixed to a final state where the air masses both have the same final temperature, T, we have

$$m_1 c_{pd}\left(T - T_1\right) + m_2 c_{pd}\left(T - T_2\right) = 0$$

Solving for T we obtain

$$T \approx \frac{m_1}{m_1 + m_2}T_1 + \frac{m_2}{m_1 + m_2}T_2$$

The total mass $m = m_1 + m_2$ remains constant during the mixing process, so the specific humidity is a mass-weighted average of q_{v1} and q_{v2}

$$q_v = \frac{m_1}{m_1 + m_2}q_{v_1} + \frac{m_2}{m_1 + m_2}q_{v_2}$$

Thus, both the temperature and specific humidity mix linearly if the heat capacity of the water vapor is neglected. Since $q_v \approx w_v$, we can also assume that the mixing ratios mix linearly. If we further assume that $w_v \approx \varepsilon e/p$, then vapor pressure mixes linearly as well.

Because of the nonlinearity of the Clausius–Clapeyron equation, adiabatic isobaric mixing results in an increase in relative humidity. This mixing process is illustrated in Figure 6.3 using a T, e diagram. If Y_1 and Y_2 are the image points for the two

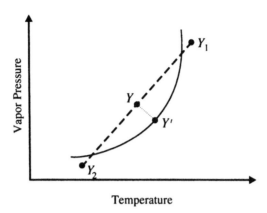

Temperature

Figure 6.3 Adiabatic isobaric mixing and condensation. Two air masses with (e, T) given by points Y_1 and Y_2 mix, resulting in a single air mass with (e, T) given by point Y. Since $\mathcal{H} > 1$ at this point, water will condense, and the temperature of the air mass will increase while the vapor pressure decreases. Condensation will continue until the temperature and vapor pressure of the air mass coincide with the saturation vapor pressure curve (point Y').

air masses, the image point for the mixture lies on a straight line joining Y_1 and Y_2. If $m_1 = m_2$, then T and e for the mixture will lie midpoint on this line. Because of the exponential relationship between e_s and T, the mixing process increases the relative humidity. In the example shown in Figure 6.3, the mixing process results in the image point Y having a relative humidity that exceeds 100%, crossing the $f = 1$ line into the liquid phase (see also Figure 4.3). Water will condense and latent heat will be released, with the final equilibrium image point at Y' on the $f = 1$ line.

The slope of the line between Y and Y' can be determined from the first law of thermodynamics for an adiabatic isobaric process in which condensation occurs (6.23):

$$dh = 0 \approx c_p \, dT + L_{lv} \, dw_s$$

Using the definition of the saturated water vapor mixing ratio, $w_s = \varepsilon e_s / p$, we can write

$$0 = c_p \, dT + \frac{L_{lv} \varepsilon}{p} \, de$$

or

$$\frac{de}{dT} = -\frac{p c_p}{\varepsilon L_{lv}} \tag{6.26}$$

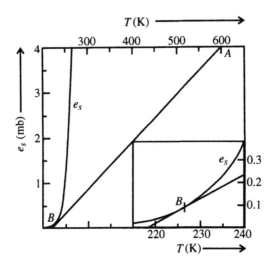

Figure 6.4 The formation of contrails by adiabatic, isobaric mixing. A jet flying at 200 mb ejects water vapor into the atmosphere at the temperature and vapor pressure represented by point A. For atmospheric temperatures less than about –47°C (226 K), the water vapor will condense, forming condensation trails. (From Ludlam, 1980.)

The value of (e, T) at Y' can be found by simultaneously solving (6.26) with the Clausius–Clapeyron equation (4.19). The amount of liquid water condensed during the mixing is

$$\Delta w_l = \frac{\varepsilon}{p} \left[e(Y) - e(Y') \right] \tag{6.27}$$

A notable example of the formation of clouds by adiabatic, isobaric mixing occurs when the exhaust gases from the combustion of fuels by an aircraft mixes with the ambient atmosphere. The trails of clouds often formed by an aircraft in flight at high altitude are referred to as condensation trails, or *contrails*. In the exhaust, the aircraft ejects heat and water vapor; the temperature of the exhaust is typically 600 K. Figure 6.4 indicates that for an aircraft flying at 200 mb, atmospheric temperatures below about –47°C will form contrails. Once contrails form, their persistence depends on the atmospheric humidity and the rate at which the exhaust trail is diffused. If the particles are ice, atmospheric humidity in excess of the ice saturation value will result in growth of the contrails.

6.5 Saturated Adiabatic Cooling

Adiabatic cooling is the most important mechanism by which moist air is brought to saturation. As described in Sections 2.1 and 2.10, adiabatic expansion in the atmosphere occurs when a dry air mass rises due to mechanical lifting (e.g., orographic, frontal), large-scale low-level convergence, turbulent mixing, and buoyancy caused by surface heating.

Recall from Section 2.4 that the first law of thermodynamics for an adiabatic process for moist air in the absence of condensation is written as (2.19b)

$$c_p \, dT - v \, dp = 0$$

from which we derived an expression for the potential temperature (2.62)

$$\theta = T \left(\frac{1000}{p} \right)^{R/c_p}$$

and the dry adiabatic lapse rate (2.68)

$$\Gamma_d = \frac{g}{c_p} \approx 10°C \text{ km}^{-1}$$

Recall that the potential temperature, θ, is conserved in reversible, dry adiabatic processes in the atmosphere.

As air expands adiabatically and cools, the relative humidity increases as the temperature and saturation mixing ratio decrease. The water vapor mixing ratio remains constant during adiabatic ascent. At some point, the relative humidity reaches 100%, and further cooling results in condensation. To determine the temperature and pressure at which saturation is reached, we logarithmically differentiate $\mathcal{H} = e/e_s$

$$d(\ln \mathcal{H}) = d(\ln e) - d(\ln e_s) \tag{6.28a}$$

Using Dalton's law of partial pressure (1.13), we have $d(\ln p) = d(\ln e)$, and we can write the first law of thermodynamics for an adiabatic process in enthalpy form (2.19b) as

$$d(\ln e) = \frac{c_p}{R_d} \, d(\ln T) \tag{6.28b}$$

Using the Clausius–Clapeyron equation (4.19), we can write

$$d(\ln e_s) = \frac{L_{lv}}{R_v T} \, d(\ln T) \tag{6.28c}$$

Incorporating (6.28b) and (6.28c) into (6.28a), we can integrate (6.28a) from the initial condition to conditions where saturation is attained, indicated by $\mathcal{H} = 1$ and $T = T_s$, where T_s is the *saturation temperature*

$$\int_{\mathcal{H}}^{1} d(\ln \mathcal{H}') = \int_{T}^{T_s} \left(\frac{c_p}{R_d} - \frac{\varepsilon L_{lv}}{R_d T} \right) d(\ln T')$$

to obtain

$$-\ln \mathcal{H} = \frac{c_p}{R_d} \ln \left(\frac{T_s}{T} \right) + \frac{\varepsilon L_{lv}}{R_d} \left(\frac{1}{T_s} - \frac{1}{T} \right) \qquad (6.29)$$

Equation (6.29) can be solved numerically to obtain T_s. An approximate but simpler equation for T_s, given initial values of T (in kelvins) and $\mathcal{H}$, is given by (Bolton, 1980)

$$T_s = \frac{1}{\dfrac{1}{T - 55} - \dfrac{\ln \mathcal{H}}{2840}} + 55 \qquad (6.30)$$

The *saturation pressure*, p_s, can be obtained from (2.22) to be

$$\ln \frac{p_s}{p} = \frac{c_p}{R_d} \ln \frac{T_s}{T}$$

or, taking anti-logs,

$$p_s = p \left(\frac{T_s}{T} \right)^{c_p / R_d} \qquad (6.31)$$

The coordinate (T_s, p_s) is known as the *saturation point* of the air mass.

During ascent, the water vapor mixing ratio, w_v, remains constant until saturation occurs. The dew-point temperature, however, decreases slightly during the ascent as pressure decreases. Recall from (6.18) that

$$d(\ln e) = \frac{L_{lv}}{R_v T_D^2} dT_D \qquad (6.32)$$

Using Dalton's law of partial pressure (1.13), we can write the hypsometric equation (1.46) as

$$d(\ln e) = -\frac{g}{R_d T} dz \qquad (6.33)$$

Combining (6.32) and (6.33), we obtain

$$\frac{dT_D}{dz} = -\frac{T_D^2 g}{\varepsilon L_{lv} T} = \frac{T_D^2 c_p}{\varepsilon L_{lv} T} \Gamma_d \qquad (6.34a)$$

For typical atmospheric values, dT_D/dz is approximately one-sixth of the dry adiabatic lapse rate. At saturation level, T becomes equal to T_D and to T_s. The *lifting condensation level*, z_s, corresponds to the level of the saturation pressure, p_s.

Using (6.34a) and the definition of the dry adiabatic lapse rate, $\Gamma_d = g/c_p$, we can write

$$\frac{d(T - T_D)}{dz} = \left(1 + \frac{T_D^2 c_p}{\varepsilon L_{lv} T}\right) \Gamma_d \qquad (6.34b)$$

When $T = T_D$, the saturation level has been reached, and a value of z_s can be determined by integrating (6.34b):

$$\int_{T_0 - T_{D0}}^{0} d(T - T_D) = \int_{0}^{z_s} \left[\left(1 + \frac{T_D^2 c_p}{\varepsilon L_{lv} T}\right) \Gamma_d\right] dz \qquad (6.34c)$$

where $T_0 - T_{D0}$ is the dew-point depression at the surface. For a parcel of air lifted from the surface, the value of z_s can be estimated from (6.34c) to be

$$z_s \approx 0.12 \left(T_0 - T_{D0}\right) \quad \text{(km)} \qquad (6.35)$$

This relation is an approximate expression of the height of the lifting condensation level achieved in an adiabatic ascent where T_0 and T_{D0} represent the initial temperature and dew-point temperature of the air mass that is being lifted. Note that z_s can be determined directly from (1.45) if p_s and T_s are known. Calculation of the lifting condensation level provides a good estimate of the cloud base height for clouds that form by adiabatic ascent.

Once saturation occurs, further lifting of the air mass results in condensation. Because of the latent heat released during condensation, the decrease of temperature with height will be smaller than that in dry adiabatic ascent. In addition, the potential temperature, θ, which was conserved in a reversible dry adiabatic ascent, is no longer conserved once condensation occurs.

A derivation of an approximate form of the *saturated adiabatic lapse rate*, Γ_s, is given here by starting with the adiabatic entropy equation (6.12) in the following approximate form:

$$0 = c_{pd}\, d(\ln T) - R_d\, d(\ln p) + \frac{L_{lv}}{T}\, dw_s \tag{6.36}$$

Using the hypsometric equation (1.46)

$$\frac{dp}{p} = -\frac{g}{R_d T}\, dz$$

and logarithmically differentiating the equation for saturation mixing ratio (4.37),

$$\frac{dw_s}{w_s} = \frac{de_s}{e_s} - \frac{dp}{p}$$

we can rewrite (6.36) as

$$-L_{lv} w_s \left(\frac{de_s}{e_s} - \frac{dp}{p} \right) = c_p\, dT + g\, dz \tag{6.37}$$

Dividing by an incremental dz and solving for $-dT/dz$, we obtain

$$-\frac{dT}{dz} = \frac{L_{lv}}{c_p}\, w_s \left(\frac{1}{e_s}\frac{de_s}{dz} + \frac{g}{RT} \right) + \frac{g}{c_p} \tag{6.38}$$

Using the chain rule, we can write the term de_s/dz as

$$\frac{de_s}{dz} = \frac{de_s}{dT}\frac{dT}{dz} \tag{6.39}$$

and substitute into (6.38) to obtain

$$-\frac{dT}{dz}\left(1 + \frac{de_s}{dT}\frac{L_{lv}}{c_p}\frac{w_s}{e_s}\right) = \frac{g}{c_p}\left(\frac{L_{lv}w_s}{RT} + 1\right)$$

Incorporating the Clausius–Clapeyron equation (4.19), solving for $dT/dz = -\Gamma_s$ and noting that $\Gamma_d = -g/c_p$ (2.68), we obtain finally

$$\Gamma_s = \Gamma_d \left(\frac{1 + \dfrac{L_{lv}w_s}{R_d T}}{1 + \dfrac{\varepsilon L_{lv}^2 w_s}{c_{pd} R_d T^2}}\right) \tag{6.40}$$

The denominator of (6.40) is larger than the numerator, and thus $\Gamma_s < \Gamma_d$. Table 6.1 shows values of Γ_s for selected values of T and p. It is seen that the temperature variation of Γ_s exceeds the pressure variation. At low temperatures and high pressures, Γ_s approaches Γ_d.

Values of Γ_s determined from (6.40) are within about 0.5% of the values determined from a more exact form of the entropy equation (6.11). Because of the approximate nature of (6.40), Γ_s is sometimes called the *pseudo-adiabatic lapse rate*.

The amount of water condensed in saturated adiabatic ascent, called the *adiabatic liquid water mixing ratio*, can be determined from the adiabatic enthalpy equation (6.7b):

$$0 = c_p\, dT - L_{lv}\, dw_l - v\, dp$$

Table 6.1 Γ_s for selected values of temperature and pressure.

T (°C)	p (hPa)		
	1000	700	500
−30	9.2	9.0	8.7
−20	8.6	8.2	7.8
−10	7.7	7.1	6.4
0	6.5	5.8	5.1
10	5.3	4.6	4.0
20	4.3	3.7	3.3

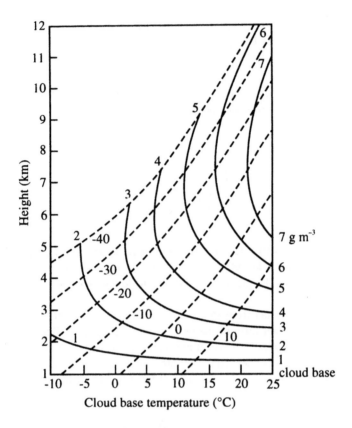

Figure 6.5 Adiabatic liquid water mixing ratio as a function of height above the cloud base and cloud base temperature. (After Goody, 1995.)

Solving for dw_l and incorporating the hydrostatic equation (1.33), we obtain

$$dw_l = \frac{c_p}{L_{lv}}\left(\frac{dT}{dz} + \frac{g}{c_p}\right) dz$$

Substituting $\Gamma_d = g/c_p$ and $\Gamma_s = -dT/dz$ yields

$$dw_l = \frac{c_p}{L_{lv}}\left(\Gamma_d - \Gamma_s\right) dz \qquad (6.41)$$

Integrating (6.41) from cloud base to height z gives the adiabatic liquid water mixing ratio at height z. Because of the complicated form of Γ_s, this equation must be integrated numerically. Integration of (6.41) shows that the adiabatic liquid water content increases with height above the cloud base and increasing cloud base temperature (Figure 6.5). Because of the variation of Γ_s with temperature, clouds with warmer bases have larger values of $\Gamma_d - \Gamma_s$ and thus larger values of the adiabatic liquid water content. The adiabatic liquid water content represents an upper bound on the liquid water that can be produced in a cloud by rising motion. Processes such as precipitation and mixing with dry air reduce the cloud liquid water content relative to the adiabatic value.

6.6 The Ice Phase

As isobaric or adiabatic cooling proceeds, the cloud may eventually cool to the point where ice crystals form. Assuming that a water cloud is present initially, then the formation of ice crystals releases latent heat during fusion. Once the cloud glaciates, it is supersaturated with respect to ice, and deposition occurs on the ice crystals, releasing the latent heat of sublimation, until the ambient relative humidity is at ice saturation. Further cooling will result in the increase of ice water content in the cloud and the release of the latent heat of sublimation into the atmosphere.

Assuming that the thermodynamic system consists of moist air plus the condensate, and that the freezing and subsequent deposition occur isobarically and adiabatically, then the enthalpy of the system will not change during this transformation. Since enthalpy is an exact differential, the enthalpy change depends only on the initial and final states (but not on the path). Consider the following path for the warming of the system associated with the phase change:

Step 1.   Water freezes at constant T_1:

$$\Delta h_1 = -L_{il} w_l \tag{6.42}$$

Step 2.   Vapor deposits on the ice at constant T_1, until the water vapor pressure reaches the saturation value over ice at T_2:

$$\Delta h_2 = -L_{iv}\left[w_s(T_1) - w_{si}(T_2) \right] = -L_{iv} \frac{\varepsilon}{p} \left[e_s(T_1) - e_{si}(T_2) \right]$$

Assuming that $(T_2 - T_1)$ is small enough to treat as a differential, we can approximate $e_{si}(T_2)$ as

$$e_{si}(T_2) = e_{si}(T_1) + \frac{L_{iv} e_{si}(T_1)}{R_v T^2} (T_2 - T_1)$$

and Δh_2 becomes

$$\Delta h_2 = -w_s L_{iv}\left[1 - \frac{e_{si}(T_1)}{e_s(T_1)}\right] + \frac{L_{iv}^2\, w_{si}(T_1)}{R_v T^2}\left(T_2 - T_1\right) \tag{6.43}$$

where w_s has been adopted in favor of e_s using $w_s \approx \varepsilon e_s/p$.

Step 3.   The system is heated from T_1 to T_2:

$$\Delta h_3 = c_p\left(T_2 - T_1\right) \tag{6.44}$$

Since $\Delta h_1 + \Delta h_2 + \Delta h_3 = 0$, we can incorporate (6.42), (6.43), and (6.44) and solve for $\Delta T = T_2 - T_1$:

$$\Delta T = \frac{L_{il}w_l + L_{iv}w_s\left(1 - \dfrac{e_{si}}{e_s}\right)}{c_p + \dfrac{\varepsilon w_l L_{iv}^2}{R_d T^2}} \tag{6.45}$$

Equation (6.45) gives the increase in temperature due to the freezing of cloud water and the subsequent deposition of water vapor onto the ice crystals. In clouds that cool by adiabatic ascent, the freezing does not occur isobarically, but gradually over a temperature interval.

Once the cloud has glaciated, further adiabatic ascent results in deposition of water vapor onto the ice crystals. Analogously to (6.40), the *ice-saturation adiabatic lapse rate* is determined to be

$$\Gamma_{si} = \Gamma_d\left(\frac{1 + \dfrac{L_{iv}w_{si}}{R_d T}}{1 + \dfrac{\varepsilon L_{iv}^2 w_{si}}{c_{pd} R_d T^2}}\right) \tag{6.46}$$

The melting process is distinctly different from the freezing process. Melting may occur as ice particles fall to temperatures that are above the melting point. In contrast to freezing, which may be distributed through a considerable vertical depth, melting of ice particles can be quite localized, occurring in a very narrow layer around the freezing point. Cooling of the atmosphere from the melting can result in an isothermal layer near 0°C. Because of their large size and density, hailstones do not melt at the freezing level in the same manner as a small ice crystal or a snowflake with a low density, but melt over a deeper layer. If atmospheric relative humidities are low in the atmosphere below the melting level, then the melting water will evaporate, cooling the hailstone and retarding the melting.

6.7 Conserved Moist Thermodynamic Variables

As shown in Section 3.1, conserved variables are commonly used in time-dependent equations. The concept of potential temperature becomes less useful when applied to a cloud, since potential temperature is not conserved during phase changes of water. Derivation of a potential temperature that is conserved in moist adiabatic ascent eliminates the need to include latent heat source terms in the time-dependent thermodynamic equation. Additionally, a potential temperature that is conserved in moist adiabatic ascent can be used to interpret graphically numerous cloud processes and characteristics (see Sections 6.8, 7.3, and 8.5).

Recall that for a reversible, adiabatic process in dry air, the entropy equation is written as (2.26b)

$$0 = c_{pd} \, d\left(\ln T\right) - R_d \, d\left(\ln p\right)$$

It was shown in Section 2.4 that integration of the above equation gives the potential temperature (2.62)

$$\theta = T \left(\frac{p_0}{p}\right)^{R_d/c_{pd}}$$

which is conserved for dry adiabatic motions.

We seek an analogous variable that is conserved for a cloud so that the variation of temperature with pressure can be determined in a saturated adiabatic process. We begin with the adiabatic form of the complete equation for the combined first and second laws for a moist air with cloud that includes both the liquid and ice phases (6.13):

$$0 = \left(c_{pd} + w_t c_l\right) d(\ln T) - R_d \, d(\ln p_d) - w_v \, d\left(\frac{A_{lv}}{T}\right) + d\left(\frac{L_{lv} w_v}{T}\right) + w_i \, d\left(\frac{A_{il}}{T}\right) - d\left(\frac{L_{il} w_i}{T}\right)$$

A conserved potential temperature for clouds will obviously be far more complex than the potential temperature derived for a dry adiabatic process, since (6.13) is considerably more complex than (2.26b). A number of different conserved potential temperatures have been derived for clouds that employ various approximate forms of (6.13).

The simplest possible case is that in which saturation conditions are maintained, ice is not present, and heat capacity of the water vapor and condensed water are neglected relative to that of dry air. Using these approximations, the entropy equation (6.13) becomes:

$$0 = c_{pd}\, d(\ln T) - R_d\, d(\ln p) + d\left(\frac{L_{lv} w_s}{T}\right) \tag{6.47}$$

Recall that we have for a dry adiabatic process from (2.63)

$$c_{pd}\, d(\ln\theta) = c_{pd}\, d(\ln T) - R_d\, d(\ln p)$$

Equating (2.63) with (6.47) yields

$$-d\left(\frac{L_{lv} w_s}{T}\right) = c_{pd}\, d(\ln\theta)$$

This expression is integrated to a height in the atmosphere where all of the water vapor has been condensed out by adiabatic cooling. The corresponding temperature is called the *equivalent potential temperature*, θ_e. Integration of

$$-L_{lv} \int_{w_s}^{0} d\left(\frac{w_s}{T}\right) = c_p \int_{\theta}^{\theta_e} d(\ln\theta)$$

yields

$$\frac{L_{lv} w_s}{T} = c_{pd}\, \ln\left(\frac{\theta_e}{\theta}\right)$$

or

$$\theta_e = \theta \exp\left(\frac{L_{lv} w_s}{c_{pd} T}\right) \tag{6.48}$$

It is easily determined that $\theta_e > \theta$, which arises from the latent heat released from the condensation of water vapor. Because of the approximations made in (6.47), the equivalent potential temperature is only approximately conserved in a saturated adiabatic process. Although approximate, (6.48) retains the essential physics of the process, whereby the condensation of water vapor provides energy to the moist air and increases its temperature relative to what the temperature would have been in dry adiabatic ascent.

An alternative but analogous potential temperature, the *liquid water potential temperature*, is derived as follows. Writing (6.47) as

$$0 = c_{pd}\, d(\ln T) - R_d\, d(\ln p) - d\!\left(\frac{L_{lv} w_l}{T}\right)$$

we can follow a procedure analogous to the derivation of θ_e and show that (Betts, 1973)

$$\theta_l = \theta \exp\!\left(-\frac{L_{lv} w_l}{c_{pd} T}\right) \tag{6.49}$$

One advantage of θ_l over θ_e is that θ_l reverts to θ, the dry potential temperature, in the absence of liquid water.

In the presence of ice, an *ice–liquid water potential temperature* can be derived from the following approximate form of (6.13):

$$0 = c_{pd}\, d(\ln T) - R_d\, d(\ln p) - d\!\left(\frac{L_{lv} w_l}{T}\right) - d\!\left(\frac{L_{iv} w_i}{T}\right)$$

to be (Tripoli and Cotton, 1981)

$$\theta_{il} = \theta \exp\!\left(-\frac{L_{lv} w_l}{c_{pd} T} - \frac{L_{iv} w_i}{c_{pd} T}\right) \tag{6.50}$$

The derivation of the ice–liquid water potential temperature implies that it is applicable only under conditions of equilibrium, since the affinity terms were not included. Since ice and liquid are both at equilibrium only at the triple point, use of the ice–liquid water potential temperature is inconsistent physically at temperatures away from the triple point. Nevertheless, the ice–liquid water potential temperature is an economical and not too inaccurate way to treat ice processes in a numerical cloud model.

The *entropy potential temperature*, θ_η, includes ice processes and is derived from the complete form of the adiabatic entropy equation (6.13) to be (Hauf and Holler, 1987):

$$\theta_\eta = T\left(\frac{p_0}{p}\right)^{R_d/(c_{pd}+w_t c_l)} \exp\left[\frac{\left(L_{iv}+A_{iv}\right)w_l}{\left[\left(c_{pd}+w_t c_l\right)T\right]} - \frac{\left(L_{il}+A_{il}\right)w_i}{\left[\left(c_{pd}+w_t c_l\right)T\right]}\right]$$ (6.51)

The entropy potential temperature is thus the most general potential temperature considered here. Unlike θ_l and θ_{il}, θ_η is applicable to nonequilibrium conditions such as subsaturated or supersaturated environments.

A major application of the conserved potential temperatures is their use as prognostic variables in cloud models (Section 8.6). Use of the more complex potential temperatures such as θ_{il} and θ_η is desirable in terms of their accuracy; however, a nontrivial calculation is required to invert (6.50) and (6.51) to obtain the physical temperature, T. When various other uncertainties are introduced into a calculation or model, the more approximate forms of the potential temperature can be justified.

Another moist thermodynamic variable that is often used is the *moist static energy*, h. It is conserved in hydrostatic saturated adiabatic processes. We start from the following adiabatic form of the first law of thermodynamics:

$$0 = \left(c_{pd}+w_t c_l\right)dT + d\left(L_{iv}w_v\right) - v\,dp$$

Using the hydrostatic equation (1.33), we may write

$$0 = \left(c_{pd}+w_t c_l\right)dT + d\left(L_{iv}w_v\right) + g\,dz \equiv dh$$

where the term $(1 + w_t)$ accounts for the contribution of the condensed water to the atmospheric density. Upon integration, the moist static energy is shown to be

$$h = \left(c_{pd}+w_t c_l\right)T + L_{lv}w_v + \left(1 + w_t\right)gz$$ (6.52)

The moist static energy is conserved for adiabatic, saturated or unsaturated transformations for a closed system in which the pressure change is hydrostatic.

It is important to note the conditions under which θ_e and the other conserved thermodynamic variables are not conserved. Examples include cases where external radiative heating or conduction takes place, since these alter the entropy. Other examples include atmospheric conditions in which latent heating occurs externally, such as the evaporation of water into air from the ocean or when precipitation falls out.

In this chapter, we have considered numerous temperatures and potential temperatures, which are defined in the context of their conservative properties regarding

Table 6.2 Conservative properties of several parameters (C=conservative; N=nonconservative).

Parameter	Isobaric cooling no condensation	Isobaric cooling with condensation	Adiabatic expansion no condensation	Adiabatic expansion with condensation
w_v	C	N	C	N
$\mathcal{H}$	N	C	N	C
T_D	C	N	N	N
θ	N	N	C	N
θ_e	N	N	C	C
η	N	N	C	C

certain moist atmospheric processes. Table 6.2 summarizes how various temperature, humidity, and other thermodynamic parameters vary in response to certain types of moist processes.

6.8 Aerological Diagrams

The principal function of a thermodynamic diagram is to provide a graphical display of a thermodynamic process. The following examples of thermodynamic diagrams have been used thus far in the text: (T, s) diagram (Section 1.9); (p, V) diagram (Sections 2.4 and 4.2); and (e, T) diagram (Sections 4.2 and 6.4). Here we consider a special class of thermodynamic diagrams called *aerological diagrams*. An aerological diagram is used to represent the vertical structure of the atmosphere and major types of processes to which moist air may be subjected, including isobaric cooling, dry adiabatic processes, and saturated adiabatic processes.

The simplest and most common form of the aerological diagram has pressure as the ordinate and temperature as the abscissa. While the temperature scale is linear, it is usually desirable to have the ordinate approximately representative of height above the surface. Thus the ordinate may be proportional to $-\ln p$ (the *Emagram*) or to p^{R/c_p} (the *Stuve diagram*). The Emagram has the advantage over the Stuve diagram in that area on the diagram is proportional to energy. Before the advent of computers, aerological diagrams were used widely in weather forecasting applications and the energy–area equivalence of the diagram was an important consideration. For the present purposes, we use the aerological diagram to illustrate certain moist atmospheric processes, and the energy–area equivalence is not an important consideration. Because of the simplicity of its construction, we use the Stuve diagram (sometimes referred to as a *pseudo-adiabatic chart*) to illustrate the utility of aerological diagrams in understanding moist thermodynamic processes.

The construction of the pseudo-adiabatic chart is illustrated in Figure 6.6 (see also Appendix E). The temperature scale is linear, while the pressure scale is proportional to p^{R/c_p}. From (2.62), it is easily seen that the dry adiabats or lines of constant potential temperature are straight lines. Pseudo-adiabats (θ_e = constant), are shown by the curved dashed lines. Lines of constant saturated water vapor mixing ratio (w_s = constant) are given by the thin solid lines in Figure 6.6. The ordinate p^{R/c_p} can be interpreted in terms of altitude, z, using (1.45). The use of the pseudo-adiabatic chart is illustrated with the following examples.

Figure 6.7 illustrates vertical profiles of temperature and dew-point temperature plotted on an aerological diagram. Such observations are obtained using balloons, aircraft or remote sensing. From the definition of dew-point temperature (6.15), it is easily seen that by reading off the saturation mixing ratio at the dew-point temperature at a given level on the diagram, one obtains the actual water vapor mixing ratio. Conversely, if the mixing ratio is given, the dew-point temperature may be read off the diagram.

The adiabatic ascent of a parcel from the surface is represented schematically in Figure 6.8. Consider a parcel with $p = p_0$, $T = T_0$, and $w_v = w_0$. The potential temperature of this parcel corresponds to the value of the dry adiabat that passes through T_0,

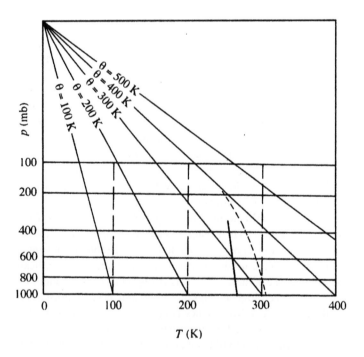

Figure 6.6 Construction of the pseudo-adiabatic chart.

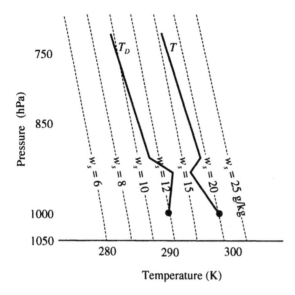

Figure 6.7 Determination of w, w_s, and T_D given the vertical profiles of temperature and dew-point temperature.

p_0. In adiabatic ascent, the parcel will be lifted dry adiabatically along an isopleth of constant θ, that passes through p_0, T_0. In this ascent, the temperature and saturation mixing ratio decrease while the actual mixing ratio remains the same. The level where the saturation mixing ratio equals the actual mixing ratio (the intersection of the constant θ line with the constant w_0 line) corresponds to T_s, p_s, z_s; the saturation temperature and pressure and the lifting condensation level. The thermodynamic properties of air that continues to ascend above the saturation point is found by following the pseudo-adiabat (line of constant θ_e) that passes through T_s, p_s. The mixing ratio of the parcel in pseudo-adiabatic ascent corresponds to the saturation mixing ratio at that level (the intersection of the pseudo-adiabat that passes through T_s, p_s with the constant mixing ratio line). The adiabatic liquid water content at a given level above the saturation point is approximated by subtracting the saturation mixing ratio from the original mixing ratio, w_0.

The equivalent potential temperature, θ_e, corresponding to T_0, p_0, is determined by following the pseudo-adiabat through T_s, p_s to very low pressure, until the pseudo-adiabat is essentially parallel to the dry adiabat. By following the dry adiabat down to a pressure of p_0 and reading off the corresponding temperature, the equivalent temperature, T_e, is obtained; by continuing to follow this dry adiabat down to $p = 1000$ mb, the equivalent potential temperature, θ_e, is obtained. The *equivalent temperature* is related to the equivalent potential temperature analogously to (2.62) as

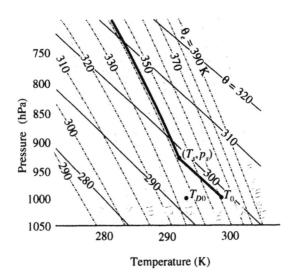

Temperature (K)

Figure 6.8 Adiabatic ascent of a parcel from p_0. The parcel initially ascends dry adiabatically along the constant potential temperature line that passes through $(T_0, 1000 \text{ hPa})$. As the parcel ascends, the saturation mixing ratio decreases while the actual mixing ratio remains the same. At the point at which the actual mixing ratio of the parcel is equal to the saturation mixing ratio, the parcel becomes saturated. Further lifting of the parcel occurs along the saturated adiabat that passes through the point, (T_s, p_s).

$$\theta_e = T_e \left(\frac{p_0}{p} \right)^{R_d/c_p}$$

The wet-bulb temperature, T_W, can be approximated by following the pseudo-adiabat that passes through p_s, T_s down to the level of p_0 and reading the corresponding temperature. By continuing to follow this pseudo-adiabat down to $p = 1000$ mb, the wet-bulb potential temperature, θ_W, is determined. Note that while the pseudo-adiabatic wet-bulb temperature is almost numerically equivalent to the adiabatic isobaric wet-bulb temperature defined in Section 6.3, they are slightly different. In the case of the pseudo-adiabatic wet-bulb temperature, water is evaporated into the air through pseudo-adiabatic descent, while water is evaporated isobarically in the atmosphere in the determination of the adiabatic isobaric wet-bulb temperature.

While aerological diagrams are useful for illustrating schematically the results of thermodynamic transformations of moist air, their use as a computational tool has been superseded by computers.

Notes

General reference sources for this chapter include *Atmospheric Thermodynamics* (1981, Chapters IV and VII) by Iribarne and Godson, *Atmospheric Convection* (1994, Chapter 4) by Emanuel, *Clouds and Storms* (1980, Chapter 3) by Ludlam, and *The Ceaseless Wind* (1986, Chapter 4) by Dutton.

A more detailed discussion of aerological diagrams is given in *Atmospheric Thermodynamics* (1981, Chapter VI) by Iribarne and Godson.

Problems

1. For a pressure of 1000 mb, determine the following. You may use the e_s table in Appendix D. Given:
a) $w_s = 5$ g kg^{-1}, find T.
b) $T = 25°C$, find w_s.
c) $T = 30°C$ and $w = 15$ g kg^{-1}, find $\mathcal{H}$.
d) $T = 20°C$ and $T_D = 15$, find $\mathcal{H}$.
e) $T = 15°C$ and $\mathcal{H} = 0.8$, find T_D.
f) $w = 20$ g kg^{-1}, find T_D.
g) $T_F = -10°C$, find T_D.

2. Consider a 1 kg parcel of moist air at $p = 1000$ mb, $T = 30°C$ and $\mathcal{H} = 0.95$. The parcel passes over a cold ocean so that the parcel cools to 25°C. Assume that only heat (no moisture) is transferred between the ocean and the parcel.
a) What is the initial vapor pressure and mixing ratio of the parcel?
b) What is the dew-point temperature?
c) How much water condenses?

3. During the formation of a radiation fog, 4000 J kg^{-1} is lost after saturation started, at 10°C. The pressure is 1000 mb. Estimate the following:
a) final temperature;
b) vapor pressure;
c) liquid mixing ratio.

4. Home humidifiers, or "swamp coolers," operate by evaporating water into the air in the house, and thereby raise its relative humidity. Consider a house having a volume of 200 m^3 in which the air temperature is initially 21°C and the relative humidity is 10%. Compute the amount of water that must be evaporated to raise the relative humidity to 60%. Assume a constant pressure process at 1010 hPa in which the heat required for evaporation is supplied by the air itself.

5. Under what atmospheric conditions can you "see" your breath? Assume that the air exhaled is at $T = 36°C$ and $\mathcal{H} = 80\%$ at a pressure of 1000 hPa. Find the maximum atmospheric temperature at which condensation will occur when the atmospheric relative humidity is
a)   90%;
b)   10%.

6. A parcel of air with $T = 25°C$ and $T_D = 20°C$ at 1000 hPa is lifted mechanically to a height of 500 mb. Assuming that only adiabatic processes occur, what will be the temperature, water vapor mixing ratio, and liquid water mixing ratio of the parcel at 500 hPa?

Chapter 7 | Static Stability of the Atmosphere and Ocean

In Chapter 1, we considered hydrostatic equilibrium in the atmosphere and ocean, whereby the gravitational acceleration is balanced by the vertical pressure gradient force. Here we examine vertical displacements in a fluid that is in hydrostatic balance. A parcel moving vertically within the fluid is subject to adiabatic expansion or compression, and hence its temperature will change. As the parcel moves vertically, it may become warmer or cooler than the surrounding fluid at a particular level. If the parcel becomes more or less dense than the surrounding fluid, it will be subject to an acceleration downward or upward. By *Archimedes' principle*, the buoyant upthrust is equal to the weight of the surrounding fluid that is displaced. Hence a parcel of the fluid that is displaced vertically is subject to a buoyancy force. If the buoyancy force acting on the displaced mass returns it to its initial position, then the fluid is statically *stable*. If the displaced mass is accelerated away from its initial position, then the fluid is statically *unstable*. If the displaced mass remains in balance with its surroundings, then the fluid is in a state of *neutral equilibrium.*

The static stability of the atmosphere is important in the explanation and prediction of cumulus convection and severe storms, rainfall, boundary layer turbulence, and large-scale atmospheric dynamics. The static stability of the ocean is important in the determination of boundary layer turbulence, internal mixing, convection, and the formation of deep water.

7.1 Stability Criteria

To understand the static stability of the atmosphere and ocean, and to determine criteria for stability, consider a small mass, or *parcel*, that is displaced vertically in a fluid at rest and in hydrostatic equilibrium. The parcel initially has the same thermodynamic state as the surrounding fluid that is at the same vertical level. Once the parcel is displaced away from its initial position, its thermodynamic state may become different from the environment at a corresponding vertical level. The variables in the parcel are denoted by a prime (e.g., T') to differentiate from those in the surrounding environment (e.g., T). The following simplifying assumptions are adopted in the *parcel method*:

1) The parcel retains its identity and does not mix with its environment.
2) The parcel motion does not disturb its environment.
3) The pressure p' of the parcel adjusts instantaneously to the ambient pressure p of the fluid surrounding the parcel.
4) The parcel moves isentropically, so that its potential temperature θ' remains constant.

The impact of these simplifications will be considered after the solution to the elementary problem has been obtained.

The fluid environment is assumed to be in hydrostatic equilibrium, which can be expressed following (1.33) as

$$0 = -g - \frac{1}{\rho}\frac{\partial p}{\partial z} \tag{7.1}$$

Consider a small displacement of the parcel in the vertical direction. From Newton's second law of motion, the acceleration of the parcel must be equal to sum of the gravitational and pressure gradient forces. We can therefore write the following expression for the acceleration of the parcel:

$$\frac{du_z'}{dt} = -g - \frac{1}{\rho'}\frac{\partial p'}{\partial z} \tag{7.2}$$

where u_z' is the vertical velocity of the parcel. From assumption 3) above and $\partial p'/\partial z = \partial p/\partial z = -\rho g$, we obtain

$$\frac{du_z'}{dt} = g\frac{\rho - \rho'}{\rho'} \tag{7.3}$$

The term on the right-hand side of (7.3) is the *buoyancy force* on the parcel. If the parcel is less dense than its surroundings, then it will accelerate upwards. The term *reduced gravity* is often used to denote the negative of the buoyancy force.

We can write (7.3) in terms of vertical density gradients by considering a small vertical displacement of the parcel from its original location. Let $z = 0$ at the initial location, where the parcel density is $\rho_0' = \rho_0$. By using Taylor's theorem, we expand the density of the parcel about the initial location

$$\rho' = \rho_0' + \left(\frac{d\rho'}{dz}\right)z + \ldots \tag{7.4}$$

where we can ignore higher-order terms involving powers of z if the vertical displacement is small. In the same way, we write for the density of the environment

$$\rho = \rho_0 + \left(\frac{d\rho}{dz}\right) z + \ldots \tag{7.5}$$

Substituting (7.4) and (7.5) into (7.3), we obtain

$$\frac{du_z}{dt} = \frac{g}{\rho_0'} \left[\left(\frac{d\rho}{dz}\right) - \left(\frac{d\rho'}{dz}\right) \right] z \approx \frac{g}{\rho_0} \left[\left(\frac{d\rho}{dz}\right) - \left(\frac{d\rho'}{dz}\right) \right] z \tag{7.6}$$

For the parcel to be stable to a vertical displacement, so that any vertical displacement would be followed by an acceleration returning the parcel to its initial position, $d\rho'$ must exceed the corresponding vertical density change of the surroundings, $d\rho$.

The *Brunt–Väisälä frequency*, N, is defined by

$$N^2 = \frac{g}{\rho_0} \left[\left(\frac{d\rho'}{dz}\right) - \left(\frac{d\rho}{dz}\right) \right] \tag{7.7}$$

and is also referred to as the *buoyancy frequency*. Substituting (7.7) into (7.6) yields

$$\frac{d^2z}{dt^2} + N^2 z = 0 \tag{7.8}$$

where we have written $d^2z/dt^2 = du_z/dt$ for the acceleration. This equation is of the form of the equation of motion for a linear harmonic oscillator. The solution to (7.8) is easily shown to be

$$z = A_1 \exp(iNt) + B_1 \exp(-iNt) \tag{7.9}$$

where A_1 and B_1 are arbitrary constants. If $N^2 < 0$, then the solution to (7.8) is

$$z = A_1 \exp\left(|N|t\right) + B_1 \exp\left(-|N|t\right) \tag{7.10}$$

The first term on the right-hand side of (7.10) implies that the displacement will increase exponentially with time, a clear case of instability. That is, for an arbitrary upward or downward displacement, the parcel will accelerate away from its inital location. If $N^2 > 0$, then the solution to (7.8) is

$$z = A_1 \cos(Nt) + B_1 \sin(Nt) \tag{7.11}$$

so that a parcel oscillates in the vertical with frequency N and period of oscillation, τ_g, given by

$$\tau_g = \frac{2\pi}{N} \tag{7.12}$$

Because the position of the parcel oscillates about its initial position when $N^2 > 0$, this is a stable situation. If $N^2 = 0$, the parcel is neutral and there is no acceleration.

From (7.8), we obtain the following criteria of static stability for a fluid with respect to small vertical displacements:

$$
\begin{aligned}
N^2 > 0&: \text{ stable} \\
N^2 = 0&: \text{ neutral} \\
N^2 < 0&: \text{ unstable}
\end{aligned}
\tag{7.13}
$$

To interpret more completely these stability criteria, we examine the buoyancy frequency, N, individually for the ocean and the atmosphere.

First, consider the application of (7.7) to the ocean. The vertical density change $d\rho'/dz$ for an ocean parcel that moves vertically and isentropically can be written as

$$\frac{d\rho'}{dz} = \left(\frac{\partial \rho'}{\partial p'}\right)\frac{dp'}{dz} + \left(\frac{\partial \rho'}{\partial T'}\right)\frac{dT'}{dz} \tag{7.14}$$

Note that the salinity of the parcel does not change during vertical motion. The vertical density change $d\rho/dz$ in the ocean environment is given by

$$\frac{d\rho}{dz} = \left(\frac{\partial \rho}{\partial p}\right)\frac{dp}{dz} + \left(\frac{\partial \rho}{\partial T}\right)\frac{dT}{dz} + \left(\frac{\partial \rho}{\partial s}\right)\frac{ds}{dz} \tag{7.15}$$

By incorporating (7.14) and (7.15) into the expression for the buoyancy frequency (7.7), we obtain

$$N^2 = g\alpha\left(\frac{dT}{dz} + \Gamma'_{ad}\right) - g\beta\frac{ds}{dz} \tag{7.16}$$

where we have used definitions of the thermal expansion coefficient α (1.31a), the saline contraction coefficient β (1.31c), and the adiabatic lapse rate Γ (2.68), and we have ignored the effect of pressure fluctuations. Expression (7.16) states that if there is no vertical salinity gradient in the environment, then the ocean is stable to vertical displacements if the lapse rate in the ocean is less than the adiabatic value. Also, if the environmental lapse rate is equal to the adiabatic lapse rate, the ocean will be unstable to vertical displacements if the salinity decreases with depth.

We can follow a similar procedure for the moist (but unsaturated) atmosphere. Using the ideal gas law and ignoring pressure fluctuations, we can write (7.7) as

$$N^2 = \frac{g}{T_0}\left(\frac{dT_v}{dz} + \Gamma_d'\right) \qquad (7.17)$$

From the definition of the virtual potential temperature θ_v in (2.67b), we can show from (2.63) that

$$\frac{1}{\theta}\frac{d\theta_v}{dz} = \frac{1}{T}\frac{dT_v}{dz} - \frac{R_d}{c_{pd}}\frac{1}{p}\frac{dp}{dz} = \frac{1}{T}\left(\frac{dT_v}{dz} - \frac{g}{c_{pd}}\right) \qquad (7.18)$$

and we can write (7.17) as

$$N^2 = \frac{g}{\theta_0}\frac{d\theta_v}{dz} \qquad (7.19)$$

The static stability criteria for moist but unsaturated atmosphere can thus be written equivalently as

$$\frac{d\theta_v}{dz} > 0 \quad \text{or} \quad -\frac{dT_v}{dz} < \Gamma_d : \quad stable$$

$$\frac{d\theta_v}{dz} = 0 \quad \text{or} \quad -\frac{dT_v}{dz} = \Gamma_d : \quad neutral \qquad (7.20)$$

$$\frac{d\theta_v}{dz} < 0 \quad \text{or} \quad -\frac{dT_v}{dz} > \Gamma_d : \quad unstable$$

The term $-dT_v/dz$ in (7.20) can be interpreted as the lapse rate of virtual temperature in the environment. In considering atmospheric static stability, it is important to retain the virtual temperature correction, since differences in the specific humidity between the parcel and the environment can be significant in determining the buoyancy force, especially in the tropics.

Static instability in the atmosphere and ocean ($N^2 < 0$) gives rise to *buoyant convection*, which refers to vertical motions induced by buoyant accelerations. In the atmosphere, convection is typically induced by heating of the Earth's surface and commonly results in the formation of cumuliform clouds (Section 8.5). In the ocean, convection is induced by an increase in surface density caused by cooling and/or an increase in salinity caused by excessive evaporation or sea ice formation; thus convection in the ocean consists of downward motions. Convection in both the atmosphere and ocean transports heat, water vapor or salinity, and momentum, and acts to reduce the static instability.

From the requirements of mass continuity (3.7), an upward moving parcel in the atmosphere must be accompanied by descending motion in the surrounding environment to replace the volume vacated by the rising parcel. If the descending air is cloud free, it will warm at the dry adiabatic rate. The increase in the temperature in the surrounding air will reduce $(T_v - T_v')$ and thus reduce $(\rho - \rho')$ in (7.3), hence reducing the buoyancy force. Analogously in the ocean, descending convective parcels will induce compensating rising motions in the surrounding water that act to diminish the buoyancy force.

7.2 Stability of a Saturated Atmosphere

Vertical displacements of air parcels frequently result in phase changes of water substance, which affect the buoyancy of the air and thus the static stability criteria. When a saturated parcel of air is displaced vertically, its temperature changes according to the saturated adiabatic lapse rate. Application of the simple parcel model to a saturated air parcel yields the following expression:

$$N^2 = \frac{g}{T_0}\left(\frac{dT_{vl}}{dz} + \Gamma_s'\right) \tag{7.21}$$

where Γ_s is the saturated adiabatic lapse rate according to (6.40). Because of the weight of the condensed water, a *liquid water virtual temperature* T_{vl} is used instead of T_v as defined by (1.25), whereby

$$T_{vl} = T\left(1 + 0.608\,w_v - w_l\right) \tag{7.22}$$

Note that if $w_l = 0$, then $T_{vl} = T_v$. Derivation of an expression analogous to (7.20) for saturated air shows that cloudy air is unstable when θ_e decreases upward and/or total water ($w_t = w_v + w_l$) increases upward.

Allowing for the possibility of condensation/evaporation during the parcel displacement leads to five possible states of static stability for moist air:

$$-\frac{dT_{vl}}{dz} < \Gamma_s : \quad absolutely\, stable$$

$$-\frac{dT_{vl}}{dz} = \Gamma_s : \quad saturated\, neutral$$

$$\Gamma_s < -\frac{dT_{vl}}{dz} < \Gamma_d : \quad conditionally\, unstable \qquad (7.23)$$

$$-\frac{dT_{vl}}{dz} = \Gamma_d : \quad dry\, neutral$$

$$-\frac{dT_{vl}}{dz} > \Gamma_d : \quad absolutely\, unstable$$

These stability criteria are illustrated in Figure 7.1. The conditional instability criterion implies that an unsaturated parcel is stable to a vertical displacement, while a saturated parcel is unstable. Figure 7.2 gives an example of conditional instability plotted on an aerological diagram. A moist but unsaturated parcel that is forced to ascend from 1000 hPa is at first colder than the environment and subject to a

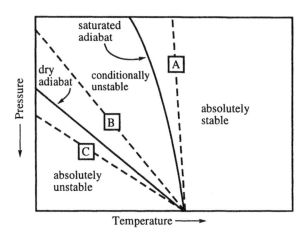

Figure 7.1 Regions of stability, instability, and conditional stability illustrated on an aerological diagram. When the environmental lapse rate is less than the saturated adiabatic lapse rate (e.g., lapse rate A), the atmosphere is absolutely stable. When the environmental lapse rate is greater than the saturated lapse rate, but less than the dry adiabatic lapse rate (e.g., lapse rate B), the atmosphere is conditionally stable. When the environmental lapse rate is greater than the dry adiabatic lapse rate (e.g., lapse rate C), the atmosphere is absolutely unstable.

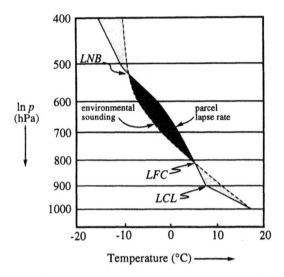

Figure 7.2 Convective instability illustrated on an aerological diagram. The dashed line represents the environment (T) and the solid line represents the parcel (T'). Below 810 mb and above 530 mb, energy is required to lift the parcel. Above 810 mb and below 530 mb, the parcel accelerates freely. The dark shaded area represents the convective available potential energy (CAPE), while the two lightly shaded areas represent the convection inhibition energy (CINE).

downward (restoring) buoyancy force. The parcel reaches its lifting condensation level (LCL) at 900 hPa and ascends along the saturated adiabat thereafter. If the parcel is lifted above 810 hPa, it becomes warmer than the environment and is subject to an upward buoyancy force. Conditional instability implies that there is a critical vertical displacement from which a parcel may change from being stable to unstable (or vice versa). The level of this critical perturbation ($p = 810$ hPa in Figure 7.2) is called the *level of free convection (LFC)*. The parcel will continue to accelerate upwards until the *level of neutral buoyancy (LNB)* is reached ($p = 530$ hPa in Figure 7.2), which denotes the level above which the parcel in saturated adiabatic ascent would become cooler than the environment.

The amount of energy available for the upward acceleration of a particular parcel is called the *convective available potential energy (CAPE)*. On a thermodynamic diagram whose area is proportional to energy (e.g., the emagram; see Section 6.8), *CAPE* is proportional to the area enclosed by the two curves that delineate the temperature of a parcel and its environment, as illustrated by the darker shaded region in Figure 7.2. The amount of *CAPE* of a parcel lifted from a height z (at or above the *LFC*) to the *LNB* is given by the vertical integral of the buoyancy force between these levels

$$CAPE(z) = \int_{z}^{LNB} g\, \frac{\rho - \rho'}{\rho'}\, dz \qquad (7.24)$$

where the units of $CAPE$ are J kg^{-1}. If the environment is in hydrostatic equilibrium we can use (1.26) and (1.33) to obtain

$$CAPE(p) = \int_{p(LNB)}^{p(z)} R_d\left(T_v' - T_v\right) d(\ln p) \qquad (7.25)$$

$CAPE$ is defined only for parcels that are positively buoyant somewhere in the vertical profile. The term *convection inhibition energy* ($CINE$) is analogous to $CAPE$ but refers to a negative area on the thermodynamic diagram.

An important simplification made in the elementary parcel theory is that the parcel does not mix with its environment. However, as a buoyant parcel ascends, mixing typically occurs through the parcel boundaries as a result of turbulent motions, which is called *entrainment*. Since the environmental air is typically cooler and drier than a rising saturated parcel, entrainment will lower the buoyancy of the parcel and reduce the amount of condensed water. Consider a mass m of saturated cloudy air which rises from a level z, entraining a mass dm of environmental air over the distance dz. The cloudy air has a temperature T', and the environmental air, T. Applying the first law of thermodynamics to the mixture $m + dm$ and assuming no heat transfer mechanisms occur other than condensation, evaporation, and mixing, we have

$$m\left(c_{pd}\, dT' - R_d T' \frac{dp}{p}\right) = -m L_{lv}\, dq_s - c_{pd}\,(T' - T)\, dm - L_{lv}\,(q_s' - q_v)\, dm \quad (7.26)$$

The first term on the right-hand side of (7.26) describes the latent heat released by the cloudy parcel in ascent; the second term denotes the heat required to warm the entrained air; and the third term describes the latent heat required to evaporate just enough water from the cloudy air to saturate the mixed parcel.

Following the procedure developed in Section 6.5 to derive an expression for the saturated adiabatic lapse rate (6.40), we use (7.26) to determine an expression for the lapse rate of a saturated parcel subject to entrainment, Γ_m:

$$\Gamma_m = \Gamma_s + \frac{\dfrac{1}{m}\dfrac{dm}{dz}\left[(T' - T) + \dfrac{L_{lv}}{c_{pd}}(q_s' - q_v)\right]}{1 + \dfrac{\varepsilon L_{lv}^2 q_s}{c_{pd} R_d T^2}} \qquad (7.27)$$

Note that (7.27) reduces to Γ_s if $dm/dz = 0$, i.e., if no entrainment takes place. For $dm/dz > 0$ and $T' > T$, then $\Gamma_m > \Gamma_s$. In effect, the mixing of cloudy air with dry environmental air reduces the density difference between the parcel and its environment, hence reducing the buoyancy force. Since the lapse rate in an entraining cloud is greater than the saturated adiabatic lapse rate, an entraining cloud achieves a smaller vertical velocity relative to that predicted by parcel theory. To assess stability criteria for an entraining cloud parcel, Γ_m can be substituted into (7.23) for the stability criteria. The process of entrainment of environmental air into vertically developing clouds is not adequately understood at present, and a complete discussion of the hydrodynamics involved in entrainment is beyond the scope of this text. However, the expressions (7.26) and (7.27) describe qualitatively the effect that entrainment has on the stability criteria.

7.3 Processes Producing Changes in Stability

The static stability of a layer in the atmosphere or ocean is modified by:

1) vertical motions in the layer; and
2) differential heating or cooling of the layer.

Vertical motions in a layer modify the layer static stability in the following way. Consider the large-scale ascent of a dry atmospheric layer (Figure 7.3), during which the mass of the layer remains constant (i.e., there is no horizontal or vertical convergence of air). From (7.18), we can write

$$\frac{1}{\theta}\frac{d\theta}{dz} = \frac{\Gamma_d - \Gamma_{env}}{T} \tag{7.28}$$

where Γ_{env} is the lapse rate of the environment and we have ignored the virtual temperature effects for dry air. By incorporating the hydrostatic equation (1.33) and the ideal gas law (1.15), we obtain

$$\frac{1}{\theta}\frac{d\theta}{dp} = -\frac{R_d}{g}\frac{\left(\Gamma_d - \Gamma_{env}\right)}{p} \tag{7.29}$$

During ascent or descent of the layer, the derivative $d\theta/dp$ remains constant, since θ is conserved in dry adiabatic motion and the mass of the layer remains constant. Hence we can write

$$\Gamma_d - \Gamma_{env} = C_1 p$$

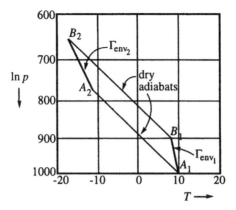

Figure 7.3 An initially stable layer A_1B_1 is made less stable as a result of dry adiabatic ascent.

where C_1 is a constant. Therefore, during the ascent of a layer, the lapse rate of the layer approaches the dry adiabatic lapse rate. Thus an initially stable layer is made less stable, while an initially unstable layer is made more stable. The reverse occurs during the descent of a layer when the pressure increases, whereby the environmental lapse rate moves further away from Γ_d.

Now consider the stability of a lifted layer when condensation occurs (Figure 7.4). Let the line A_1B_1 represent the lapse rate of the layer initially, and A_2B_2 after lifting adiabatically to saturation. After lifting, it is seen that the layer is unstable relative to the saturated adiabat. By considering numerous examples such as shown in Figure 7.4, it can be inferred that:

i)   the saturated layer will be stable if $d\theta_e/dz > 0$;
ii) the saturated layer will be neutral if $d\theta_e/dz = 0$;
iii) the saturated layer will be unstable if $d\theta_e/dz < 0$.

The above stability criteria follow from the fact that $d\theta_e/dz$ does not change sign during lifting. This analysis of the changes in stability associated with lifting a layer of moist air is relevant in determining whether unstable conditions and deep convection are likely to occur during orographic or frontal lifting, for example.

In addition to the large-scale ascent and descent of a layer, the static stability of a layer in the atmosphere or ocean can be modified as result of vertical turbulent mixing (see Section 3.6). Consider two isolated masses m_1 and m_2 at the pressure levels p_1 and p_2 with temperatures T_1 and T_2 (Figure 7.5). Ignore for now any gradients in salinity or water vapor mixing ratio that might influence stability. To examine the thermodynamics of the mixing process, consider an idealized process whereby these two masses are both brought adiabatically to an intermediate pressure level where

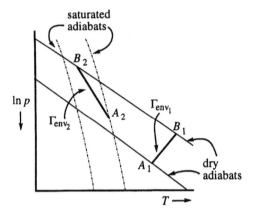

Figure 7.4 Destabilizing an initially stable atmospheric layer. The initially stable and unsaturated inversion layer A_1B_1 is lifted adiabatically. If the bottom of the layer reaches saturation before the top of the layer (as, for example, in an inversion layer in which the mixing ratio decreases with height), further lifting will destabilize the layer. This occurs because the bottom of the layer cools at the much slower saturated adiabatic lapse rate, while the top of the layer continues to cool at the faster dry adiabatic lapse rate.

they are mixed. The potential temperature of the mixture is a weighted mean potential temperature (see also Section 6.4) of the two masses. Finally, the two masses return to their original pressure level, both masses having the same potential temperature. If the entire layer $\Delta p = p_2 - p_1$ mixes vertically as a result of the motion of turbulent eddies (Section 6.4), then the potential temperature of the layer will become constant, with a value corresponding to the mass-weighted potential temperature of the layer. Thus vertical mixing acts to destabilize a layer that is initially stable, and stabilize a layer that is initially unstable.

The influence of vertical mixing on the stability of the atmosphere is modified if condensation occurs during the mixing process. After mixing, a layer will be characterized by a weighted mean total water mixing ratio and a weighted mean liquid water potential temperature (6.50). If the average mixing ratio line intersects the average potential temperature line (Figure 7.6), then from that level upwards, condensation occurs and the final temperature distribution follows the saturated adiabat. The level of intersection is called the *mixing condensation level (MCL)*. As a result of vertical mixing under conditions when a cloud forms in the layer, the stability of the layer will be modified towards the dry adiabat below the cloud and towards the saturated adiabat in the cloud layer.

In addition to changes in stability associated with large-scale and turbulent vertical motions, changes in stability can also occur by differential heating/cooling in an atmosphere or ocean layer. The presence of boundary layer clouds will radiatively cool the top of the boundary layer, contributing to destabilization of the atmospheric

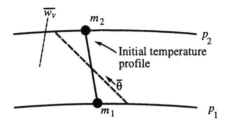

Figure 7.5 Vertical mixing of air parcels, m_1 and m_2, without condensation. Two air parcels, initially at different pressure levels, mix at an intermediate pressure level. The potential temperature of the mixture is a mass-weighted average of the individual parcels' potential temperatures. Mixing of an entire layer results in a constant potential temperature $\overline{\theta}$ throughout the layer. This destabilizes an initially stable layer and stabilizes an initially unstable layer. Because the dry adiabat corresponding to $\overline{\theta}$ does not intersect the average mixing ratio line, $\overline{w}_v$, the mixing process is dry adiabatic and no condensation occurs.

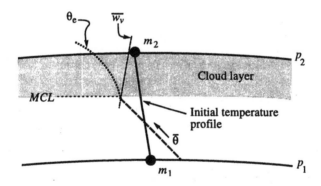

Figure 7.6 Vertical mixing of air parcels, m_1 and m_2, with condensation. If the mixing of two air parcels results in an average potential temperature, $\overline{\theta}$, that intersects the average mixing ratio line, $\overline{w}_v$, then from the level of intersection upwards, condensation will occur and the final temperature distribution will follow a saturated adiabat, θ_e. The lapse rate below the cloud layer moves towards the dry adiabatic lapse rate, while the lapse rate within the cloud layer moves towards the saturated adiabatic lapse rate.

boundary layer. Differential heat advection over a layer in the atmosphere or ocean can also modify the static stability.

Consider Figure 7.7, which represents the coupled atmospheric and oceanic boundary layers in a state that is initially neutral. Net heating of the air/sea interface will increase the surface temperature; this will destabilize the atmosphere ($N^2 < 0$) and stabilize the ocean ($N^2 > 0$). Net cooling of the air sea interface will decrease the

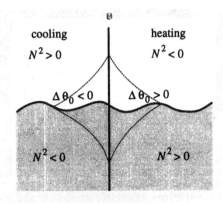

Figure 7.7 Heating of the ocean surface ($\Delta \theta_0 > 0$), represented schematically by the dotted line to the right of the initial potential temperature curve, destabilizes the lower atmosphere while stabilizing the upper ocean.   The reverse occurs for surface cooling ($\Delta \theta_0 < 0$).

surface temperature, stabilizing the atmosphere ($N^2 > 0$) and destabilizing the ocean ($N^2 < 0$). Modifications to ocean stability also occur by surface exchanges of fresh water and salt. Precipitation, melting of sea ice, and river runoff contribute to the stabilization of the upper ocean ($N^2 > 0$), while evaporation and the freezing of sea-water destabilize the upper ocean ($N^2 < 0$).

Notes

A treatment of the stability of both the atmosphere and ocean is given by Gill (1982, Chapter 3) *Atmosphere-Ocean Dynamics.*

A detailed discussion of atmospheric stability is given by Emanuel (1994, Chapter 6) *Atmospheric Convection* and Iribarne and Godson (1981, Chapter IX) *Atmospheric Thermodynamics.*

Problems

1. In an unstable layer of air near the ground, the temperature is found to decrease linearly at a rate of 2.5°C per 100 m. A parcel of air at the bottom of the layer with $T = 280$ K is given an initial upward velocity of 1 m s^{-1}. Assuming the parcel ascends dry adiabatically, what will be its height and speed after one minute?

2. Calculate the period of oscillation for a thick atmospheric layer is isothermal, with $T = -3°C$. A dry parcel is given an upward impetus and begins to oscillate about its original position.

3. Suppose that the environmental lapse rate is dry adiabatic, with a temperature of 280 K at 900 hPa, and a relative humidity of 50%. Consider a parcel of saturated air at 900 hPa at 280 K, initially at rest. If this parcel is given an upward displacement, it will be positively buoyant and will continue to ascend. Neglecting entrainment and aerodynamic resistance, calculate the parcel's upward velocity at 700 hPa, assuming the following:
a) elementary parcel theory without including the virtual temperature correction;
b) elementary parcel theory including the virtual temperature correction;
c) parcel theory with a correction for the weight of condensed water, assuming full adiabatic water content.

Chapter 8 | Cloud Characteristics and Processes

The most distinctive feature of the Earth when viewed from space is the presence of clouds covering approximately half of its surface area (Figure 8.1). The latent heat released in clouds is an important source of energy for scales of motion ranging from the global atmospheric circulation, hurricanes and mid-latitude cyclones, to individual storms. Clouds are dominant components in the Earth's hydrological cycle, transporting water vertically and horizontally and removing water from the atmosphere

Figure 8.1 View of the Earth from satellite.

through precipitation. Clouds are associated with some of the most damaging weather in the world: torrential rains, severe winds and tornadoes, hail, thunder and lightning, and snow storms. Clouds are a major factor in determining the Earth's radiation budget, by reflecting shortwave radiation and emitting longwave radiation. The interactions of clouds with radiation determine the amount of radiation that reaches the Earth's surface and thus influence surface evaporation and evapotranspiration rates. Clouds are a major element of surface weather and are the prime determinant of atmospheric visibility. Clouds are also important in atmospheric chemistry because they play an active role in many chemical reactions and transport chemicals through updrafts and scavenging associated with precipitation.

A major challenge in understanding and modeling clouds is the broad range of spatial scales involved. The scales range from the micron scale of individual cloud drops, to the scale of an individual cloud (kilometers), up to the scale of the largest cloud systems (1000 km). Our present computational capability allows only a small range of spatial scales to be simulated explicitly in a single model. Processes on the other scales are either specified or parameterized.

In this chapter, the characteristics and classification of clouds are discussed. We describe precipitation and cloud radiative interactions: these are the two principle irreversible processes associated with clouds and are the primary thermodynamic effects that clouds have on their environment. Finally, we introduce the problem of parameterization of cloud processes in large-scale models.

8.1 Cloud Classification and Characteristics

Cloud classification schemes were introduced in the early 19th century based upon the physical appearance of clouds from the perspective of a surface observer. Such schemes allowed for uniform identification of clouds around the world by different observers. It has been argued that alternative cloud classifications based either upon physical principles (e.g., cloud motions or cloud-radiative characteristics) would be more useful to atmospheric science. However, no such alternative cloud classification has achieved any widespread acceptance; thus the standard morphological cloud classification of the World Meteorological Organization is used here.

The morphological cloud classification scheme is based on three cloud characteristics: cloud shape, cloud height, and whether or not the cloud is precipitating. There are three principal cloud shapes:

1) curly or fibrous clouds are known as *cirrus* clouds;
2) layered or stratified clouds are known as *stratus* clouds;
3) lumpy or heaped clouds, increasing upward from a horizontal base, are known as *cumulus* clouds.

Clouds are also distinguished by the heights above ground level at which they form:

1) high clouds whose bases are higher than 6 km in the tropics and 3 km in the polar regions (prefix: *cirro*);
2) middle clouds whose bases lie between 2 and 8 km in the tropics and 2 and 4 km in the polar regions (prefix: *alto*);
3) low clouds whose bases lie below 2 km;
4) clouds of vertical development.

The prefix *nimbo* or the suffix *nimbus* indicates the presence of rain.

Using this basic framework, the cloud classification is based on ten main cloud groups called *genera*. The definitions of the ten genera are as follows (from the *International Cloud Atlas*):

Cirrus (Ci). Detached clouds that are white and have a fibrous (hair-like) appearance or a silky sheen. These clouds appear in the form of delicate filaments, patches or narrow bands.

Cirrocumulus (Cc). Thin, white patch of cloud without shadows. The clouds are composed of very small elements in the form of grains or ripples that are merged or separate, and more or less regularly arranged. Most of the elements have an apparent width of less than 1° when viewed from the surface.

Cirrostratus (Cs). Transparent, whitish cloud veil of fibrous (hair-like) or smooth milky appearance, totally or partly covering the sky, and generally producing halo phenomena.

Altocumulus (Ac). White or gray cloud which occurs as a layer or patch, generally with shading. The clouds are composed of laminae, rounded masses, or rolls, which are sometimes partly fibrous or diffuse and which may or may not be merged. Most of the regularly arranged elements have an apparent width of between 1° and 5° when viewed from the surface.

Altostratus (As). Grayish or bluish cloud layer of striated, fibrous or uniform appearance. The layer has parts thin enough to reveal the sun at least dimly, as through ground glass. Altostratus does not show halo phenomena.

Nimbostratus (Ns). Gray cloud layer, often dark, rendered diffuse by more or less continuously falling rain or snow but not accompanied by lightning, thunder, or hail. Nimbostratus is thick enough to blot out the sun.

Stratocumulus (Sc). Gray or whitish patch or layer of cloud which almost always has dark parts, composed of tessellations, rounded masses, rolls, etc., which are non-fibrous and which may or may not be merged. Most of the regularly arranged small elements have an apparent width of more than 5° when viewed from the ground.

Stratus (St). Generally gray cloud layer which may produce drizzle, ice prisms or snow grains. When the sun is visible through the cloud, its outline is clearly discernible. Stratus clouds do not produce halo phenomena except possibly at very cold temperatures. Sometimes stratus clouds appear in the form of ragged patches.

Cumulus (Cu). Detached clouds, generally dense and with sharp outlines, developing vertically in the form of rising mounds, domes or towers, of which the bulging upper part often resembles a cauliflower. The sunlit parts of these clouds are typically brilliant white, while their base is relatively dark and nearly horizontal.

Cumulonimbus (Cb). Heavy, dense clouds, with considerable vertical extent, in the form of a huge tower. At least part of their upper portion is usually smooth, fibrous, or striated and is nearly always flattened; this part often spreads out in the shape of an anvil or vast plume. Under the base of this cloud, which is generally very dark, there are frequently low ragged clouds and precipitation.

Fog is not included as a genus in this cloud classification scheme. *Fog* is composed of very small water drops (sometimes ice crystals) in suspension in the atmosphere and it reduces the visibility at the surface to less than 1 km. It will be shown in Section 8.4 that fog may be considered as a stratus cloud whose base is low enough to reach the ground.

Besides their morphological appearance to a ground observer, the different cloud genera are associated with different characteristic values of cloud temperature and phase, amount of condensed water, cloud vertical velocities and turbulence, and cloud time scales. These cloud characteristics determine whether or not the cloud will precipitate and the form and amount of precipitation, and also the influence of the cloud on the radiation balance.

Satellite views of the Earth reveal organized cloud patterns, some of which extend over distances of hundreds to thousands of kilometers (Figure 8.1). In the mid-latitudes, organized cloud patterns are commonly associated with frontal systems. Mesoscale convective complexes are found commonly over warm land surfaces and over tropical oceans. Even within large-scale cloud systems, clouds exhibit fine detail in their horizontal structure.

Because they are most closely coupled to the ocean surface, the cloud types stratus, stratocumulus, cumulus, and cumulonimbus are described more completely in Sections 8.4 and 8.5.

8.2 Precipitation Processes

The nucleation and diffusional growth of cloud particles was discussed in Chapter 5. In Table 5.5, it was shown that diffusional growth of water drops is not sufficient to grow particles that are large enough to precipitate over the lifetime of many clouds that are observed to precipitate. Diffusional growth of ice crystals can occur sufficiently rapidly in a water-saturated environment to form particles that are large enough to have a significant fall speed. However, many clouds are observed to form precipitation-sized particles in warm clouds on time scales as short as 10–20 minutes.

The mechanisms that form precipitation-sized particles can be divided into warm-cloud and cold-cloud processes, the distinction arising from the absence or presence of ice particles. In both warm- and cold-cloud rain production, the collision and coalescence of cloud particles to form larger cloud particles, also called *accretional growth*, is an essential element. The following cloud particle interactions can occur, giving rise to precipitation-sized particles: collision and coalescence between water drops; collection of water drops by ice crystals; and aggregation of ice particles. A collision between cloud particles is not sufficient for aggregation; once the particles collide, they must coalesce or "stick" together.

Central to understanding precipitation and also the collision and coalescence process is the concept of particle terminal velocity. When the gravitational force between a cloud particle and the earth is balanced by the frictional force of the particle as it falls through the air, the speed at which the particle is falling is called the *terminal velocity*. For a small spherical liquid drop,[1] we may approximate the terminal velocity, u_T, as

$$u_T = k_1 r^2 \tag{8.1a}$$

with $k_1 = 1.19 \times 10^6$ cm^{-1} s^{-1}. This quadratic dependence of fall speed on size for drops with $r < 30$ μm is called *Stokes' law*. Stokes' law does not hold for larger particles, since the shape of larger drops is deformed as they fall and the frictional force becomes more complex. Experiments with falling drops have provided the following approximations for larger drops to be

$$u_T = k_2 r \tag{8.1b}$$

with $k_2 = 8 \times 10^3$ s^{-1}. This equation is valid for particles in the size range 40 μm $< r <$ 0.6 mm. For the largest category of particles, 0.6 mm $< r <$ 2 mm, we have

$$u_T = k_3 r^{1/2} \tag{8.1c}$$

where $k_3 = 2.01 \times 10^3 (\rho_0/\rho_a)^{1/2}$ cm$^{1/2}$ s^{-1} and ρ_0 is a reference density of 1.2 kg m^{-3}. The terminal velocity of cloud drops as a function of particle radius is shown in Figure 8.2. Cloud drops with r < 50 μm do not have an appreciable fall speed. The maximum raindrop size is about $r = 3$ mm; larger raindrops break up owing to aerodynamical forces on the drop.

The fall speed of ice particles is more difficult to determine theoretically than that for liquid cloud drops, because of the complex shapes and variable density of ice

---

[1] The gravitational force on a sphere is given by $\mathcal{F}_g = (4/3)\,\pi r^3 g(\rho_l - \rho_a)$ and the frictional force exerted on a sphere of radius r is given by $\mathcal{F}_R = (\pi/2)\,r^2 u_T^2 \rho_a C_D$, where C_D is the drag coefficient characterizing the flow. Using the Reynolds number, $Re = 2\rho_a u_T r/\mu$, with μ the dynamic viscosity of the fluid, we can write $\mathcal{F}_R = 6\pi\mu r u_T (C_D Re/24)$. For very small Reynolds number ($r < 30$ μm), $(C_D Re/24) = 1$, and $u_T = (2/9)\,r^2 g \rho_l/\mu = k_1 r^2$.

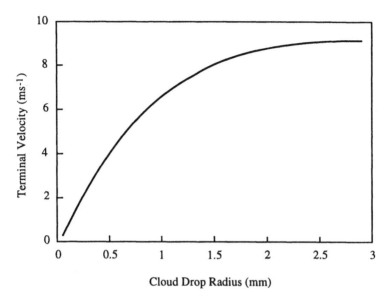

Figure 8.2 Terminal velocity of cloud drops as a function of drop radius. (Data from Gunn and Kinzer, 1949.)

crystals. Figure 8.3 shows the observed fall speeds of different ice crystal types. Graupel are the fastest falling ice particles, with single (unaggregated) crystals having fall speeds less than 1 m s^{-1}.

Figures 8.2 and 8.3 show that cloud particles fall at varying speeds, depending on their size, shape, and density. In general, larger particles fall faster than small particles and may collide with a smaller, slower particle that is in its path (Figure 8.4). A larger particle will not necessarily capture a smaller particle in its path, due to inertial and aerodynamic forces.

A *collection efficiency*, $\mathcal{E}$, is defined as the probability that a collision will occur between two particles located at random in the volume swept out by the faster falling particle, and that the particles "stick" or coalesce upon colliding. Figure 8.5 presents the collision efficiencies for small collector drops. Note that the collision efficiency in Figure 8.5 is virtually identical to the collection efficiency since the coalescence probability is unity when the drops are all smaller than 100 μm. For any size of collector drop R, the collision efficiency is small for small values of r/R, since small drops have little inertia and are easily deflected by the flow around the collector drop. As r/R increases, the inertia of the drops increases, accounting for an increase in the collision efficiency. As r/R approaches unity, values of $\mathcal{E}$ may exceed unity as the trailing drop can be attracted into the wake of a drop falling close by (but not directly in the swept out volume) at nearly the same speed.

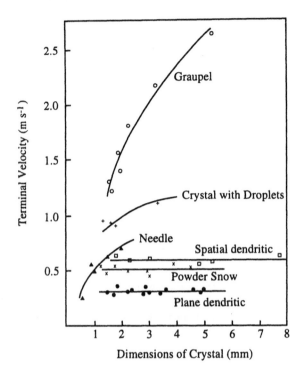

Figure 8.3 Observed terminal velocities of ice particles as a function of crystal type and size. (From Fletcher, 1962.)

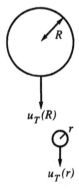

Figure 8.4 Collision geometry for a collector drop of radius R falling with speed $u_T(R)$ through a population of smaller drops of radius r, falling with a speed $u_T(r)$.

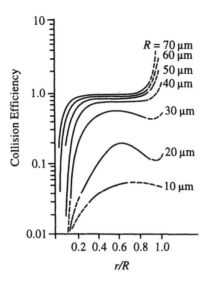

Figure 8.5 Collision efficiencies for collector drops of radius R and drops of radius r. (From Klett and Davis, 1973.)

The collection efficiency associated with ice particles is more complex than that for interactions between liquid water drops. The term *riming* is used to describe the capture of supercooled drops by an ice particle. *Aggregation* describes the collection of ice particles by other ice particles to form snowflakes. Far less is known about the efficiency of ice particle collection than is known about the corresponding collection by water drops. Determination of the collision efficiency is made difficult by the complex aerodynamics of falling ice particles. If an ice and liquid particle do collide, the coalescence efficiency is believed to be unity because freezing of the drop is likely to occur on contact. The coalescence efficiency in aggregation is greatest for crystals that have open crystal structures such as dendrites, where the dendritic branches can "hook" onto each other. Observations suggest that significant aggregation is possible only at temperatures warmer than $-10°C$.

The growth rate of a particle by collision and coalescence is illustrated by examining a continuous collection process. A particle of radius R falling through a population of smaller particles (see Figure 8.4) will sweep out drops of radius r from a volume in unit time as

$$\pi \left(R + r\right)^2 \left[u_T\left(R\right) - u_T\left(r\right)\right] n\left(r\right) \mathcal{E}\left(R, r\right)$$

where $\mathcal{E}\left(R,r\right)$ is the collection efficiency and $n(r)$ is the particle size distribution (5.41). The total rate of increase in volume of the collector particle can be obtained by integrating over all particle sizes

$$\frac{dV}{dt} = \int_0^R \pi (R+r)^2 \frac{4}{3}\pi r^3 \,\mathcal{E}(R,r)\, n(r)\left[u_T(R) - u_T(r)\right] dr$$

This equation can be written in terms of a time derivative of radius:

$$\frac{dR}{dt} = \frac{\pi}{3}\int_0^R \left(\frac{R+r}{R}\right)^2 \left[u_T(R) - u_T(r)\right] n(r)\, r^3 \,\mathcal{E}(R,r)\, dr \qquad (8.2)$$

If $R \gg r$ and $u_T(R) \gg u_T(r)$, we can simplify (8.2) to be

$$\frac{dR}{dt} \approx \frac{\rho_a \overline{\mathcal{E}}}{4\rho_l}\, w_l\, u_T(R) \qquad (8.3)$$

where $\overline{\mathcal{E}}$ is an average collection efficiency and w_l is the liquid water mixing ratio

$$w_l = \frac{\rho_l}{\rho_a}\int_0^\infty \frac{4}{3}\,\pi\, n(r)\, r^3\, dr \qquad (8.4)$$

with units mass of liquid water per mass of dry air. For a particle to reach a size large enough to precipitate out of the cloud, its terminal velocity u_T must exceed the updraft velocity within the cloud.

Since collection efficiency increases with the radius of the collecting drop, and the terminal velocity increases with radius, rate of growth by collection proceeds more and more rapidly as drop size increases. Figure 8.6 compares the condensational growth rate with the accretional growth rate as a function of radius. The condensational growth rate decreases with radius (following (5.26)) while the accretional growth rate increases with radius. For radii greater than about 25 μm, the rate of growth by accretion exceeds that by condensation. Figure 8.6 shows a "gap" between approximately 10 and 25 μm where drop growth rates by both mechanisms are very slow. For precipitation to form as rapidly as is observed, drops must somehow reach a size of 25 μm more rapidly than would occur through simple diffusional and accretional growth. Modifications to diffusional growth were described in Section 5.4, whereby processes such as turbulence and entrainment broaden the drop spectra beyond that predicted by simple diffusional growth and result in larger drops.

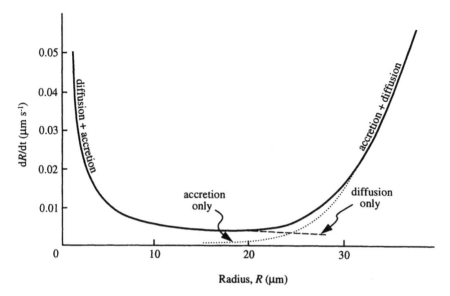

Figure 8.6 Drop growth rate by condensation and accretion. The dashed line represents growth by diffusion only, and the dotted line represents growth by accretion only, while the solid curve represents the combined growth rate. Condensational growth rate decreases with increasing radius, while accretional growth rate increases with increasing radius.

Modification of the continuous collection model can also help explain more rapid drop growth in the gap between 10 and 25 μm. In the continuous collection model described above, it is assumed that the collector particle collides in a continuous and uniform fashion with smaller cloud particles that are distributed uniformly in space. In reality, collisions are individual events that are statistically distributed in space and time. This has given rise to the *stochastic collection* model that accounts for the probabilistic aspects of collision and coalescence. Using the stochastic model, some drops are "statistically favored" for rapid growth. The stochastic collection model is illustrated using the following example (Figure 8.7). Consider 100 drops, all having the same radius. Assume that 10% of the drops undergo a collision in one second. After two seconds, one statistically favored drop triples its mass, while 81 of the drops remain at their initial size. Stochastic processes are of particular importance for the first 20 collisions or so, because they allow the largest drops to get past the "gap" shown in Figure 8.6. After this point, there is a sufficient number of large drops and the collection becomes essentially continuous.

The formation of precipitation in the atmosphere can be classified as either warm-cloud or cold-cloud processes. Warm-cloud processes produce precipitation solely

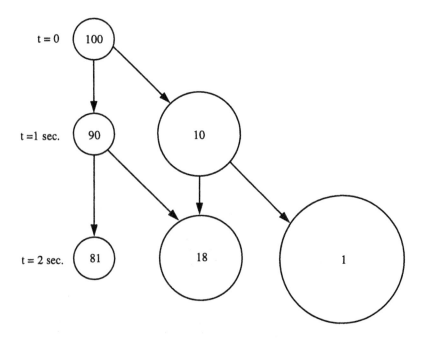

Figure 8.7 Illustration of stochastic collection model. At $t = 0$, a population of 100 drops all have the same initial radius. If 10% of the drops undergo collision and coalescence, then after one second, 10 of the drops will have grown larger, while 90 of the drops remain at the initial size. In the next second, if 10% of the drops in each category undergo collision, then one of the larger drops will undergo a second collision, while nine of the larger drops retain their size. Of the 90 that did *not* collide in the first second, nine will collide in the next second, while 81 will not. Thus in two seconds, there will be 81 drops that have not collided at all, 18 drops that have collided once, and one statistically favored drop that will have collided twice. (After Berry, 1967.)

through condensation and collection between water drops. Observations show that rain can develop in warm cumulus clouds over a time period as short as 15 minutes. Cold-cloud processes (commonly referred to as the *Bergeron mechanism*) produce precipitation through rapid diffusional growth of ice crystals in an initially water-saturated environment, then subsequently by aggregation processes. The difference between the rate of precipitation initiation by the warm-cloud and cold-cloud processes is illustrated in Figure 8.8 by comparing the growth rate of an ice crystal with that of a large drop. Initially, the ice crystal grows rapidly by diffusion. The growth of the drop is impeded initially by its small collection efficiency, but its growth rate increases once the drop has grown sufficiently so that its collection efficiency is no longer small.

Depending on cloud temperature, the amount of condensed water, and the drop sizes, precipitation may be initiated in less than 30 minutes by either the cold-cloud or

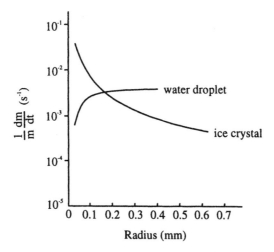

Figure 8.8 Growth rates for ice crystals and water drop. Initially, the ice growth rate exceeds the water drop growth rate. However, once the water drop grows to a sufficient size, its collection efficiency is no longer small, and its growth rate increases rapidly. (From Rogers, 1976.)

warm-cloud process. Development of precipitation particles to significant size by either the cold-cloud or warm-rain process depends on aggregation processes.

Not all clouds form precipitation-size particles. Precipitation formation is favored in clouds with a large condensed water content (typically arising from adiabatic cooling) and broad drop spectra. The dynamics of cloud motions therefore play an important role in determining whether or not a cloud precipitates. Cumuliform clouds are favored for precipitation development, because of strong updraft velocities that produce a substantial amount of condensed water. Low-level stratiform clouds rarely produce more than drizzle, since they rarely have a large amount of condensed water or the cold temperatures needed to initiate ice crystal processes.

Precipitation efficiency is a concept that describes how effectively a cloud converts condensed water into precipitation. Consider a cloud that forms via adiabatic cooling. The adiabatic liquid water mixing ratio, given by (6.41)

$$dw_l = \frac{c_p}{L_{lv}} \left(\Gamma_d - \Gamma_s \right) dz$$

can be differentiated with respect to time to yield

$$\frac{dw_l}{dt} = \frac{c_p}{L_{lv}} \left(\Gamma_d - \Gamma_s \right) u_z \tag{8.5}$$

which gives the rate of condensation at level z. The *liquid water path,* $\mathcal{W}_l$, is defined as the vertical integral of the liquid water mixing ratio:

$$\mathcal{W}_l = \int_{z_b}^{z_t} \rho_a w_l \, dz \tag{8.6}$$

with units kg m^{-2}. If all of the adiabatic liquid water were to fall out of the cloud, the depth of the adiabatic precipitation, P_{ad}, would be

$$P_{ad} = \frac{\mathcal{W}_l}{\rho_l} \tag{8.7}$$

where $\mathcal{W}_l$ here is the adiabatic liquid water path. Taking the time derivative of (8.7) and incorporating (8.5) and (8.6) gives

$$\dot{P}_{ad} = \frac{\rho_a}{\rho_l} \int_{z_b}^{z_t} \frac{dw_l}{dt} \, dz = \frac{\rho_a}{\rho_l} \int_{z_b}^{z_t} \frac{c_p}{L_{lv}} \left(\Gamma_d - \Gamma_s \right) u_z \, dz$$

where $\dot{P}_{ad}$ is therefore the *adiabatic precipitation rate* in units m s^{-1}. A *precipitation efficiency* can then be defined as the ratio of the actual precipitation rate to the adiabatic precipitation rate. Even in cumulonimbus, precipitation efficiency typically does not exceed 0.3.

Particles of solid or liquid water falling through the air are called *hydrometeors*. The term hydrometeor typically refers to precipitation particles that reach the ground, whereas the term *virga* denotes precipitation that does not reach the ground, evaporating the in the region below cloud base. Hydrometeors are classified as follows.

Drizzle consists of small water drops with diameter less than 500 μm. Drizzle is typically rather uniform precipitation that falls from warm stratus clouds.

Rain is precipitation of liquid water in which the drops are larger than drizzle. Rain may form from either warm or cold clouds.

Diamond dust consists of small ice crystals with a low fall speeds. Diamond dust commonly forms by isobaric cooling of lower tropospheric air during winter in the polar regions.

Snow is precipitation of solid water in the form of ice crystals that are larger than diamond dust and may be aggregated into large particles at temperatures greater than –10°C. Snow usually develops in clouds that form by adiabatic cooling.

Sleet consists of generally transparent, solid grains of ice that are less than 5 mm in diameter that have formed from the freezing of raindrops or the refreezing of melted snowflakes when falling through a below-freezing layer of air near the Earth's surface.

Graupel are solid precipitation particles with the appearance of pellets that form by the riming of ice crystals less than 5 mm in diameter.

Hail is precipitation of solid water in the form of irregular lumps of ice that have diameters exceeding 5 mm. Hail forms by riming in thunderstorms with vigorous updrafts.

Freezing rain is supercooled rain that falls in liquid form but freezes on impact to form a coating of glaze upon the ground and on exposed objects.

Rain, snow, sleet, and freezing rain can in principle form in the same cloud. The state of the precipitation once it reaches the ground depends on the temperature of the subcloud layer (Figure 8.9). If the subcloud layer is above freezing (Figure 8.9d),

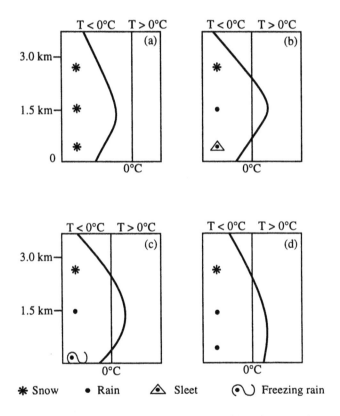

Figure 8.9 The temperature of the subcloud layer determines the type of precipitation that reaches the surface. In (a), the air temperature is below freezing at all levels, and ice crystals formed within the cloud reach the ground as snow. In (b), a region of above-freezing air causes the ice crystals to melt before they enter a deep, subfreezing layer near the surface, where they refreeze and reach the ground as sleet. In (c), the subfreezing surface layer is shallow, and the liquid refreezes on contact with the surface rather than during the descent. In (d), the above-freezing layer between about 2.5 km and the surface is sufficiently deep to allow the ice crystals to melt completely and reach the surface as rain.

then any solid precipitation particles will melt before reaching the ground. If the subcloud layer remains below freezing (Figure 8.9a), then solid precipitation particles will reach the ground as snow. If a temperature inversion is present in the subcloud layer (Figure 8.9b and c), solid precipitation particles will melt, but may become supercooled (freezing rain) or partially refreeze as the particle falls into a subfreezing layer above the surface.

The actual rainfall rate can be related to the size spectra of the rain drops. Based upon a large number of observed rainfall drop size spectra, Marshall and Palmer (1948) suggested the following drop size distribution for rainfall, $N(d)$:

$$N(d) = N_1 \exp\left(-\Lambda d\right) \tag{8.8}$$

where d is the drop diameter and $N_1 = 0.08$ cm^{-4} is a constant. The slope factor, Λ, depends only on rainfall rate and is given by

$$\Lambda(\dot{P}) = 41\, \dot{P}^{-0.21}$$

where Λ has units cm^{-1} and $\dot{P}$ is in mm hr^{-1}. Observations show that not all raindrop size distributions have this simple exponential form, but observations from many different regions have found an exponential form when sufficent individual samples are averaged. For snow, the slope factor, Λ, has been related to precipitation rate by Gunn and Marshall (1958):

$$\Lambda(\dot{P}) = 25.5\, \dot{P}^{-0.48}$$

and

$$N_1 = 0.038\, \dot{P}^{-0.87}$$

where $\dot{P}$ is given as the water-equivalent depth of the accumulated snow.

8.3 Radiative Transfer in a Cloudy Atmosphere

The interactions of clouds with radiation are important for the Earth's energy balance, the surface heat budget, and for the dynamics of clouds and larger-scale motions. A complete discussion of radiative transfer in a cloudy atmosphere is beyond the scope of this text. Here we present a simple framework for understanding cloud-radiative interactions in the context of cloud microphysical characteristics.

To understand the interactions of a cloud with radiation, we first examine the interaction of single spherical particle with electromagnetic radiation. Consider an isolated particle that is irradiated by an incident, plane electromagnetic wave (Figure 8.10).

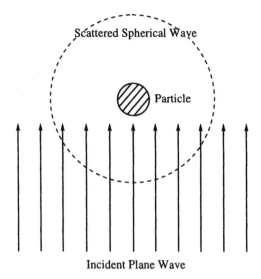

Figure 8.10 A plane electromagnetic wave is incident on an isolated particle. The presence of the particle, with electric and magnetic properties that differ from the surrounding medium, distorts the plane wave. The particle both absorbs and scatters some of the plane wave energy, thus diminishing the amplitude of the wave. The sum of the absorption and the scattering by the particle is called the extinction.

The plane wave preserves its character only if it propagates through a homogenous medium. The presence of the particle, with different electric and magnetic properties from the surrounding atmosphere, distorts the wave. The plane wave may be diminished in amplitude (absorption) and by the creation of a spherical wave (scattered energy) that travels outward from the particle. The total energy lost by the plane wave corresponds to extinction.

To determine how a spherical particle interacts with a stream of radiation of a specified wavelength, there are two parameters that must be specified. The first of these parameters is the *complex index of refraction, n*. The index of refraction is the ratio of the speed of light in a vacuum to the speed of light through the material. The index of refraction depends on wavelength, and consists of a real (n_{re}) and imaginary (n_{im}) component

$$n = n_{re} + i n_{im}$$

The imaginary component corresponds to absorption and the real component to scattering. Figure 8.11 shows the real and imaginary refractive indices for water and ice as they vary with wavelength. For both water and ice, $n_{im} \approx 0$ for wavelengths in the visible part of the spectrum (0.4–0.8 μm), implying no absorption of visible radiation

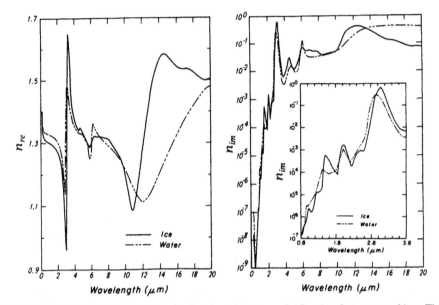

Figure 8.11 Real and imaginary components of the index of refraction for water and ice. The real refractive index of water varies greatly from that of ice for wavelengths greater than about 10 μm. In the visible part of the spectrum (0.4–0.8 μm), $n_{im} \approx 0$, indicating that visible radiation is not absorbed by clouds. In the near infrared region ($\gtrsim 1\mu$m), however, significant absorption occurs by both water and ice. (From Liou, 1992.)

by clouds. At wavelengths greater than about 1 μm (in the near infrared portion of the spectrum), values of n_{im} lead to significant absorption of radiation by both liquid water and ice clouds. Note the significant differences in values of n_{im} for liquid water versus ice at wavelengths 1.6 and 3.7 μm; these differences have been proposed as a way of discriminating ice clouds from water clouds via satellite remote sensing.

The second parameter that must be specified is the dimensionless *size parameter*, x

$$x = \frac{2\pi r}{\lambda} \qquad (8.9)$$

The size parameter is the ratio of the particle circumference to the wavelength. Table 8.1 gives values of the size parameter for different sizes of cloud particles at two different wavelengths: one each in the shortwave and longwave portions of the spectrum. For cloud drops ($r \approx 10 \mu$m), the size parameter is much greater than unity in the visible portion of the spectrum, while it is of order unity in the longwave portion of the spectrum.

Table 8.1 Values of the size parameter, x, for varying values of drop radius, r, and wavelength, λ.

r (μm)	λ (μm)	
	0.5	10
10	125.7	6.3
100	1,256.6	62.8
1,000	12,566.4	628.3

Mie theory solves Maxwell's equations of electromagnetic radiation for a sphere, given the complex index of refraction at the wavelength under consideration and the size parameter. Mie's solution is in terms of an expansion in spherical harmonics and Bessel functions,[2] and is beyond the scope of this text. Principal output from the Mie solution includes the dimensionless scattering and extinction efficiencies. The *scattering efficiency*, Q_{sca}, and the *extinction efficiency*, Q_{ext}, represent the fractional area of the incident beam that is removed by scattering and the combination of scattering and absorption (extinction). An *absorption efficiency*, Q_{abs}, is determined from $Q_{abs} = Q_{ext} - Q_{sca}$.

Figure 8.12 illustrates the extinction efficiency for a water drop in air. For small values of x, Q_{ext} is seen to increase approximately as x^4, which is the *Rayleigh scattering* regime. Q_{ext} attains a maximum value near $x = 2\pi$ (or $r = \lambda$). At higher values of x, Q_{ext} oscillates, asymptoting to $Q_{ext} \approx 2$ for very large values of x; this is the *geometric optics* regime. Values of $Q_{ext} > 1$ imply that the sphere scatters and/or absorbs more radiation than it intercepts. This seeming paradox results from energy diffracted about the sphere in addition to energy redirected by reflection inside the sphere.

To determine the radiative interactions with a spectrum of particle sizes, the Mie coefficients are determined from the integral cross-sectional area of the drops, scaled by the respective efficiency:

$$\sigma_{ext} = \int_0^\infty n(r)\, \pi\, r^2 Q_{ext}(x)\, dr \tag{8.10a}$$

$$\sigma_{sca} = \int_0^\infty n(r)\, \pi\, r^2 Q_{sca}(x)\, dr \tag{8.10b}$$

---

[2] See Liou (1980) and Goody and Yung (1989) for Mie's solution.

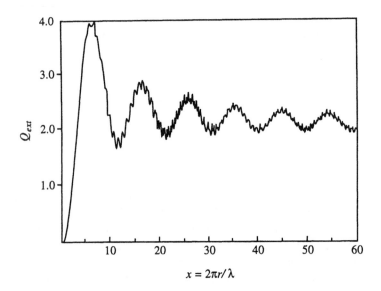

Figure 8.12 Extinction efficiency for a water drop in air ($n = 1.33$) calculated for $\lambda = 0.5\ \mu$m as a function of size parameter, x.

where σ_{ext} and σ_{sca} are the volume extinction and scattering coefficients, respectively, $\sigma_{abs} = \sigma_{ext} - \sigma_{sca}$, and $n(r)$ is the drop size distribution. The Mie cross sections vary with wavelength, associated with variations in the refractive index that determine the scattering and extinction efficiencies. The optical depth through a cloud of spherical particles is determined from:

$$\tau_{ext} = \int \sigma_{ext}\, dz \qquad (8.11a)$$

$$\tau_{sca} = \int \sigma_{sca}\, dz \qquad (8.11b)$$

where τ_{ext} and τ_{sca} are the extinction and scattering optical depths, respectively, and $\tau_{abs} = \tau_{ext} - \tau_{sca}$. Analogous to the volume absorption coefficient in Section 3.3, the Mie volume scattering and extinction coefficients have units m^{-1}.

In the geometric optics limit ($x \gg 1$), it was shown in Figure 8.12 that $Q_{ext} \approx 2$. Substituting this value into (8.10a) and using (8.6), we obtain from (8.11a)

$$\tau_{ext} = \frac{3 \mathcal{W}_l}{2 \rho_l \overline{r_e}} \tag{8.12}$$

where we have introduced the *effective radius*, r_e, defined as

$$r_e = \frac{\displaystyle\int_0^\infty r^3 n(r)\, dr}{\displaystyle\int_0^\infty r^2 n(r)\, dr} \tag{8.13}$$

and the notation $\overline{r_e}$ in (8.12) refers to the value of r_e averaged over the depth of the cloud. The effective radius can be thought of as an average drop radius that is weighted by the drop cross-sectional area. Since the cross-sectional area varies as r^2, larger drops have a greater weight in determining the effective radius, and thus $r_e > \overline{r}$ ($\overline{r}$ as defined in problem 5.9). From (8.12) it is seen that the optical depth increases with increasing liquid water path. For the same liquid water path, the optical depth will increase for smaller particles. Smaller particles are associated with a greater optical depth because they have a greater cross-sectional area per mass of condensed water. The approximate optical depth in (8.12) is valid for $x \gg 1$; for cloud particles ($r \sim 10$ μm) the geometric optics regime is applicable in the shortwave spectrum ($\lambda < 4\ \mu m$).

When the shortwave optical depth exceeds unity, multiple scattering occurs as an individual photon is scattered repeatedly by successive drops. Because of multiple scattering, the irradiances cannot be determined simply by incorporating the optical depth into a form of the radiative transfer equation such as Beer's law (3.31). By solving a form of the radiative transfer equation that includes a scattering source term (multiple scattering), we can determine the reflectivity, transmissivity, and absorptivity of the clouds, as well as vertical profiles of irradiances and heating rates. Figure 8.13 shows the variation of cloud shortwave reflectivity and absorptivity with $\mathcal{W}_l$ and r_e. Both cloud reflectivity and absorptivity increase with increasing $\mathcal{W}_l$. For a given value of $\mathcal{W}_l$, the reflectivity decreases as drop size increases, while the variation of cloud absorptivity with r_e varies according to the value of $\mathcal{W}_l$: for $\mathcal{W}_l \approx 20$ g m^{-2}, absorptivity increases as drop size increases. The values shown in Figure 8.13 are for a solar zenith angle $Z = 60°$. Cloud reflectivity increases with increasing solar zenith angle, while cloud absorptivity decreases slightly at high solar zenith angles.

Note that Figure 8.13 shows the broadband cloud reflectivity and absorptivity, integrated over the entire shortwave spectrum. Clouds show a strong spectral variation in their radiative properties, arising particularly from variations in the imaginary part of

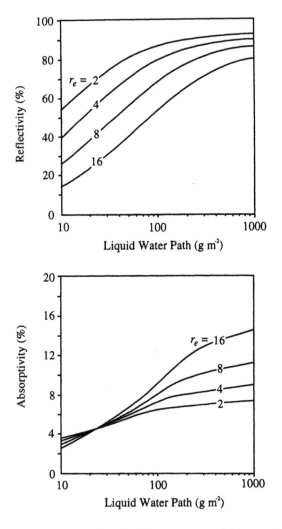

Figure 8.13 Cloud shortwave reflectivity and absorptivity as a function of liquid water path W_l and effective radius r_e for a solar zenith angle of 60°. (From Slingo, 1989.)

the refractive index of water. It is seen from Figure 8.11 that at wavelengths less than about 0.7 μm, the imaginary part of the refractive index is nearly zero and thus absorption of radiation by clouds effectively vanishes at these wavelengths. On the other hand, at wavelengths greater than 0.7 μm, the imaginary part of the index of refraction is large, and significant absorption by clouds occurs. As a result of spectral variations of the refractive index of water, clouds also show greatest shortwave reflectivity at wavelengths less than 0.7 μm.

In the longwave portion of the spectrum, $x \approx 1$ and the full Mie solution must be employed. Because longwave radiation is diffuse and there is not a preferred scattering angle, scattering can be accounted for by the *diffusivity factor,* $\beta_{sca} \approx 1.66$, that effectively accounts for the increased optical depth associated with scattering. The longwave absorption optical depth at wavelength λ can be determined following (3.25) as

$$\tau_{abs} = \beta_{sca}\, k_{abs}\, \mathcal{W}_l \tag{8.14}$$

where $\mathcal{W}_l$ is the liquid water path (8.6) and k_{abs} is the mass absorption coefficient, and the subscript λ is dropped for convenience. Values of the mass absorption coefficient can be parameterized from Mie calculations to be of the form

$$k_{abs} = a + \frac{b}{r_e} \tag{8.15}$$

where a and b are constants that vary with wavelength. Incorporating (8.15) into (8.14) gives

$$\tau_{abs} = \beta_{sca}\left(a + \frac{b}{r_e}\right)\mathcal{W}_l \tag{8.16}$$

The longwave transmissivity of a cloud can be written using (3.28) as

$$\mathcal{T} = \exp\left[-\beta_{sca}\left(a + \frac{b}{r_e}\right)\mathcal{W}_l\right] \tag{8.17}$$

Since longwave scattering is minimal, we have from (3.16) and (3.19)

$$\epsilon = 1 - \mathcal{T} = 1 - \exp\left[-\beta\left(a + \frac{b}{r_e}\right)\mathcal{W}_l\right] \tag{8.18}$$

The transmissivity decreases with increasing values of $\mathcal{W}_l$, while the emissivity increases with increasing values of $\mathcal{W}_l$. For a constant value of $\mathcal{W}_l$, the transmissivity increases for larger values of r_e and the emissivity decreases. Because water drops are such efficient absorbers/emitters of longwave radiation, values of $\mathcal{W}_l > 50$ gm^{-2} cause a cloud to emit nearly as a black body (Figure 8.14). The emissivity of a cloud is always slightly less than unity because of a small amount of reflection.

Figure 8.15 shows vertical profiles of shortwave and longwave fluxes observed in a stratocumulus cloud deck. The radiative heating rate corresponding to these fluxes can be determined from the gradient in net flux profiles following (3.34). The gradient of net longwave flux is largest near the cloud top, in the layer about 50 m below the top, where the liquid water path is typically high, and where the longwave cooling rate can reach as high as 10°C hr^{-1}. Clouds that have a smaller amount of liquid water near the cloud top have the longwave cooling distributed over a greater depth, and have lower values of peak longwave cooling rate. Since the cloud base is typically cooler than the surface, radiative heating commonly occurs near the base. In the interior of a thick cloud, the heating rates are nearly zero. The magnitude of the solar heating within the cloud relative to the longwave heating depends strongly on the solar zenith angle. Solar heating typically occurs over a greater depth in the cloud than does longwave heating.

Thus far, we have examined extinction by spherical particles in liquid water clouds. The radiative properties of ice clouds are more difficult to determine, since extinction by nonspherical ice particles is complicated. A commonly used approximation for ice crystals is to use an effective radius of the ice crystal size distribution corresponding to spheres of equal surface area. Values of the ice crystal optical depth can then be determined using Mie theory, and derived parameters such as emissivity and reflectivity can be calculated using (8.14)–(8.18) and substituting $\mathcal{W}_i$ (*ice water path*) for $\mathcal{W}_l$. Note that the longwave mass absorption coefficient in (8.16) is different for ice particles than for the liquid particles because of their different indices of refraction (see Figure 8.11). Figure 8.16 shows values of the longwave emissivity and shortwave reflectivity for ice clouds. Comparison of the values in Figure 8.16 with those in Figures 8.13 and 8.14 shows that the values of emissivity and reflectivity are smaller than those found for liquid water clouds. This is explained primarily by the smaller amounts of condensed ice and the larger sizes of the ice crystals.

The discussion above has related the radiative characteristics of the cloud to their microphysical characteristics. However, we have deliberately simplified the problem. Radiative transfer in the presence of ice clouds remains an active area of research, whose major challenge is to relate ice crystal habit, spatial orientation, and size spectra to the radiation field. The results shown in Figures 8.13–8.16 have been derived by assuming the clouds are horizontally homogeneous and of infinite horizontal extent. Real clouds have finite horizontal extent and horizontal inhomogeneities within individual cloud elements; these inhomogeneities alter both the shortwave and longwave properties of the clouds. Effects of finite geometry and horizontal inhomogeneities on the radiative transfer are simulated using three-dimensional radiative transfer models or using Monte Carlo techniques that trace individual photons.

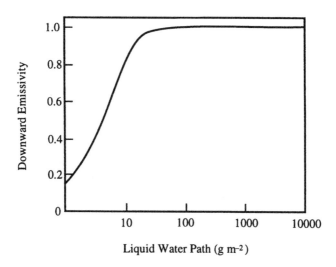

Figure 8.14 Downward emissivity as a function of liquid water path. The emissivity approaches unity for large $\mathcal{W}_l$. (After Stephens, 1978.)

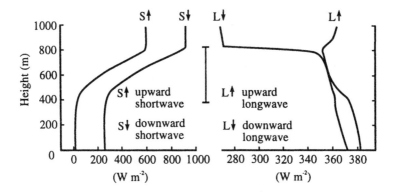

Figure 8.15 Shortwave and longwave fluxes in a stratocumulus deck. The vertical bar shows the extent of the cloud layer. (After Nicholls, 1984.)

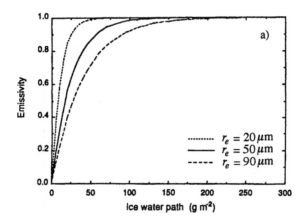

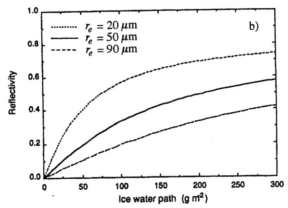

Figure 8.16 Cirrus radiative properties for varying values of particle effective radius as a function of ice water path: a) infrared emissivity; b) shortwave reflectivity for solar zenith angle $Z = 30°$. (From Ebert and Curry, 1992.)

8.4 Fogs, Stratus Clouds, and Stratocumulus Clouds

Fogs, stratus clouds, and stratocumulus clouds are considered collectively because they all occur in the atmospheric boundary layer (within approximately 2 km of the surface) and have some common microphysical characteristics and formation mechanisms. This class of clouds form over both continental and marine locations; here the focus is on those that form in marine locales.

Fogs, stratus clouds, and stratocumulus clouds are not typically associated with precipitation, although they often produce drizzle, or diamond dust in the polar regions. The importance of these clouds lies in their effect on the Earth's radiation budget, arising from the large spatial extent of the clouds.

8.4.1 Fog and stratus

Fog differs from cloud only in that the base of fog is at the Earth's surface, while cloud bases are above the surface. By international convention, fog reduces visibility below 1 km; if the visual range is greater than 1 km, then the condensate is called *mist*. *Visual range* is the greatest distance at which it is just possible to see with the unaided eye a prominent dark object against the sky at the horizon during daytime. Visual range decreases as liquid water content increases and as drop size decreases.

The physical mechanisms that result in the formation of fog involve three primary processes:

1) cooling of the air to its dew point (isobaric cooling, Section 6.2);
2) addition of water vapor to the air (Section 6.3);
3) vertical mixing of moist air parcels with different temperatures (Section 6.4).

Most fogs are influenced by more than one of these processes, although one mechanism may dominate its formation.

Radiation fog is formed primarily by isobaric cooling. Strong radiative cooling of the Earth's surface occurs at night under conditions of clear skies and light winds. The air just above the surface cools by the transfer of longwave radiation, losing energy to the colder surface below as well as cooling to space. If the radiative cooling of the air proceeds to the dew-point temperature, then a fog forms. If the fog becomes optically thick, the level of maximum longwave cooling occurs at the fog top, allowing the fog to propagate upwards. Due to purely radiative processes, the interior of a thick fog will tend to become isothermal. However, the fog-top cooling destabilizes the lapse rate within the fog. This destabilization increases turbulent mixing, which acts to enhance the fog development if the air above the fog is moist; otherwise turbulent mixing can act to dissipate the fog. Radiation fogs dissipate when the radiative cooling is reduced, most commonly by the daytime solar heating, or by the advection of a cloud overhead which reduces the fog's longwave cooling to space.

Radiation fog occurs frequently over land at night. Over the open ocean, nighttime longwave cooling of the surface typically does not reduce the sea surface temperature by more than a degree and thus radiation fogs rarely form. However, over sea ice during the polar night, radiation fogs are very common and may reach great depths. During the polar night, radiation fogs are not subject to the ordinary diurnal cycle, and a fog can form and persist for periods of days to weeks. Consider the simulation shown in Figure 8.17 of a radiation-advective fog over the Arctic sea ice during winter. Warm, moist air is advected over the sea ice surface. The air above the

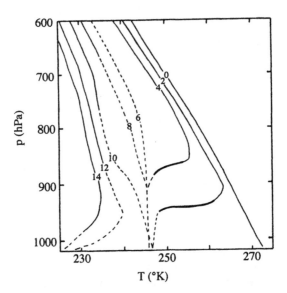

Figure 8.17 Evolution of a radiation-advective fog over the Arctic sea ice during winter. At day 0, the temperature corresponds to a maritime polar air mass that is advected over the Arctic Ocean. Profiles are given every other day for a two-week period indicated by the numbers on the curves. Thin lines correspond to clear air; heavy lines correspond to water drops or mixed phase; and dashed lines correspond to ice crystals. (From Curry, 1983.)

surface cools radiatively and also by eddy diffusion, and the fog initiates as a thin liquid layer above the surface, which increases rapidly with depth. As the fog cools below –15°C, the ice phase is initiated and the cloud becomes mixed phase; at temperatures below –22°C, the cloud is completely crystalline. After the first few hours and for the first five days, the fog is divided into an upper liquid layer and a lower crystalline layer. After six days of cooling the fog is completely crystalline and the fog has reached a maximum height of nearly 3 km. Gravitational fallout of the ice particles (diamond dust) acts to deplete the fog of condensate, and by the end of two weeks, there is only a thin layer of ice crystals remaining, just above the surface.

When cold air streams over a water surface that is much warmer than the air temperature, the surface water evaporates at a rapid rate and fills the air above the surface with fog. Such fogs, called *steam fogs*, are commonly observed when cold wintertime continental air masses are advected over the ocean or other bodies of water, off the coasts of high-latitude continents and at the sea ice margin. Rather extreme temperature differences between the air and surface are required for steam fog to form, since the cold air above the warm water becomes unstable as a result of surface heating, and the moisture is carried away from the surface and mixed with the air aloft. High relative humidity in the cold air is also favorable for steam fog formation. Steam fog is shallow and resembles irregular tufts of whirling smoke emanating from the water.

Mixing fog occurs over a cold ocean surface if there is a large surface evaporation flux and a small surface sensible heat flux. Vertical mixing of near saturated surface air with the warmer air above the surface can produce condensation following Section 7.3. Such fogs are common off the west coast of continents at subtropical latitudes and over high-latitude seas.

Rain fog occurs when rain that was formed in a warmer layer aloft falls through a layer of colder air near the ground. Rain fog occurs most frequently in association with weather fronts that have marked temperature contrasts (Figure 8.18). Rain evaporates rapidly in the subsaturated, cold air, increasing the water vapor content of the cold air and decreasing its temperature (see Section 6.3). If the cold air reaches saturation, its temperature will be at the wet-bulb temperature, T_W. If the saturation vapor pressure of the falling rain exceeds $e_s(T_W)$, then the rain will evaporate further, causing condensation to occur and a fog to form.

Stratus clouds form via the same mechanisms as fog, the only difference being the location of the cloud base. Fog can be transformed into stratus, as the lower portion of the fog evaporates and the upper part rises. Stratus may be transformed into fog in the following way. Longwave cooling at the cloud top decreases the stability of the layer (Section 7.3). This will increase turbulent mixing in the layer, causing a turbulent transport downward of cool air and cloud drops. Evaporation of drops beneath the cloud base causes cooling and an increase in humidity, further lowering the cloud base.

Perhaps the most unusual stratus clouds are those that form over the Arctic Ocean during summer, when the sea ice is melting. Monthly average low-cloud amounts are

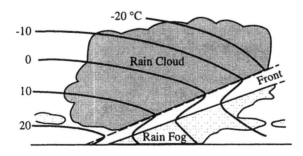

Figure 8.18 When rain falls through a layer of cold air near the ground, it evaporates rapidly in the subsaturated air. This decreases the temperature and increases the water vapor content in the layer, creating an environment in which *rain fog* is likely to form.

nearly 70% over the central Arctic Ocean for the months of May through September. Relatively warm air from the south that is advected over the ice pack is modified by turbulent and radiative processes. The most striking feature of the summertime Arctic stratus is the presence of multiple cloud layers. It is common for Arctic stratus to occur in a number of well-defined layers separated by intervening clear regions that are several hundred meters thick. As many as five simultaneous layers have been reported. Several theories have been proposed for the cloud layering:

1) persistent solar heating in the cloud interior that results in evaporation;
2) advection of an upper cloud layer over a surface fog;
3) relative humidity variations in the overlying air, with different dew-point depressions, causing condensation to occur via isobaric cooling at different vertical levels that are separated by clear interstices.

8.4.2 Stratocumulus

Marine stratocumulus clouds occur frequently at subtropical latitudes to the west of continents, in association with the subtropical anticyclones. In middle and high latitudes, large regions of stratocumulus occur where cold air from the continent streams over a warm underlying ocean. Stratocumulus clouds form via two basic mechanisms:

1) transformation of a fog or stratus layer by turbulent mixing;
2) shallow convection initiated by heating of the lower atmosphere by a warm ocean surface.

Turbulent motions in the atmospheric boundary layer are generated by wind shear and buoyancy fluxes. The upward and downward motions associated with a turbulent eddy (Figure 3.7) are associated with adiabatic cooling and warming. These vertical motions tend to produce an atmospheric boundary layer that is well mixed vertically (Section 7.3), with constant potential temperature and water vapor mixing ratio throughout the atmospheric boundary layer. If the upward motions in the eddies result in the top of the eddy rising above the lifting condensation level, then small cloud elements form. Cloud-top radiative cooling then promotes further turbulent mixing and condensation. The depth of the stratocumulus cloud-top mixed layer is typically determined by a balance between large-scale sinking motions (*subsidence*), which tends to make the layer shallower, and cloud-top entrainment, which commonly causes the depth of the mixed layer to increase.

Figure 8.19 illustrates the vertical structure of a stratocumulus cloud deck that was observed over the North Sea. Note that over the depth of boundary layer, θ_e and $w_t = w_v + w_l$ are nearly constant, indicating that the layer is well mixed, with a cloud base at about 400 m. The liquid water content increases with height above the cloud base to a maximum value of about 0.6 g kg^{-1}, corresponding nearly to the adiabatic

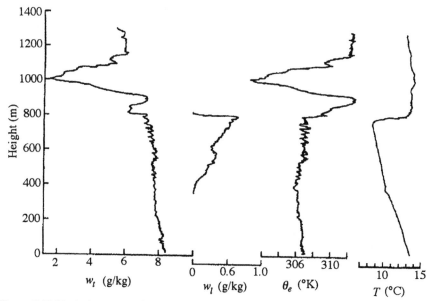

Figure 8.19 Vertical structure of a stratocumulus cloud deck observed over the North Sea (55 °N). Over the depth of the boundary layer, the total water mixing ratio and the potential temperature are nearly constant, indicating a well-mixed layer. The cloud base is at about 400 m. (After Nicholls, 1984.)

liquid water content. Just below the cloud top, the liquid water content is much less than the adiabatic value because of cloud-top entrainment of dry air.

While stratocumulus clouds do not rain, they frequently produce drizzle. Drizzle can influence the stability of a cloud layer by altering the vertical distribution of latent heating. As drizzle settles from the top of the cloud layer, it removes water which cannot subsequently be evaporated at that level; as a consequence, drizzle contributes to net latent heating in the upper part of the cloud layer. However, as the drizzle settles into the subcloud layer, it evaporates, cooling the subcloud layer. As a result of this process, the layer between the heating aloft and cooling below is stabilized, whereas the shallow layers above the heating zones or below the evaporatively cooled zone are destabilized. This process tends to form two shallow unstable layers that are decoupled by an intermediate stable layer (Figure 8.20). Other processes that may contribute to the decoupling of a stratocumulus-topped mixed layer include decreased surface buoyancy fluxes, solar heating, and entrainment of warmer and drier air. In some cases, the decoupling of the mixed layer results in the destruction of the cloud layer. In other cases, the decoupling leads to the buildup of conditional instability in the subcloud layer, and cumulus clouds rise out of the subcloud layer into the stratocumulus deck. In this case, the cumulus clouds supply the stratocumulus layer with moisture being lost by entrainment and drizzle.

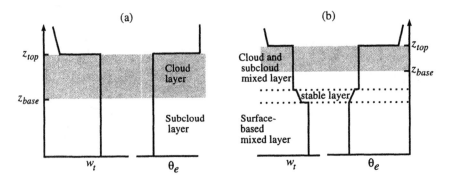

Figure 8.20 Idealized boundary-layer profiles of total water content and equivalent potential temperature. In (b), the cloud and subcloud layers are decoupled from the surface mixed layer by a stable intermediate layer. Decoupling may occur for a number of reasons, including the fallout of drizzle from the upper cloud layer, and its subsequent evaporation in the subcloud layer; a decrease in surface buoyancy fluxes; solar heating; and entrainment of warmer, drier air. (After Turton and Nicholls, 1987.)

As the boundary layer is heated from below, it can become conditionally unstable and sometimes result in the formation of a broad field of small cumulus clouds rather than stratocumulus. The difference between stratocumulus and shallow cumulus is the relative strength of the heating and mixing in the boundary layer. Entrainment across the top of the cloudy mixed layer has also been hypothesized to contribute to the breakup of stratocumulus into cumulus clouds. When the jump in equivalent potential temperature, $\Delta\theta_e$, at the top of the cloudy mixed layer is negative and exceeds a certain magnitude, a parcel of air entrained into the mixed layer from above and mixed with the cloudy air becomes denser than its surroundings. Because of the negative buoyancy, the air from aloft would be rapidly mixed through a portion of the cloud, leaving patchy, dissipating stratocumulus. Whether or not a stratocumulus deck will break up into a cumulus cloud field depends on a complex interplay between cloud-top entrainment and heating and moistening from the ocean surface.

8.5 Cumuliform Clouds

Cumuliform clouds form by the local ascent of warm, buoyant air parcels. These clouds occur over all regions of the ice-free global ocean, with highest frequency in the tropics. When compared with the stratiform clouds examined in Section 8.4, cumuliform clouds are associated with stronger vertical air motions and often with intense precipitation.

While there are many species of cumuliform clouds, their forms can be grouped into the following categories:

- *Fair weather cumulus*: individual elements have horizontal and vertical scales of approximately 1 km; these clouds do not precipitate.
- *Towering cumulus*: attain widths and depths of several kilometers; these frequently precipitate.
- *Cumulonimbus*: widths of tens of kilometers and may extend vertically to the tropopause, where their tops spread out and form an anvil-shaped cloud; associated with heavy precipitation, lightning, thunder, and sometimes hail.
- *Mesoscale convective complexes*: aggregation of cumulonimbus clouds that extends over hundreds of kilometers, produces large amounts of rain, and develops large circulation patterns in addition to the convective-scale air motions.

The depth and structure of a cumuliform cloud depends on the fluxes of heat and moisture from the surface, the stability of the atmosphere, and the large-scale vertical velocity. Examination of a typical vertical temperature profile in a convective environment provides insight into the characteristics of cumuliform clouds (Figure 8.21). A superadiabatic layer just above the surface, typically with a depth of 30–100 m, results in the buoyant production of thermal eddies. From this level up to the cloud

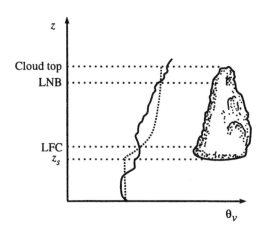

Figure 8.21 Typical temperature profiles in a convective environment. The solid profile represents the environmental temperature; the dashed profile corresponds to the temperature within the cloud. The cloud base forms near the lifting condensation level, z_s. Near the cloud base, the temperature increases more rapidly with height in the cloud than in the surroundings, resulting in a relatively large temperature difference between the environmental temperature and the cloud interior temperature. A cloud that reaches the level of free convection (LFC) will accelerate upwards until it reaches the level of neutral buoyancy (LNB), where the environmental temperature is equal to the interior cloud temperature.

base, the atmospheric boundary layer is well mixed, which is reflected by a constant value of θ_v. The cloud base is found approximately at the lifting condensation level, z_s (6.35). To interpret the vertical structure of the cloud and temperature above cloud base, it is useful to refer to Section 7.2 and Figure 7.2. If the buoyant acceleration of the cloud parcel from the surface instability is sufficient to reach the level of free convection (LFC), then the cloud will accelerate and grow vertically. If the cloud does not reach the LFC, then the cloud will remain shallow and undeveloped (e.g., fair weather cumuli). A cloud that does reach the LFC will continue to accelerate upwards until it reaches the level of neutral buoyancy (LNB). Although the cloud parcel acceleration is zero at the LNB, it may still have inertia and penetrate above the level of neutral buoyancy so that an adiabatic cloud might have its top at some distance above the level of neutral buoyancy. However, in the presence of entrainment, cloud tops rarely reach as high as the level of neutral buoyancy. The vertical velocity at the cloud base is typically about 3 m s⁻¹, and may increase with height at a rate of about 4 m s⁻¹ km⁻¹, reaching a maximum below the cloud top.

8.5.1 Non-precipitating cumulus

A developing cumulus cloud is characterized by a rising thermal tower whose outline is sharply defined by protuberances with a cauliflower appearance. The protuberances continually emerge from the top of the cloud. After reaching a peak height, a tower subsides and the protuberances become less pronounced or disappear altogether. The cloud edges become tenuous and the tower evaporates completely. An individual cumulus tower goes through a life cycle of growth and decay over a period of minutes for fair weather cumulus and about an hour for towering cumulus. In the absence of precipitation, the condensation and subsequent evaporation of a cumulus cloud results in no net latent heating of the atmosphere. Cumulus clouds with diameters less than 5 km and a depth of less than 1 km are not commonly observed to precipitate.

The vertical variation in cloud liquid water content shows an increase from cloud base to within a few hundred meters of the cloud top, where it rapidly falls to zero. Figure 8.22 illustrates the vertical variation in liquid water content from cloud base to cloud top in cumulus clouds in terms of the ratio of the observed liquid water content to the adiabatic value (6.41). The ratio w_l/w_l^{ad} represents the departure of the liquid water mixing ratio from the adiabatic value due primarily to the effects of entrainment. The liquid water content is seen to depart significantly from the adiabatic values within 500 m of the cloud base. The relative importance of air entrained into cumuli from the sides versus the cloud top has not been fully resolved.

Figure 8.23 shows observations of horizontal variability associated with cumulus clouds. Extreme variations in the vertical velocity are seen, from updrafts of about 6 m s⁻¹ to downdrafts of more than 4 m s⁻¹ in less than 50 m across the cloud. Upon entering the cloud, the liquid water content jumps from zero to substantial values. Within a cloud element, the liquid water content fluctuates slightly in response to vertical velocity fluctuations while maintaining high values on average.

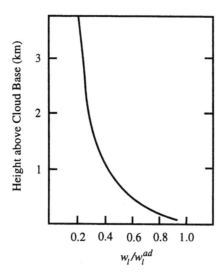

Figure 8.22 Observed variation with height above the cloud base of liquid water mixing ratio, w_l as a fraction of the adiabatic value, w_l^{ad} in large cumulus clouds. At the cloud base, the ratio is typically near unity, but decreases rapidly with height in the cloud due primarily to entrainment. (After Ludlam, 1980.)

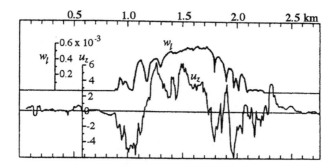

Figure 8.23 Liquid water concentration, w_l and vertical air speeds, u_z, obtained during horizontal traverses of cumulus clouds in aircraft. The water concentration measurements were taken near 1.25 km above the cloud base and 250 m from the cloud top. Vertical air speed measurements were obtained at 1 km above the cloud base. (Adapted from Ludlam, 1980.)

8.5.2 Precipitating cumuliform clouds

When a large reservoir of *CAPE* (7.25) is present in the atmosphere, convection can penetrate to great heights, sometimes exceeding the height of the tropopause. Deep convective clouds have a longer lifetime than shallow cumulus clouds, allowing time for precipitation particles to form by either the warm-rain or cold-rain mechanisms. Precipitating cumuliform clouds range from isolated precipitating cumulus towers, cumulonimbus, to the squall lines and mesoscale convective complexes resulting from the clustering of cumuliform clouds. Figure 8.24 illustrates a mature cumulonimbus cloud, and the regions of updraft and downdraft within the cloud.

Formation of precipitation changes dramatically the thermodynamics of a cumuliform cloud. In Section 7.2, we showed that the presence of condensed water affects cloud buoyancy and thus vertical accelerations by generating a downward-directed drag force equivalent to the weight of the suspended water. Hence, an immediate consequence of precipitation is that it unloads an updraft from the weight of the condensed water. The intense precipitation of cumulonimbus and mesoscale convective complexes can influence the dynamics of the storm. Precipitation particles can initiate downward acceleration by dragging air downward. Evaporation and melting of the precipitation particles cool the air and hence contribute to downdraft formation and maintenance. As the downdraft air approaches the Earth's surface, it spreads laterally and can undercut the surrounding warm, moist air. Depending on the environmental wind field, evaporation of precipitation can constructively or destructively contribute to the subsequent propagation of the cloud. If precipitation falls into the inflow flank of the cloud

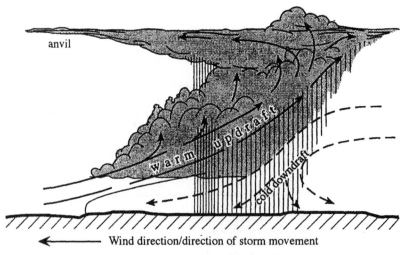

Figure 8.24 A cumulonimbus cloud in its mature stage. (After Goody, 1995.)

updraft, the parent cumulus tower will dissipate. If the precipitation falls downstream of the inflow, the cold air mass can undercut the warm moist air and lift it to the lifting condensation level or level of free convection and initiate new convective towers.

When showers form in cumulus towers which have risen well above the 0°C level, the ice phase is initiated and buoyancy is enhanced (see Section 6.6). Latent heat released during the freezing of supercooled drops and the subsequent vapor deposition growth of ice particles can augment cloud buoyancy production. Glaciation of a growing cumulus cloud can promote the vertical growth of convective towers several kilometers above a nearby cloud that is nonglaciated. If the warm rain process has been suppressed, glaciation will initiate the formation of precipitation, further increasing the cloud buoyancy by diminishing the water loading through precipitation. Glaciation in a cumulus cloud is evidenced by the appearance of a characteristic *anvil,* which is a smooth spreading layer of ice crystals at the top of the cloud (Figure 8.24). The anvil may spread with the wind and extend for hundreds of kilometers. Dissipation of the cloud begins once the updraft has been cut off from the source of warm, moist air. Precipitation and the associated downdraft gradually decrease, finally ending about 20 minutes after the updraft has ceased. Because sublimation is slow at the cold temperatures of the upper troposphere, anvil clouds may persist for many hours after the parent cloud has dissipated.

The typical thunderstorm is a complex of individual convective cells, each with a lifetime of 30 to 40 minutes; this complex is called a *multi-cell thunderstorm.* Multicell thunderstorms may occur in large groups and complexes, sharing a common shield of anvil cloud. If these cloud clusters have a minimum duration of six hours and a cold anvil cloud spreading over an area of at least 100 km in one direction, they are referred to as mesoscale convective complexes. Mesoscale convective complexes form frequently over land and over the equatorial oceans. As a result of the circulations generated by the mesoscale convective complex, stratiform precipitation contributes approximately 50% to the overall rainfall. The large extent of the anvil cloud and the large amount of condensed water results in a substantial perturbation to surface radiation balance.

8.6 Parameterization of Cloud Microphysical Processes

Cloud models are used to test our understanding of cloud processes and to predict the future state of a cloud system. To simulate the formation, evolution, and dissipation of a cloud or system of clouds and their interaction with the large-scale environment, thermodynamic, microphysical, and dynamical processes must be included. Because of the complexity of the equations, analytical solutions do not exist and numerical solutions must be obtained. The equations developed previously for cloud particle nucleation, diffusional growth, the formation of precipitation-sized particles, and the complexity of the thermodynamics and the small-scale motions associated with cloud formation and dissipation processes are too complex for routine inclusion in three-dimensional models of the atmosphere. Therefore, it is necessary to employ parameterizations of these processes that are aimed at capturing in a few simple formulae the essential physics embodied by the complete equations.

In this section we illustrate the fundamentals of parameterizing cloud microphysical processes. The emphasis here is not on the numerical and computational details, but rather on the general construction of a one-dimensional numerical cloud model. The equations required for a one-dimensional numerical cloud model consist of Newton's second law of motion applied to air, the conservation of mass (3.8), the first law of thermodynamics (3.6), and conservation equations for atmospheric water (e.g., (3.57)). For simplicity, consider only a warm cloud (no ice particles) with a specified velocity field. Individual equations for water vapor mixing ratio (w_v), cloud liquid water mixing ratio (w_l), and precipitation mixing ratio (w_R) are included. Neglecting any source terms that are not related directly to condensation and precipitation and ignoring partial cloudiness, the equations may be written as

$$\frac{\partial \theta}{\partial t} + u_z \frac{\partial \theta}{\partial z} = \frac{\theta}{T} \frac{L_{lv}}{c_p} \left(C_{vl} - E_{Rv} \right) \tag{8.19}$$

$$\frac{\partial w_v}{\partial t} + u_z \frac{\partial w_v}{\partial z} = -C_{vl} + E_{Rv} \tag{8.20}$$

$$\frac{\partial w_l}{\partial t} + u_z \frac{\partial w_l}{\partial z} = C_{vl} - A_{lR} - K_{lR} \tag{8.21}$$

$$\frac{\partial w_R}{\partial t} + u_z \frac{\partial w_R}{\partial z} + \frac{\partial}{\partial z} \left(w_R u_T \right) = A_{lR} + K_{lR} - E_{Rv} \tag{8.22}$$

where C_{vl} is the condensation of cloud water, E_{Rv} is the evaporation of rainwater. A_{lR} is the autoconversion, which is the rate at which cloud water content decreases as particles grow to precipitation size by coalescence and/or vapor diffusion. K_{lR} is the collection of cloud water, which is the rate at which the precipitation content increases as a result of collection of cloud drops by raindrops. The term $\partial(w_R u_T)/\partial z$ is the fallout of the rainwater. The model therefore produces rain in the following way: cloud water appears by condensation, C_{vl}; once sufficient cloud water is produced, then microphysical processes lead to production of rain by autoconversion, A_{lR}; once rain is present, the amount of rain can be increased through A_{lR} and or K_{lR}; and once sufficient rainwater is produced, then the raindrops fall and some of the rain may evaporate in the subcloud layer, E_{Rv}.

To calculate the sources and sinks of cloud water and rainwater so as not to require detailed knowledge of the cloud drop and raindrop spectra, parameterizations must be used that relate these sources and sinks to the model prognostic variables in (8.19)–(8.22). Parameterizations of these terms have been developed using calculations from more complex models that include the evolution of drops of different sizes. This type of approximation, referred to as *bulk microphysics parameterization,* has been included in most of the simpler cloud models and in increasing numbers of global numerical weather prediction and climate models.

The term C_{vl}, which describes the condensation and evaporation of cloud water, is determined following Kessler (1969) to be

$$C_{vl} = \delta \frac{dw_s}{dt}$$

where $\delta = 1$ if the air is saturated and $\delta = 0$ if the air is unsaturated, w_s is the saturation mixing ratio, and dw_s/dt is the rate at which vapor is being converted to liquid (e.g., in moist adiabatic ascent or by radiative cooling).

The autoconversion term can be parameterized as (following Kessler, 1969)

$$A_{lR} = a_1 \left(w_l - w_{lo} \right)$$

where w_{lo} is a threshold value of liquid water mixing ratio for the onset of production of raindrops by collision and coalescence among cloud particles. Values of w_{lo} commonly range from 0.4 to 1.0 g kg^{-1}, with different values chosen for different cloud types and CCN activity spectra. a_1 is an efficiency factor for accretion, where $a_1 = 0$ if $w_l < w_{lo}$ and $a_1 > 0$ if $w_l > w_{lo}$ with typical values of $a_1 \approx 10^{-4}$ s^{-1}.

The collection term, K_{lR}, can be parameterized as (following Lee, 1989)

$$K_{lR} = k_1 \, w_l^{1.029} \, w_R^{1.042}$$

where $k_1 \approx 7.57$ s^{-1} is an efficiency factor for collection.

Evaporation of rain, E_{Rv}, is parameterized to be a function of the rain water content and the relative specific humidity of the air (following Lee, 1989):

$$E_{Rv} = 0.00136 \, w_R^{0.42} \left(w_s - w_v \right)^{0.746}$$

where E_{Rv} is given in s^{-1}.

The terminal velocity of the rain particles is given as a function only of w_R by (Manton and Cotton, 1977)

$$u_T = 2.13 \left(\frac{\rho_l}{\rho_a} \right)^{1/2} \left(g w_R \right)^{1/2}$$

where ρ_l is the density of liquid water, ρ_a is the density of air, and u_T has units of m s^{-1}.

Addition of the ice phase to the bulk microphysical parameterizations increases substantially the complexity of these parameterizations. Inclusion of the ice phase is handled typically by adding three additional prognostic equations, for cloud ice, snow, and graupel. A bulk microphysical model with ice must include additional conversion terms such as depositional nucleation, contact nucleation, secondary ice crystal production, freezing, melting, and sublimation.

Increasingly sophisticated bulk microphysical parameterizations continue to be developed. Particularly when the ice phase is included, many of the parameterizations require that arbitrary assumptions be made and introduce a level of uncertainty when these parameterizations are applied to a broad spectrum of cloud types. In a cloud model or a large-scale model of the atmosphere, cloud microphysical processes may have a strong impact on the cloud-scale motions and on larger-scale motions; these variations in the velocity field can then feed back onto the cloud microphysical processes.

Here we have illustrated the concept of cloud bulk microphysical parameterization, which is only one aspect of cloud parameterization needed for large-scale atmospheric models. The scale of a single grid cell in a large-scale atmospheric model is typically on the order of 100 km in the horizontal and 1 km in the vertical. Individual clouds are almost always smaller than this grid scale. It remains a major challenge in atmospheric modeling to parameterize the cloud fractional coverage at different vertical levels, which is needed to calculate the atmospheric radiative fluxes, and to parameterize the transport of heat, moisture, and momentum associated with cumulus convection.

Notes

The *International Cloud Atlas* (1987; Vol. II) by WMO contains many plates that illustrate cloud visual characteristics of the different cloud types.

Storm and Cloud Dynamics (1989) by Cotton and Anthes provides an extensive discussion of the topics discussed in this chapter.

Radiation and Cloud Processes in the Atmosphere (1992) by Liou is a comprehensive treatment of cloud radiative processes.

More extensive treatments of precipitation processes are given in *Cloud Microphysical Processes* (1997) by Pruppacher and Klett and *A Short Course in Cloud Physics* (1989) by Rogers and Yau.

Cloud Dynamics (1993) by Houze provides a thorough treatment of the dynamics of the different cloud types.

Extensive discussions of convective clouds are given in *Atmospheric Convection* (1994; Chapters 7-9) by Emanuel and in *Clouds and Storms* (1980; Chapters 7-8) by Ludlam.

Problems

1. An air parcel initially at $p = 1000$ hPa, $T = 20°C$ and $w_v = 10$ g kg^{-1} is forced upward by a mountain which has a top at 750 hPa. The lapse rate of the environment is 7°C km^{-1} and the relative humidity of the environment is 80%. Assume that the precipitation efficiency of the cloud is 50%. After the cloud precipitates, the air then descends down the other side of the mountain, to the initial pressure $p = 1000$ hPa. What will the temperature and water vapor mixing ratio be after it descends? Note: the strong dry wind in the lee of the mountain, called a *foehn* or *chinook*, is often attributed to this mechanism.

2. Calculate the time required for the diameter of a spherical snowflake to increase from 1 mm to 1 cm if it grows by aggregation as it falls through a cloud of small ice crystals present in an amount 1 g kg^{-1}. You may assume that the collection efficiency is unity, that the density of the snowflake is 100 kg m^{-3}, and the difference in the fall speeds of the snowflake and ice crystals is constant and equal to 1 m s^{-1}.

3. In a small cumulus cloud with base at 1 km and top at 2 km, the liquid water content increases linearly with height to maximum value at cloud top of 1 g kg^{-1}.
a) A drop of 100 μm diameter starts to fall from the top of the cloud. What will be its size when it leaves the cloud base? Assume that the collection efficiency is 0.8 and that there is no vertical air velocity.
b) If all of the condensed water falls out as rain, how much rain would be received at the Earth's surface (in cm)?

4. Calculate the distance through which a water drop, initially 1 mm in radius, falls before evaporating completely in an environment of $\mathcal{H} = 70\%$ and $T = 278$ K.

5. Cloud drop size distributions are determined from aircraft measurements using an instrument called the Forward Single Scattering Probe (FSSP), which determines drop sizes and concentrations by using optical techniques. The FSSP typically gives counts of particles in 15 different-size bins, with $\Delta r = 1.5$ μm. These counts can then be converted into concentrations (# cm^{-3}) by taking into account aircraft velocity, etc. One drop size distribution, $n(r)$, given below, which was determined from the FSSP during the Arctic Stratus Experiment from the middle of a stratus deck. The values of the drop radii given at the top of the table represent the average radius of the 1.5 mm bin.

r (μm):	1.6	3.1	4.7	6.3	7.8	9.4	11.0	12.5	14.1	15.7
$n(r)$:	94	103	105	94	23	23	0.6	0.1	0	0

a) Compute the following for the above drop size distribution:
  N: total number concentration of drops (# cm^{-3})
  w_l: liquid water mixing ratio (g kg^{-1})
  $\bar{r}$: mean drop size radius (μm)

r_e: drop equivalent radius (μm)

σ_{ext}: extinction cross section for solar (shortwave) radiation (m^{-1})

b) Assuming that the cloud is 300 m deep and that the microphysical properties are vertically homogeneous, compute $\mathcal{W}$, the liquid water path (g m^{-2}).

c) For shortwave radiation, compute τ_{ext}, the cloud optical depth for shortwave radiation.

6. The cloud in 5 has the following characteristics: Cloud base: 705 m; Cloud top: 1026 m; Cloud temperature: 277 K (cloud is isothermal); Surface albedo: 0.58

An aircraft flies above the cloud and below the cloud, with both upward and downward facing solar and infrared radiometers. The following fluxes are measured from the aircraft:

	Solar flux (W m^{-2})	
	Downward	Upward
Cloud top	595	399
Cloud base	337	187

	Longwave Flux (W m^{-2})	
	Downward	Upward
Cloud top	235	328
Cloud base	326	331

Determine the following:

a) Solar transmissivity, reflectivity, and absorptivity for the cloud layer.

b) Since the surface albedo (reflectivity) is 0.58, what you calculated in 6a has a contribution from surface reflectivity. The (reduced) reflectivity, $\mathcal{R}^\circ$, and the transmissivity, $\mathcal{T}^\circ$, of the cloud layer, corresponding to the situation of the surface albedo equal to zero, can be determined from the *interaction principle*:

$$\mathcal{R}^\circ = \frac{\mathcal{R} - \alpha_0\, \mathcal{T}^2}{1 - \alpha_0^2\, \mathcal{T}^2}$$

$$\mathcal{T}^\circ = \left(1 - \alpha_0\, \mathcal{R}^\circ\right) \mathcal{T}$$

where α_0 is the surface albedo and $\mathcal{R}$ and $\mathcal{T}$ are determined in 6a.

c) Infrared emissivity for the entire cloud layer.

d) Solar heating rate for the cloud layer.

e) Longwave heating rate for the cloud layer.

Chapter 9 | Ocean Surface Exchanges of Heat and Fresh Water

Heating and cooling at the ocean surface determine the sea surface temperature, which is a major determinant of the static stability of both the lower atmosphere and the upper ocean. The surface heat fluxes at the air/sea interface are central to the interaction and coupling between the atmosphere and ocean. The processes that determine energy transfer between the surface and atmosphere include: net surface radiation flux; the surface turbulent sensible heat and latent heat fluxes; heat transfer by precipitation; and storage and transport of energy below the ocean surface.

The ocean surface salinity budget plays an important role in determining the stability of the upper ocean. The saline surface water in the high-latitude North Atlantic Ocean is a key factor that allows surface water to sink deep into the ocean. On the other hand, fresh surface water acts to stabilize the mixed layer in the Arctic Ocean and the tropical western Pacific.

The ocean surface heat and salinity fluxes, when combined, determine the ocean surface buoyancy flux.

9.1 Ocean Surface Energy Budget

The surface energy budget represents the sum of heat fluxes at the air/sea interface, including fluxes from both the air and ocean sides of the interface. Because the entire ocean mixed layer is active in transferring heat to the interface, the formulation of the surface energy budget commonly includes the heat budget of the entire ocean mixed layer.[1] From the illustration in Figure 9.1, we can write the surface energy budget equation for the ocean[2] as

$$F_{Q0}^{net} - F_{Q0}^{adv} - F_{Q0}^{ent} = F_{Q0}^{rad} + F_{Q0}^{SH} + F_{Q0}^{LH} + F_{Q0}^{PR} \tag{9.1}$$

---

[1] For simplicity, we assume here that all pentrating solar radiation is absorbed in the ocean mixed layer. Heat exchange processes occurring at the ocean side of the air/sea interface are discussed in detail in Sections 11.1–11.3.

[2] Here we consider only the air/ocean interface; the surface energy balance at the ice/ocean interface is examined in Section 10.6.

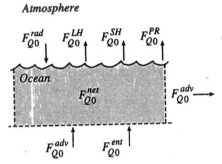

Figure 9.1 Terms in the upper ocean heat budget (see text; equation 9.1). A term is positive if it adds heat to the upper ocean, and negative if it removes heat. The shaded region represents the ocean mixed layer.

where the notation F_Q denotes the flux density of energy in W m^{-2}. The term F_{Q0}^{rad} represents the net surface radiation flux, F_{Q0}^{SH} refers to the surface turbulent flux of sensible heat, F_{Q0}^{LH} is the surface turbulent flux of latent heat, and F_{Q0}^{PR} is the heat transfer by precipitation, F_{Q0}^{adv} represents the transport of energy into or out of the ocean mixed layer via fluid motions, F_{Q0}^{ent} represents the transport of energy into or out of the ocean mixed layer via entrainment and/or molecular diffusion at the base of the ocean mixed layer, and F_{Q0}^{net} is the ocean heat storage term. The sign convention used in (9.1) is that a term adding heat to the mixed layer is positive and a term removing heat from the mixed layer is negative. The remainder of this section addresses the evaluation of the terms on the right-hand side of (9.1).

9.1.1 Surface radiation flux

The net surface radiation flux, F_{Q0}^{rad}, is the sum of the net solar and longwave fluxes at the surface

$$F_{Q0}^{rad} = \left(1 - \alpha_0\right) F_0^{SW} + F_0^{LW} - \epsilon_0 \sigma T_0^4 \tag{9.2}$$

where F_0^{SW} is the downwelling solar radiation flux at the surface, α_0 is the shortwave surface *albedo* (reflectivity), F_0^{LW} is the downwelling infrared radiation flux at the surface, T_0 is the surface temperature, and ϵ_0 is the surface longwave emissivity.

The surface downwelling solar and infrared radiation fluxes, F_0^{SW} and F_0^{LW}, depend on: the amount of radiation incident at the top of the atmosphere; the atmospheric temperature profile; the optical depth of all gaseous constituents in the atmosphere; the optical depth of the atmospheric aerosol; and the optical depth and fractional area coverage of cloud. Given these inputs and the surface temperature and albedo, radiative transfer equations developed from the principles given in Sections

3.3 and 8.3 can be solved numerically to determine values of F_0^{SW} and F_0^{LW}. Numerous radiative transfer models have been developed of varying degrees of complexity. Because of the complexity of solving the radiative transfer equations numerically, and because frequently the observations are inadequate to provide all of the input variables needed for the model, some simple empirical formulations have been developed to determine F_0^{SW} and F_0^{LW} that depend only upon readily available conventional surface observations, such as fractional cloud cover and the surface air temperature.

Under cloudless conditions, the downwelling surface shortwave radiation flux varies with the amount of radiation received at the top of the atmosphere and with variations primarily in aerosol and water vapor content. A useful expression for determining F_0^{SW} under cloudless conditions ($F_0^{SW, clr}$) is given by (Zillman, 1972)

$$F_0^{SW, clr} = S \cos^2 Z \left[1.085 \cos Z + (2.7 + \cos Z)\, e_a + 0.01 \right]^{-1} \qquad (9.3)$$

where e_a is the near-surface atmospheric vapor pressure in bars. The *solar constant*, S, is defined as the amount of solar radiation received per unit time and per unit area, perpendicular to the sun's rays at the top of the atmosphere, at the mean Earth–sun distance. The solar constant has been monitored by satellite and is found to be about 1370 ± 4 W m^{-2}. The *solar zenith angle*, Z, is defined as the angle between the vertical direction and the direction of the incoming solar beam.[3] The presence of clouds reduces the surface solar radiation flux relative to clear-sky values because clouds reflect and absorb solar radiation. The effects of clouds can be included following Reed (1977):

$$F_0^{SW} = F_0^{SW, clr} \left[1 + 0.0019\,(90 - Z) - 0.62\,A_c \right] \qquad (9.4)$$

where A_c is the fractional area of the sky covered by clouds. Equation (9.4) has been found to overestimate F_0^{SW} as much as 6% under some circumstances and should not be applied to values of $A_c < 6\%$.

A portion of the radiation that reaches the sea surface is reflected back into the atmosphere. The shortwave surface albedo, α_0, over the ice-free[4] ocean varies with the solar zenith angle, cloud characteristics, surface wind speed, and the presence of impurities in the water. The dependence of the ocean surface albedo on the solar zenith angle and cloud cover is shown in Figure 9.2. When the sun is nearly overhead

---

[3] See Section 12.1 for further discussion of the solar constant and the solar zenith angle, including mathematical expressions.

[4] The albedo of the ice-covered ocean is addressed in Section 10.5.

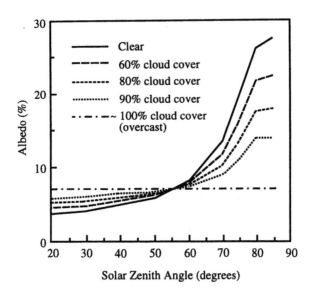

Figure 9.2 The dependence of ocean surface albedo on cloud cover and solar zenith angle. (Data from Mirinova, 1973.)

(low solar zenith angle), the surface albedo varies between 5% and 8%, depending on the cloud cover. When the solar zenith angle is large, the surface albedo increases substantially, particularly for clear skies. Since clouds scatter solar radiation very effectively, solar radiation beneath a cloud becomes increasingly diffuse, and the surface albedo becomes insensitive to solar zenith angle. The surface albedo of water varies spectrally, with higher values in the visible wavelengths (0.4 to 0.8 μm) than in the near-infrared wavelengths (0.8 to 4.0 μm).

The downwelling infrared radiation at the surface under cloudless skies, $F_0^{LW,clr}$, has been parameterized by (Swinbank, 1963):

$$F_0^{LW,clr} = \sigma T_a^4 \left(1 - 9.365 \times 10^{-6} T_a\right) \tag{9.5}$$

where T_a is the near-surface atmospheric temperature. Values of F_0^{LW} under cloudy conditions are larger than those under clear conditions, because clouds emit strongly in the infrared. For cloudy sky conditions, F_0^{LW} can be determined following Zillman (1972):

$$F_0^{LW} = F_0^{LW,clr} + 0.96\, A_c\, \sigma T_a^4 \left(1 - 9.2 \times 10^{-6} T_a^2\right) \tag{9.6}$$

Longwave radiation is absorbed and emitted in the top 1 mm of the ocean. The ocean surface emits nearly as a black body, with a surface emissivity $\epsilon_0 = 0.97$.

Although expressions like (9.3)–(9.6) are simple to use and are used frequently by oceanographers, their accuracy is significantly less than calculations made using a complete radiative transfer model.

9.1.2 Surface turbulent heat fluxes

The terms F_{Q0}^{SH} and F_{Q0}^{LH} in (9.1) refer to the surface turbulent fluxes of sensible and latent heat, respectively. These terms can be written as

$$F_{Q0}^{SH} = -\rho c_{pd} \left(\overline{w'\theta'}\right)_0 \tag{9.7}$$

$$F_{Q0}^{LH} = -\rho L_{lv} \left(\overline{w'q_v'}\right)_0 \tag{9.8}$$

following the notation in Section 3.6. Note that over a snow- or ice-covered ocean surface, L_{iv} should be substituted for L_{lv} in (9.8).

The covariances $\left(\overline{w'\theta'}\right)_0$ and $\left(\overline{w'q_v'}\right)_0$ can be determined from high-frequency measurements of the vertical velocity, potential temperature, and specific humidity. However, such measurements are rarely available; hence, it is desirable to estimate the covariances in terms of parameters that are routinely measured or included in numerical models. The most common method of estimating the surface turbulent fluxes is through the use of bulk aerodynamic formulae. *Bulk aerodynamic formulae* are based on the premise that the near-surface turbulence arises from the mean wind-shear over the surface, and that the turbulent fluxes of heat and moisture are proportional to their gradients just above the ocean surface.

Central to the bulk aerodynamic formulae are the *aerodynamic transfer coefficients*. The aerodynamic transfer coefficient for momentum (also referred to as the *drag coefficient*) is denoted by C_D and relates the vertical flux of horizontal momentum at the surface (or the *surface stress*) to the square of the velocity difference between the atmosphere just above the surface (nominally at $z=10$ m) and the surface:

$$u_*^2 = \overline{uw_0}^2 + \overline{vw_0}^2 = C_D \left(u_a - u_0\right)^2 \tag{9.9}$$

where u_* is called the *friction velocity*. The subscript 0 refers to the value at the ocean surface, and the subscript a refers to a reference level in the atmosphere, which is typically 10 m. The term $u_a = \sqrt{u_x + u_y}$ refers to the wind speed at the atmospheric reference level and the value of u_0 is the component of ocean surface velocity along the wind direction. The term u_0 is commonly neglected relative to u_a but should not be neglected in regions of the ocean where surface currents are strong and winds are weak.

The bulk aerodynamic formulae for the sensible and latent heat fluxes at the surface are written analogously as

$$F_{Q0}^{SH} = \rho c_p C_{DH} \left(u_a - u_0\right) \left(\theta_a - \theta_0\right) \tag{9.10}$$

$$F_{Q0}^{LH} = \rho L_{lv} C_{DE} \left(u_a - u_0\right) \left(q_{va} - q_{v0}\right) \tag{9.11}$$

where C_{DH} and C_{DE} are the aerodynamic transfer coefficients for temperature and humidity, respectively. The term θ_0 is the potential temperature corresponding to the interfacial temperature at the ocean surface. The term q_{v0} is the saturation specific humidity of the seawater at the interfacial temperature at the ocean surface, which is determined from the definition of a saturation specific humidity, q_s, following (4.39) and Raoult's law (4.48)

$$q_{v0} = q_s\left(T_0, p_0\right) \left(1 - \frac{i n_{solt}}{n_{H_2O}}\right) \approx 0.98\, q_s \tag{9.12}$$

where the approximate factor 0.98 holds for a salinity of 35 psu.

The key aspect of evaluating (9.10) and (9.11) is to determine values of the aerodynamic transfer coefficients. Wind blowing over the surface generates eddies whose size depends on the roughness of the surface. The surface roughness of the ocean, z_0, arises from viscous effects and surface waves. The generated turbulence also depends on the static stability. Under ordinary conditions, C_{DH} and C_{DE} are nearly equal. Various methods have been used to evaluate the aerodynamic transfer coefficients; here we present a simple technique for their evaluation that includes the essential physics of the problem without doing a detailed derivation.[5] Assuming that $C_{DH} = C_{DE}$, we can write the following expression:

$$C_{DH} = \frac{k^2}{\left(\ln \dfrac{z_a}{z_0}\right)^2} f\left(Ri_B\right) \tag{9.13}$$

where $k = 0.4$ is the *von Karman constant*. An expression for the ocean surface roughness length that includes viscous and gravity wave effects is given by:

$$z_0 = 0.11 \frac{\nu}{u} + 0.16 \frac{u_*}{g} \tag{9.14}$$

---

[5] Derivations of expressions for the drag coefficients are found in Geernaert and Plant (1990); Kraus and Businger (1994); Kantha and Clayson (1999).

where v is the molecular viscosity and is on the order of 1.5×10^{-5} m^2 s^{-1} and u_*, the friction velocity, is given by

$$u_*^2 = \frac{k^2 \left(u_a - u_0 \right)^2}{\left(\ln \dfrac{z_a}{z_0} \right)^2} f(Ri_B)$$

(9.15)

The *bulk Richardson number*, Ri_B, is the square of the ratio of the buoyancy frequency to the wind shear

$$Ri_B = \left(\frac{N}{\dfrac{\partial u}{\partial z}} \right)^2 \approx \frac{g}{\theta_0} \frac{z_a \left(\theta_{va} - \theta_{v0} \right)}{\left(u_a - u_0 \right)^2}$$

(9.16)

If the atmosphere just above the surface is statically stable, then $Ri_B > 0$; if unstable, the $Ri_B < 0$; if neutral, then $Ri_B = 0$. The stability parameter $f(Ri_B)$ is determined to be (Louis, 1979)

$$f(Ri_B) = \frac{1}{1 + 15 \sqrt{1 + 5Ri_B}} \qquad \text{for } Ri_B > 0$$

(9.17)

$$f(Ri_B) = 1 - \frac{15 Ri_B}{1 + 75 \left(\dfrac{k^2}{\left(\ln \dfrac{z_a}{z_0} \right)^2} \right) \sqrt{1 + \dfrac{z}{z_0} |Ri_B|}} \qquad \text{for } Ri_B < 0$$

(9.18)

Evaluation of C_{DH} using (9.13)–(9.18) is shown in Figure 9.3. C_{DH} is relatively small for statically stable conditions, and relatively large for statically unstable conditions. The magnitude of the transfer coefficient decreases with increasing values of z/z_0.

A problem in the use of (9.9) and (9.16) arises in the limit $(u_a - u_0) = 0$, where from (9.10) and (9.11), the turbulent fluxes should be zero. However, significant turbulent fluxes can arise from buoyant (or *free*) convection even if the wind speed is zero. In the free convection limit, the surface sensible heat flux is given by (Louis, 1979)

$$\left(\overline{w'\theta'} \right)_0 = \left(\frac{g z_a}{\theta_0} \right)^{1/2} \frac{\left(\theta_a - \theta_0 \right)^{3/2}}{3.9 \left(\dfrac{z_a}{z_0} \right)^{1/2}}$$

(9.19)

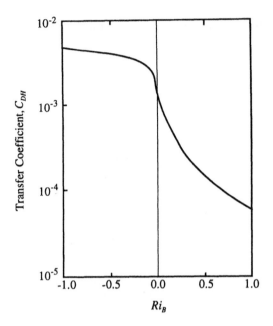

Figure 9.3 Transfer coefficient for heat as a function of the bulk Richardson number, with $z/z_0 = 5500$. As the stability increases, the transfer coefficient decreases. The magnitude of the heat transfer is thus inversely proportional to the degree of stability.

9.1.3 Heat flux from precipitation

The term F_{Q0}^{PR} in (9.1) is the heat flux at the surface due to rain or snow. Heat transfer by precipitation occurs if the precipitation is at a different temperature than the surface. Based on the analysis in Section 6.3, if a falling raindrop is in thermal equilibrium with its surroundings, then the temperature of the raindrop at a given height will be the same as the wet-bulb temperature of the atmosphere at that height. Assuming that such an equilibrium exists, we can write

$$F_{Q0}^{PR} = \rho_l \, c_{pl} \, \dot{P}_r \left(T_{wa} - T_0 \right) \tag{9.20}$$

where ρ_l and c_{pl} refer to the liquid water values appropriate to the rain and $\dot{P}_r$ is the rainfall rate in m s^{-1}. Values of F_{Q0}^{PR} are greatest for large rainfall rates and for large differences between the atmospheric wet-bulb temperature and sea surface temperature. Except for rare circumstances, $T_{wa} < T_0$, and the heat flux from rain cools the ocean. During heavy rainfall events, values of F_{Q0}^{PR} may be the largest term in the surface energy budget; however, when is averaged over longer time scales, the contribution of this term to the surface energy budget is quite small and is commonly neglected.

In the presence of snowfall the term F_{Q0}^{PR} is more complex, since the ocean must provide latent heat to melt the snow. Hence we have the following expression for the heat flux associated with snowfall into the ocean:

$$F_{Q0}^{PR} = \rho_{\jmath}\, c_{p_{\jmath}}\, \dot{P}_{\jmath} \left(T_{la} - T_0\right) - \rho_{\jmath}\, L_{il}\, \dot{P}_{\jmath} \tag{9.21}$$

where the subscript $\jmath$ refers to snow, $\dot{P}_{\jmath}$ is the snowfall rate, and T_{la} is the ice-bulb temperature (6.25) of the atmosphere just above the surface. Taking the ratio of the two terms of the right-hand side of (9.21), we obtain

$$\frac{c_{p_{\jmath}}\left(T_{la} - T_0\right)}{L_{il}} = 0.0063\left(T_{la} - T_0\right)$$

Even for extreme values such as $T_{la} - T_0 = 15°C$, the latent heat term is an order of magnitude larger than the sensible heat term in (9.21).

9.1.4 Variation of surface energy budget components

Here we consider the annual cycle of the three primary components of the surface energy budget (net radiation, sensible heat flux, and latent heat flux) for four regions of the global ocean (Figure 9.4):

1. Tropical western Pacific Ocean (1°S, 150°E);
2. Subtropical eastern Pacific Ocean (25°N, 135°W);
3. Gulf Stream (38°N, 71°W);
4. East Greenland Sea (70°N, 2°E).

Most of the latitudinal variation in net surface radiation over the ocean arises from the latitudinal changes of solar radiation and from cloudiness variations. Because of the large cloudiness over the tropical western Pacific Ocean, the net surface radiation is lower than one might expect for the tropics. The largest annual cycle of net surface radiation flux is seen in the high-latitude oceans and in the Gulf Stream. In the winter, the net surface radiation over the East Greenland Sea reaches as low as −50 W m⁻².

Although the sea surface temperatures are warm in the tropical oceans, there is little surface sensible heat flux in the tropics because values of $(T_a - T_0)$ are not large and wind speeds are generally small. The largest values of surface sensible heat flux are seen along the western coastal boundaries of mid-latitude oceans during winter

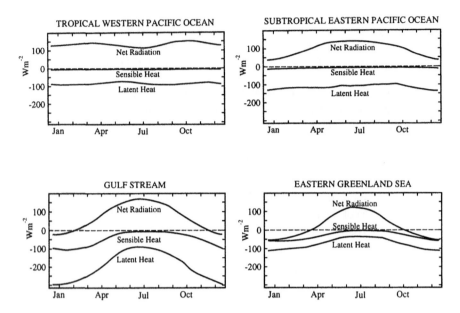

Figure 9.4 The annual cycle of net surface radiation, sensible heat flux, and latent heat flux over four regions of the global ocean.

(e.g., the Gulf Stream). Northward western boundary currents carry warm water poleward, which comes into contact with the cold dry air advected off the continents during winter. Relatively large values of surface sensible heat flux also occur in the East Greenland Sea over most of the annual cycle, as cold, dry air is continually advected off the Greenland continent. Low values of surface sensible heat fluxes are seen in the subtropical eastern Pacific Ocean. This is a result of the locally cool sea surface temperatures that arise from upwelling which occurs along the eastern margins of oceans, bringing cold water to the surface.[6]

Analogously to the surface sensible heat flux, the largest values of the surface latent heat flux occur over the Gulf Stream, with correspondingly smaller values in the subtropical eastern Pacific. Everywhere (except occasionally in the East Greenland Sea during winter), the surface latent heat flux exceeds the sensible heat flux. The *Bowen ratio*, B_0, is defined as the ratio of the surface sensible heat flux to the latent heat flux

$$B_0 = \frac{F_{Q0}^{SH}}{F_{Q0}^{LH}} \qquad (9.22)$$

---

[6] Note that a reversal of this occurs during El Niño, when the upwelling ceases and ocean surface temperatures along the eastern margins of the ocean may increase by up to 6°C.

From Figure 9.4, we can infer that values of the Bowen ratio range from 0.05 in the tropical western Pacific to 0.8 during winter in the East Greenland Sea. This latitudinal gradient of the Bowen ratio is a consequence of the decrease in sea surface temperature with latitude and the corresponding latitudinal variation in surface saturation vapor pressure, which varies exponentially with sea surface temperature following the Clausius–Clapeyron equation (4.23).

The net heat flux into the ocean is positive in the tropical oceans and along the eastern ocean boundaries. In the tropics, this arises primarily from the flux of solar radiation, while along the eastern ocean boundaries, the positive net heat flux arises principally from reduced latent heat loss. Large negative values are seen over the Gulf Stream and in the high-latitude oceans. Warm water is transported northward in the Gulf Stream, which results in a large heat loss to the atmosphere primarily through the latent heat flux. In the high latitude oceans, the combination of sensible and latent heat loss overcomes the relatively weak net radiation flux, to cool the ocean.

9.2 Ocean Surface Salinity Budget

The processes that contribute to the ocean surface salinity budget include: precipitation, evaporation, formation and melting of sea ice, river runoff, and storage and transport below the ocean surface. Analogously to the surface energy budget equation (9.1), we can write a general ocean surface salinity budget for either the air/ocean interface or the ice/ocean interface as

$$F_{s0}^{net} - F_{s0}^{adv} - F_{s0}^{ent} = \left(-\rho_l \dot{P}_r - \rho_{\jmath} \dot{P}_{\jmath} + \dot{E}_0 - \dot{R} \right) s_0 + \rho_i \frac{dh_i}{dt} \left(s_0 - s_i \right) \qquad (9.23)$$

where the subscripts 0 and i refer to the ocean surface value and the sea ice value, respectively. The terms in (9.23) are the flux density of salt at the surface, in units kg psu m^{-2} s^{-1}. A positive term denotes an increase in surface salinity. Following (9.1), F_{s0}^{net} is the ocean storage term and F_{s0}^{adv} and F_{s0}^{ent} represent the transport of salt into the ocean mixed layer via fluid motions and turbulence, respectively.

The first two terms on the right-hand side of (9.23) represent the freshening associated with rain and snowfall. Precipitation acts as a negative flux of salt, since the near surface ocean water is diluted by the precipitation as if there were a loss of salt. Because of the momentum of rain as it reaches the ocean surface, some of the drops submerge into the ocean, depending on the size (and thus terminal velocity) of the drops. The raindrops that are not submerged remain to form a freshwater "skin." The surface salinity depression associated with rainfall normally does not exceed 5 psu. Snowflakes do not submerge into the ocean because of their low density.

The term $\dot{E}_0$ in (9.23) is the evaporative flux of water from the ocean surface. Evaporation of water from the ocean increases the concentration of salt in the ocean and thus the salinity. The evaporative flux of water $\dot{E}_0$ is given by

$$\dot{E}_0 = \rho \left(\overline{w' q_v'} \right)_0 = -\frac{F_{Q0}^{LH}}{L_{lv}} \qquad (9.24)$$

and can be evaluated following Section 9.1.2.

The term $\dot{R}$ in (9.23) arises from the transport of freshwater from river runoff into the ocean. Typically, about 40% of the precipitation that falls on a continent is transported into the global ocean through river runoff. River runoff acts analogously to precipitation by diluting the ocean water and acting as negative salt flux. River runoff influences the surface salinity directly only in coastal regions.

In (9.23) the term h_i denotes sea ice thickness, ρ_i the density of sea ice, and s_i the salinity of sea ice. The term dh_i/dt reflects the growth or melting of sea ice. Because growing sea ice rejects the salt back into the melt (Section 4.6), sea ice freezing acts effectively as salt source for the ocean mixed layer. Melting sea ice freshens the ocean mixed layer, and thus acts as a negative salt flux.

Over most of the global ocean, away from the coast and from the high-latitude regions that are influenced by sea ice and snowfall, we can write a simplified version of (9.23) as

$$F_{s0}^{net} - F_{s0}^{adv} - F_{s0}^{ent} = \left(-\rho_l \dot{P}_r + \dot{E}_0 \right) s_0 \qquad (9.25)$$

Figure 9.5 shows the zonally-averaged ocean surface water balance. The evaporative flux shows a general decrease with latitude away from the equator, analogous to the surface latent heat flux described in Section 9.4.4. Precipitation exceeds evaporation in the equatorial regions, and thus there is a net freshening of the ocean surface. In the subtropical latitudes, from approximately 15° to 40°, the evaporation term dominates the surface salinity budget, and there is a net positive surface salinity flux. In the middle latitudes, there is also an excess of precipitation over evaporation in the zonally-averaged surface salinity budget. However, the zonal average hides important meridional differences and differences between ocean basins. For example, comparing the values of the Atlantic and Pacific Oceans (Table 9.1) shows that the net surface salinity flux is positive (salinating) in the Atlantic and negative (freshening) in the Pacific. This difference between the two basins arises primarily from differences in precipitation. This is believed to occur because of water vapor transport from the Atlantic to the Pacific across Central America and because of the relatively narrow width of the Atlantic basin.

Global river runoff is shown in Figure 9.6. Most of the river runoff occurs in the Northern Hemisphere, because of the larger land mass. The Amazon River provides the largest source of river runoff. The large amount of river runoff into the Arctic Ocean results in an upper ocean salinity of around 30 psu. Most of the fresh-water input to the Southern Ocean comes from glaciers via iceberg calving, basal melting under ice shelves, and wall melting (see Section 10.9).

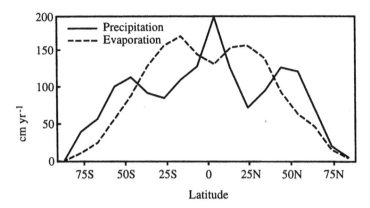

Figure 9.5 Annual zonal mean water balance at the ocean surface. (Data from Baumgartner and Reichel, 1975.)

Figure 9.6 Global river runoff. Segment areas are proportional to the annual volume flow. Most of the runoff occurs in the Northern Hemisphere because of its greater land mass. The largest volume of fresh-water input comes from the Amazon River. Circles on the coast of Antarctica indicate runoff from glaciers. (From Woods, 1984.)

Table 9.1 Precipitation and evaporation rates (mm yr^{-1}) for four ocean basins. (Data from Baumgartner and Reichel, 1975.)

	$\dot{P}$	$\dot{E}_0$
Arctic Ocean	97	53
Atlantic Ocean	761	1133
Indian Ocean	1043	1294
Pacific Ocean	1292	1202
All oceans	1066	1176

9.3 Ocean Surface Buoyancy Flux

The net fluxes of heat and salinity at the ocean surface produce an *ocean surface buoyancy flux*, F_{B0}, written as

$$F_{B0} = g \left(\frac{\alpha}{c_{p0}} F_{Q0}^{net} - \beta F_{s0}^{net} \right) \tag{9.26}$$

where c_{p0} is the specific heat of the surface water and F_{B0} has units kg m^{-1} s^{-3}, corresponding to $(g\rho u_i)$. Comparison of (9.26) with (7.16) and consideration of the instability criterion, $N^2 < 0$, implies that a negative value of F_{B0} meets the instability criterion, and would lead to sinking motion in the ocean.

By incorporating (9.1), (9.20), (9.23), and (9.24) into (9.26), and ignoring terms associated with advection, entrainment, sea ice, and snowfall, we obtain

$$F_{B0} = \frac{g\alpha}{c_{p0}} \left[F_{Q0}^{rad} + F_{Q0}^{SH} - \rho L_{lv} \dot{E}_0 + c_{pl} \dot{P}_r \left(T_{wa} - T_0 \right) \right] - g\beta \left(\dot{E}_0 - \rho_l \dot{P}_r \right) s_0 \tag{9.27}$$

We see from (9.27) that precipitation and evaporation influence the ocean surface buoyancy flux through contributions to both the net heat flux and net salt flux terms.

Evaporation increases the buoyancy flux both by cooling and by increasing salinity. Taking the ratio of the cooling term to the salinity term for evaporation, we obtain from (9.27)

$$\frac{\alpha L_{lv}}{\beta c_{p0} s_0} \qquad (9.28)$$

The value of the ratio varies with the temperature and salinity of the seawater, through the term s_0 explicitly and implicitly through variations in α and β. Table 1.4 shows that α varies by more than an order of magnitude over the range of temperatures in the global ocean. The ratio in (9.28) under conditions typical of the tropical oceans ($T = 30°C$, $s = 35$ psu) is approximately 8.0 and under conditions typical of the high-latitude oceans ($T = 0°C$, $s = 35$ psu) is approximately 0.6. The cooling effect of evaporation thus dominates the buoyancy flux at tropical latitudes, while the increase in salinity dominates at polar latitudes. At all latitudes, the cooling associated with evaporation reinforces its effect on salinization in determining the surface buoyancy flux.

We can conduct a similar analysis for the effect of precipitation on the ocean surface buoyancy flux. Precipitation decreases the buoyancy flux by freshening the ocean. The term associated with the sensible heat flux of rain typically has a cooling effect on the ocean (if $T_{wa} < T_0$) and thus increases the buoyancy flux. By taking the ratio of the heating term to the salinity term for precipitation, we obtain an expression analogous to (9.28):

$$-\frac{\alpha c_{pl} \left(T_{wa} - T_0\right)}{\beta c_{p0} s_0} \qquad (9.29)$$

The negative sign indicates that the heating effect counteracts the salinity effect on the surface buoyancy flux. Under conditions typical of the tropical oceans, this ratio is about -0.06. At high latitudes, because of the reduced value of α at cold ocean temperatures, the effect of cooling on the buoyancy flux is further diminished relative to the freshening effect of rainfall. Thus the freshening effects of rain on the buoyancy flux dominate the cooling effects of rain at all latitudes, with the cooling effect slightly counteracting the freshening effect.

When the effects of snowfall are considered, the latent heat required to melt the snow once it reaches the ocean must be included. Incorporating (9.21) into (9.27), we obtain the following ratio analogous to (9.29):

$$-\frac{\alpha \left[c_{pl} \left(T_{la} - T_0\right) + L_{il}\right]}{\beta c_{p0} s_0} \qquad (9.30)$$

For $T_{la} - T_0 = 10°C$ and $T_0 = 0°C$, the ratio is -0.35. Thus the latent heat flux associated with snowfall counteracts significantly the effect of freshening on the ocean surface buoyancy flux, although freshening still dominates the effect of snowfall on the surface buoyancy flux.

For an ocean surface covered with sea ice, we obtain the following expression for the surface buoyancy flux from (9.21), (9.23), and (9.27), where the surface is now the ice/ocean interface:

$$F_{B0} = \frac{g\alpha}{c_{p0}} \left(F_0^{SW} \, \mathcal{T}_i + \rho_i L_{il} \frac{dh_i}{dt} \right) - g\beta\rho_i \frac{dh_i}{dt} (s_0 - s_i) \tag{9.31}$$

where $\mathcal{T}_i$ is the transmissivity of the sea ice to solar radiation (see Section 3.3). The only heat flux terms that influence the ice/ocean surface are the penetration of solar radiation beneath the ice and the latent heat associated with freezing or melting ice. Ice growth releases latent heat to the ocean, and also acts as a salinity source. Thus the heating and salinity terms have opposing effects on the buoyancy flux. The ratio of the heating and salinity terms associated with ice growth can be written as

$$- \frac{\alpha L_{il}}{\beta c_{p0} (s_0 - s_i)} \tag{9.32}$$

For conditions typical of polar oceans, the ratio is about −0.1, indicating that the salinity term associated with ice growth dominates the latent heating term in determining the ocean surface buoyancy flux.

9.4 Air Mass and Upper Water Mass Modification

The exchange of heat and moisture between the atmosphere and the ocean modifies the temperature and static stability of both the lower atmosphere and upper ocean, as well as the humidity of the lower atmosphere and the salinity of the upper ocean.

An *air mass* is a widespread body of air whose properties were established while it was situated over a particular region of the Earth's surface (*air mass source region*). An air mass is approximately homogeneous in its horizontal extent, particularly with reference to temperature and moisture distribution. The stagnation or long-continued motion of air over a source region permits the vertical temperature and moisture distribution of the air to reach relative equilibrium with the underlying surface. As an air mass moves away from the source region, it undergoes modification in response to the altered underlying surface.

The most widely accepted air mass classification scheme is the *Bergeron classification*. In this system, air masses are first designated by the thermal properties of their source region: *tropical* (*T*), *polar* (*P*), and *arctic* or *antarctic* (*A*). The tropical and polar air masses are delineated by the location of the polar front, which arises

from the thermal contrasts between the tropical and polar air masses. Arctic or antarctic air masses are poleward of the polar air masses. Oftentimes, polar air masses
are not distinguished from arctic air masses. For characterizing the moisture distribution, air masses are distinguished as *continental* (*c*) and *maritime* (*m*) source regions.
Thus we have the following air mass designations: *continental tropical* (*cT*); *maritime tropical* (*mT*); *maritime polar* (*mP*); *continental polar* (*cP*); *maritime arctic* (*mA*);
and *continental arctic* (*cA*).

Further classification according to whether the air is *cold* (*k*) or *warm* (*w*) relative
to the surface over which it is moving indicates the low-level stability of the air and
the type of modification from below. When a colder air mass is advected over a
warmer surface, the lower atmosphere will become statically unstable, a condition
favorable for convective and turbulent motions (Figure 9.7). In contrast, the modification of a warm air mass that is advected over a cold surface will become stabilized,
diminishing the vertical turbulent exchange of heat and moisture. In this case, the
primary energy exchange between the atmosphere and surface is via radiative transfer. The modification of a cold air mass occurs much more rapidly than the modification of a warm air mass.

Water masses are identified by their potential temperature and salinity. A water
mass with characteristic temperature and salinity values forms when the upper ocean
is subject to specific meteorological influences over a significant period of time. The
existence of large volumes of seawater of characteristic temperature and salinity indicates that the seawater within the mass originated in the same locale in response to the
same heat and fresh-water fluxes across the air/sea interface. The name of a water
mass is identified with its oceanic source region (e.g., Mediterranean Water, South
Atlantic Central Water).

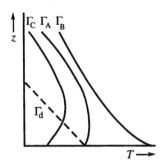

Figure 9.7 Heating and cooling of an air mass as it is advected over a warmer or cooler
surface. Initially, the air mass has a lapse rate, Γ_A, that is less than dry the adiabatic lapse rate
(dashed line). Heating from the surface increases the lapse rate to Γ_B, so that the lower
atmosphere becomes statically unstable. If this same air mass is then advected over a cooler
surface so that its lapse rate becomes Γ_C, it will once again be stable to vertical motions.

Upper water masses are generally considered to include both the mixed surface layer and the upper part of the permanent thermocline. Sinking of an upper water mass, induced by static instability or convergence, produces an *intermediate water mass* or *deep water mass*, depending on the depth to which it sinks. Intermediate and deep water masses maintain their identities for a very long time, since at these depths the only heat exchange possible between water masses is diffusive and turbulent. Intermediate and deep water masses form at high latitudes in response to the surface cooling and salinization associated with evaporation and sea ice formation. Intermediate and deep water masses are transported by ocean currents far from their source of origin. For example, Antarctic Bottom Water is found in the deepest ocean throughout the Southern Hemisphere and into the Northern Hemisphere.

Figure 9.8 shows a plot of the vertical T, s structure in different regions of the world ocean. In the upper 1000 m, the water mass structure is dominated by exchanges that occur across the air/sea interface. The high salinity of the Mediterranean Water is associated with the large amount of evaporation that occurs during winter when the

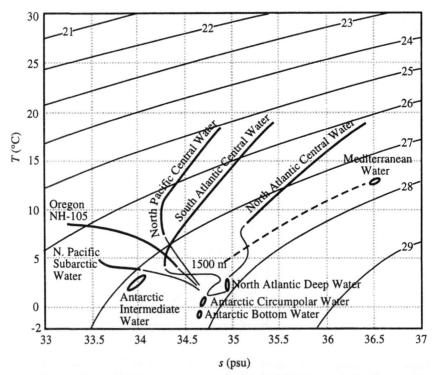

Figure 9.8 T,s curves illustrating the various types of water mass structures found in different regions of the world ocean. Broad lines represent characteristics of the upper ocean layer, from 100–1000 m; thin lines represent the deep water structure. (From Neshyba, 1987. ©John Wiley & Sons, Inc. Reprinted with permission.)

cold, dry wind from the continent blows over the relatively warm sea. This increases the density of the surface water to such an extent that convection occurs down to the sea floor at 2000 m depth. The Mediterranean Water leaves the Strait of Gibraltar at depth, where it mixes intensely with the Atlantic Water. Although the Mediterranean Water is continually modified by mixing, it can be recognized throughout much of the Atlantic Ocean by its signature values of high temperature and salinity. The North Atlantic Central Water is much more saline that the South Atlantic Central Water because of the high surface evaporation and relatively low precipitation in the North Atlantic. The complex shape of the South Atlantic T, s curve arises because its water column intersects layers of Antarctic Intermediate Water at about 1000 m and North Atlantic Deep Water at about 3000 m.

Notes

The components of the surface heat budget are discussed in more detail in *Atmosphere-Ocean Interaction* (1994) by Kraus and Businger and in *Small-scale Processes in Geophysical Flows* (1999, Chapter 4) by Kantha and Clayson.

A description of the ocean surface heat and salinity budgets and also the ocean surface buoyancy flux is given by Gill (1982, Chapter 2) *Atmosphere-Ocean Dynamics*.

Byers (1959, Chapter 14) *General Meteorology* provides a thorough discussion of air masses.

Tomczak and Godfrey (1994, Chapter 5) *Regional Oceanography: An Introduction* provides a thorough description of water masses.

Problems

1. While flying over the tropical ocean, an aircraft measured a raindrop number concentration of $N(d) = 100$ m^{-3} mm^{-1} for raindrops of a diameter $d = 2.2$ mm.

a) Assuming a Marshall–Palmer size distribution for the raindrops (8.8), determine $\dot{P}$.

b) Assuming that the rain reaches the surface with a temperature at the wet-bulb temperature of the air, calculate the sensible heat flux at the ocean surface associated with the rain, F_{Q0}^{PR}, in units of W m^{-2}, assuming a wet-bulb temperature of the air to be $T_{w0} = 26°C$ and the surface temperature is $T_0 = 30°C$.

c) Compare this value of F_{Q0}^{PR} with the magnitude of the terms shown in Figure 9.4.

2. The following data were obtained from the tropical western Pacific Ocean (1.7°S, 156°E) on November 15, 1992. At noon on this date, the solar zenith angle is $Z = 17.1°$. No clouds were present ($A_c = 0$ and $\dot{P}_r = 0$). Values of T_a, q_a, and u_a were obtained at a height of 15 m.

Time	T_0 (°C)	T_a (°C)	q_a (g kg^{-1})	u_a (m s^{-1})
Noon	29.4	28.1	18.8	4.2
2 am	29.0	28.0	18.3	4.0

Estimate the net heat flux into the ocean at noon and at 2 a.m.

3. Using the data in **2**, determine the ocean surface buoyancy flux for noon and 2 a.m.

Chapter 10 | Sea Ice, Snow, and Glaciers

Variations in snow and ice cover play a crucial role in climate via the following physical processes:

1) The presence of ice and snow causes a much larger portion of the incoming solar energy to be reflected back to space.
2) The presence of sea ice reduces the turbulent transfers of sensible and latent heat and momentum between the ocean and atmosphere, thus insulating the ocean at high latitudes.
3) Because of the latent heat associated with melting and freezing, sea ice and snow act as thermal reservoirs that delay the seasonal temperature cycle.
4) The formation and melting of sea ice alters the ocean surface buoyancy flux, thus influencing thermohaline circulations.
5) Glacial runoff from Antarctica is a major source of fresh water for the southern hemisphere oceans.
6) The volume of glacial ice is the primary determinant of global sea level.

10.1 Large-scale Morphology of Sea Ice

At its maximum seasonal extent, sea ice covers approximately 8% of the surface area in the Southern Hemisphere and 5% in the Northern Hemisphere. In the Southern Hemisphere, sea ice forms a seasonally varying ring around the Antarctic Continent with relatively small meridional variations. Most of the Antarctic sea ice is seasonal, with 80% of the ice disappearing by the end of the austral summer (Figure 10.1a). In the Northern Hemisphere, a perennial sea ice cover exists in the Arctic Ocean, which is essentially a landlocked ocean basin (Figure 10.1b). Strong meridional variations in the Arctic Ocean sea ice cover arise from the complex configuration of the Northern Hemisphere land masses and from variations in ocean currents. In winter, the sea ice extends as far south as 45°N to the coast of Japan (145°E), while the warm Atlantic water flowing northward keeps the sea ice free as far north as 80°N near Spitsbergen (10°E).

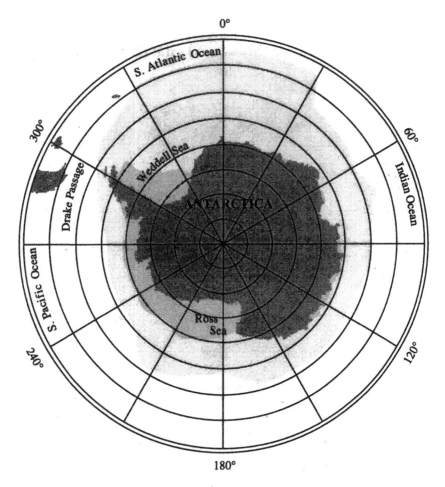

Figure 10.1a Seasonal sea ice extent for the Southern Hemisphere. Medium shading indicates summer ice cover, and light shading indicates the maximum (winter) extent of the ice cover.

Sea ice is in almost continual motion, driven by winds and ocean currents. Ice motion in the Arctic Ocean consists of anticyclonic (clockwise) motion in the Amerasian Basin and a southward drift towards the Fram Strait in the Eurasian Basin. Average ice displacements in the Arctic Ocean are 7 km day^{-1}. On an annual basis, roughly 10% of the Arctic Ocean sea ice is exported through the Fram Strait into the Greenland Sea, where it eventually melts. Ice motion in the Southern Hemisphere occurs in a clockwise direction around the continent. Horizontal gradients in ice

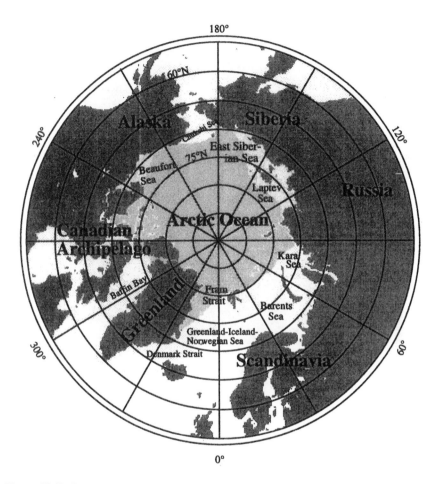

Figure 10.1b Seasonal sea ice extent for the Northern Hemisphere. Medium shading indicates summer ice cover, and light shading indicates the maximum (winter) extent of the ice cover.

motion give rise to ice divergence and convergence. Ice divergence is defined as $\nabla \cdot u_i$ where u_i is the ice drift velocity.

Diverging motions cause the ice to break apart and expose the underlying water. Cracks in the sea ice formed by this mechanism are called *leads*, which are tens to hundreds of meters in width and may extend lengthwise for many kilometers (Figure 10.2). During winter, when leads rapidly freeze over, the area of open water in the

Figure 10.2 A lead in the sea ice, caused by diverging horizontal motion, and a pressure ridge, caused by converging ice motion. (Photograph courtesy of K. Steffen.)

Arctic Ocean rarely exceeds 0.5%. During late summer in the Arctic, the lead fraction may reach 20%, when melting processes act in conjunction with ice dynamics to increase the area of open water. In the Southern Hemisphere, where the ice is not constrained by land, the amount of open water within the sea ice averages 20 to 40% throughout most of the year, indicating that the Antarctic sea ice undergoes substantial divergence. Converging and shearing ice motions cause the ice to deform and pile up. *Pressure ridges* are linear features up to several kilometers in length, consisting of a *keel* beneath the ice and a *sail* on the surface of the ice (Figure 10.2). Keels are usually less than 20 m thick, and sails commonly do not exceed 10 m in thickness. Ridge frequency in the Arctic Ocean is generally in the range of 2–6 km^{-1}. Because of the divergent nature of the velocity field in the Antarctic, ridges are uncommon except within the Weddell Sea (Figure 10.1a), which is a semi-enclosed ocean basin.

Large regions of open water within the ice pack that are quasi-stationary and often recur on an annual basis are called *polynyas*. Polynyas are typically rectangular or elliptical in shape and range in size from a few hundred meters to hundreds of kilometers. The following mechanisms contribute to the formation of polynyas: oceanic heat may enter the region in sufficient quantities to locally prevent ice formation; and ice that does form within the polynya region may be removed continually by winds. Polynyas are active sites for brine formation and influence the large-scale water mass modification as the brine is mixed and advected beyond the polynya region. Polynyas are important biological habitats for large mammals.

Sea ice can be divided into a number of different zones. The principal distinction is between the *perennial sea ice zone* and the *seasonal sea ice zone*. The zones of perennial ice (or multi-year ice) correspond to the medium-gray shaded portions in Figure 10.1 and consist of ice that survives the summer melt season. Zones of seasonal sea ice, which consist of ice that does not survive the summer melt, are indicated in Figure 10.1 by the light-gray shaded portions. In the central Arctic Ocean the thickness of the perennial ice averages 3 to 4 m. Near the Canadian Archipelago, the ice is significantly thicker as a result of extensive ridging. The residence time of perennial ice floes in the Arctic Ocean is determined primarily by the length of time before the ice is exported through the Fram Strait, and ranges from two years to several decades. Perennial ice in the Southern Hemisphere occurs primarily in the Weddell Sea but has a residence time of only a few years. Ice in the zone of seasonal sea ice is generally less than 2 m thick. Approximately 80% of the highly mobile Antarctic sea ice is seasonal. The *marginal ice zone* is located at the boundary between the ice and the open ocean. In the region of this boundary, penetration of surface waves into the ice pack breaks the ice into numerous small floes, the size of the floes increasing with distance from the ice edge. The marginal ice zone includes a region of approximately 100 km in width that is affected by the presence of the open ocean. The marginal ice zone is associated with strong horizontal gradients in surface fluxes of heat and momentum as well as cloudiness and upper ocean characteristics.

The extent and thickness of sea ice are determined by a complex interplay of thermodynamic and dynamic processes. It is the thermodynamic processes that are the focus of this chapter. Because of the economic, scientific, and military importance of the Arctic Ocean and its relative accessibility, we have far more information about the Arctic sea ice than about the Antarctic sea ice. Hence this chapter focuses predominantly on the Arctic Ocean sea ice.

10.2 Ice Thickness Distribution

The Arctic Ocean sea ice cover is particularly heterogeneous, with variations in sea ice thickness ranging from zero in leads to as much as 40 m in some pressure ridges, over horizontal scales of order 100 m. Properties of sea ice that are strongly dependent on the ice thickness distribution include compressive strength, rate of growth, surface temperature, surface albedo, turbulent heat exchange with the atmosphere, and salt and heat exchange with the ocean. The evolution of the thickness distribution of sea ice arises from both thermodynamic processes (freezing and melting) and dynamic processes (sea ice drift and deformation). Freezing and melting are primarily influenced by radiation and heat exchanges with the atmosphere and ocean, while sea ice drift and deformation are mainly influenced by winds and ocean currents. Thermodynamic processes tend to equalize the thickness of different ice floes (e.g., rapid growth of thin ice), while mechanical processes tend to seek the extremes in ice thickness (e.g., formation of ridges and leads).

The *ice thickness distribution*, $g(h_i)$, is defined such that

$$\int_{h_1}^{h_2} g(h_i)\, dh_i \tag{10.1}$$

is the fractional area covered by ice in the thickness interval $h_1 \le h_i \le h_2$ and $g(h_i)$ has units m^{-1}. The average ice thickness, $\overline{h_i}$, is determined by integrating over the ice thickness distribution:

$$\overline{h_i} = \int_0^{h_{max}} h_i\, g(h_i)\, dh_i \tag{10.2}$$

The ice thickness distribution provides a statistical representation of the ice thickness characteristics of a region that is sufficiently large so that the average ice thickness does not vary for small changes in the integration area.

Observations of the ice thickness distribution have been observed by under-ice sonar observations from submarines or ocean moorings. Figure 10.3 shows seasonally averaged ice thickness distributions obtained in the Beaufort Sea (Figure 10.1b). The general progression of the average ice thickness and the ice thickness distribution in response to the seasonal cycle of summer warming and winter cooling is depicted in this figure. The average ice thickness is greatest in spring. During summer, the average ice thickness decreases as the solar insolation increases, and the amount of open water increases. During autumn, the open water freezes and a substantial amount of thin ice forms. During winter, a small amount of open water and new ice production is apparent, with a maximum around 1 m, representing the growth of the new ice that was formed in autumn. The continued growth of first-year ice is also evident in spring.

Associated with an ice thickness distribution are variations in ice age, snow cover, albedo, surface temperature, and many other parameters. The exchange of heat between sea ice and the atmosphere and ocean is thus complicated substantially by the presence of the ice thickness distribution.

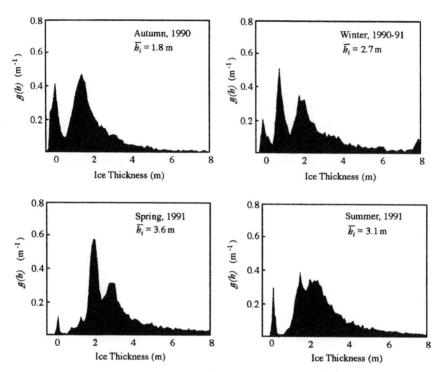

Figure 10.3 Time evolution of the ice thickness distribution in the southern Beaufort Sea, 1990–1991. In autumn 1990, two modes are present: open water and thin ice; and ice 1–2 m thick. Following this is a growth season which lasts about 250 days and then a decay season which lasts about 110 days. (Personal communication Moritz, 1999.)

Because of the virtual absence of multi-year ice and ridges in the Antarctic Ocean sea ice, the Antarctic ice thickness distribution is not nearly as complex as that for the Arctic Ocean. The Antarctic sea ice consists primarily of individual floes of first-year ice, with little variation in ice thickness. Only in the Weddell Sea, where there are ridges and multi-year ice, is there a complex ice thickness distribution.

10.3 Evolution of the Salinity Profile in Sea Ice

The presence of brine pockets in the sea ice has a profound influence on the physical, thermodynamic, and mechanical properties of sea ice. As described in Section 5.7, brine pockets are trapped in the sea ice during the growth process. The amount of salt entrapped initially during growth increases with the salinity of the seawater and with the growth rate of the sea ice. A typical salinity value for newly-formed ice in the Arctic is 14 psu, compared with approximately 30 psu for the upper Arctic Ocean.

Figure 10.4 shows idealized salinity profiles in Arctic sea ice of various ages and thicknesses. As ice thickens with time, salinity decreases. By the time sea ice has reached a thickness of 3 m, there is no brine near the surface, and the salinity near the base of the ice has decreased to a value of about 4 psu. The time evolution of the salinity profiles shown in Figure 10.4 implies a migration of brine within the sea ice, resulting in an overall decrease with time in the amount of brine.

Four mechanisms have been proposed for the transport of brine in sea ice:

1) *Brine pocket migration* occurs in the presence of a temperature gradient in the sea ice. Melting on the warm side of the brine pocket and freezing on the cool side causes the pocket to move in a direction of increasing temperature in the ice. However, this mechanism is too slow to account for significant brine transport.

2) *Brine expulsion* results from pressure buildup in freezing brine pockets. This pressure buildup can split the ice along basal planes, allowing brine to migrate. This mechanism is important primarily during the early stage of growth when the growth rate is most rapid. This mechanism explains the "u-shaped" salinity profiles apparent in ice less than 40 cm (Figure 10.4), whereby brine is expelled upwards in the upper portion of the ice that is cooling most rapidly.

3) *Gravity drainage* transports brine from the ice to the underlying seawater through the influence of gravity. Brine pockets form in vertical strings (Figures 5.13, 5.14), and depending on ice thickness, as many as 50 to 300 brine channels per square meter may exist on the underside of the ice. Gravity drainage is the principal mechanism that causes salinity to decrease in *first-year ice* (sea ice of not more than one winter's growth) whose depth exceeds 40 cm.

4) *Meltwater flushing* occurs during the summer melt season. As the ice warms, brine volume increases (Figure 10.5) and the ice becomes increasingly porous. Melting of the snow cover and the surface ice allows relatively fresh water to percolate into the ice through the brine channels, thus flushing the ice of its brine. Meltwater flushing explains the characteristic decrease of salinity near the surface in *multi-year ice* (ice generally thicker than 1 m that has survived a melt season). Successive melt seasons progressively flush the ice of its brine.

The salinity of sea ice therefore evolves as follows. Substantial amounts of brine are entrapped initially in rapidly growing ice. As the ice continues to grow by freezing at the bottom, the amount of salt entrapped diminishes gradually as the growth rate decreases. However, cooling of the upper part of the ice results in brine expulsion that allows the salinity in the lower part of the ice to remain high. By the time the ice thickness approaches 40 cm, brine channels are developed and gravity drainage promotes desalinization. At the onset of melting, flushing by fresh water occurs. Ice that has survived a summer melt season thus has substantially lower salinity than does first-year ice.

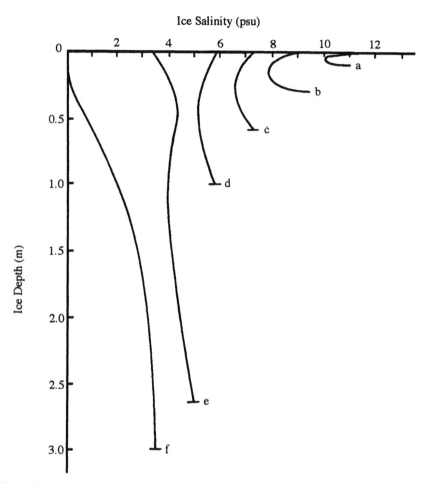

Figure 10.4 Idealized salinity profiles in Arctic sea ice of various thicknesses. Curves a–d are profiles of first-year ice. Curves e and f are profiles of multi-year ice: curve e profiles the salinity beneath hummocks and other elevated areas, while curve f profiles the salinity for ice beneath low areas. (From Maykut, 1985.)

The brine volume of sea ice has a substantial impact on its bulk thermal, mechanical, and optical properties. A very rapid increase in brine volume at high temperature is seen in Figure 10.5. The sharp increase of brine volume at temperatures greater than −5°C reduces the tensile strength of the ice, a fact well known to polar explorers. The brine volume at the bottom of growing sea ice varies from 8 to 40%. The brine volume for multi-year sea ice is typically about 10%, increasing to values as high as 40 to 50% during summer sea ice melt.

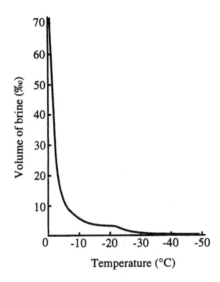

Figure 10.5 Relative volume of brine in standard sea ice. (After Assur, 1958.)

10.4 Thermal Properties of Sea Ice

In describing the thermal properties of sea ice, we refer specifically to the thermal conductivity, the specific heat capacity, and the latent heat of fusion.

Relative to pure ice, the thermal properties of sea ice are influenced primarily by its brine volume, which in turn is a function of the salinity and temperature of the ice. The density and thermal conductivity of sea ice are also influenced by the presence of air bubbles. Air bubbles accumulate in multi-year ice as a result of the freeze–thaw cycles, with a nominal air bubble volume of 15% for multi-year ice.

The density of sea ice, ρ_i, is given by (Schwerdtfeger, 1963):

$$\rho_i = \left(1 - v_a\right)\left[1 - \left(0.00456\frac{s_i}{T_i}\right)\right]\rho_0 \tag{10.3}$$

where v_a is the fractional volume of air bubbles, T_i is in °C, s_i is in psu and $\rho_0 = 917.8$ kg m^{-3} is the density of pure ice. Since air bubble volume and salinity vary with ice age, the sea ice density is therefore a function of ice age. Air bubble volume can reach substantial amounts in multiyear ice due to the melting process, reducing the mean ice density (857–922 kg m^{-3}) relative to that of first-year ice (910–930 kg m^{-3}).

In pure ice, the thermal conductivity depends only on temperature, and is approximated by (Nazintsev, 1964):

$$\kappa_o = 2.22 \left(1 - 0.0159\, T\right) \qquad (10.4)$$

where T is in °C and κ_o is in W m^{-1} K^{-1}. At $T = 0$°C, the thermal conductivity of ice is approximately four times the thermal conductivity of liquid water. Relative to pure ice, the conductivity of sea ice is influenced primarily by brine and secondarily by solid hydrates and air bubbles. An approximate expression for the thermal conductivity of sea ice, κ_i, is given by Untersteiner (1961):

$$\kappa_i = \kappa_o + b\, \frac{s_i}{T_i} \qquad (10.5)$$

where s_i is in psu, T_i is in °C, and $b = 0.1172$ W m^{-1} psu^{-1}. There is little temperature or salinity dependence of κ_i for temperatures colder than -6°C. Values of thermal conductivity vary by 10–20% in multi-year ice and up to 50% in young ice.

The specific heat of pure ice is a weak function of temperature, with a typical value $c_o = 2113$ J kg^{-1} K^{-1}. The specific heat of sea ice is complicated by the fact that sea ice is a multi-component and multi-phase system. As the temperature of the sea ice changes, the relative amounts of the solid ice, solid salts, and brine change (as indicated in Figure 4.8). Thus an effective specific heat of sea ice, c_i, is the sum of the heat required to warm the ice, solid salts, and brine plus that needed for phase transformation of the brine pockets. An approximate expression for the effective specific heat of sea ice is given by (Ono, 1967):

$$c_i = c_o + a T_i + b\, \frac{s_i}{T_i^2} \qquad (10.6)$$

where s_i is in psu, T_i is in °C, $a = 7.53$ J kg^{-1} K^{-2} and $b = 0.018$ MJ K kg^{-1} are constants. Figure 10.6 shows that the variations in c_i are very large for temperatures greater than -4°C. As the ice temperature approaches the melting point, it becomes increasingly difficult to change T_i because of increasing brine volume (Figure 10.5). Since the contribution of air to the specific heat of snow is negligibly small, the specific heat of dry snow is determined by the pure ice and has a typical value of 2113 J kg^{-1} K^{-1}.

The latent heat of fusion of pure ice, $L_{il}(T, 0)$, is the amount of heat required to melt 1 kg of ice at 0°C. The concept of latent heat of fusion of sea ice, $L_{il}(T, s)$ is complicated by the internal melting and freezing that occur over a wide temperature range. The amount of heat necessary to melt sea ice is given by Ono (1967):

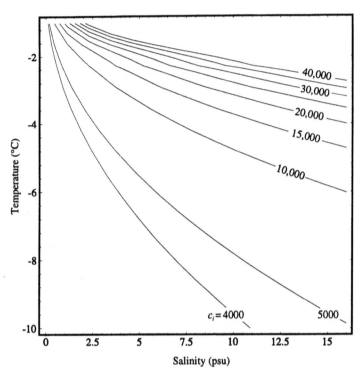

Figure 10.6 Variation of c_i with temperature and salinity. Values of c_i are in J kg^{-1} K^{-1}.

$$L_{il}(T,s) = \left[1 - s_i\left(0.000532 - \frac{0.04919}{T_i}\right)\right] L_{il}(T,0) \qquad (10.7)$$

where $L_{il}(T, 0) = 0.3335$ MJ kg^{-1}, T_i is in °C and s_i is in psu. The term $L_{il}(T,s)$, represents the amount of heat that must be conducted away from the growing interface to form 1 kg of salty ice. Note that latent heat released by freezing in the brine pockets is accounted for in c_i and is not included in the latent heat.

10.5 Optical Properties of Sea Ice and Snow

The *optical properties* of a substance determine how it interacts with solar radiation. The albedo of sea ice and snow is an important component of the surface energy budget (9.1), determining the reflection of solar radiation. The high albedo of polar surfaces results in the reflection of a substantial portion of the incident solar radiation. The albedo of snow and sea ice shows a strong variation with wavelength across the shortwave spectrum, with the albedos in the visible portion of the spectrum (0.4 to 0.8 μm) being much larger than the albedos in the near infrared portion of the spectrum (0.8 to 4.0 μm). In addition to varying spectrally, surface albedos can vary with solar zenith angle and cloud cover. The variation of surface albedo with cloud cover arises from the selective absorption of near infrared radiation by clouds (Section 8.3), which acts to enrich the stream of radiation reaching the surface in the visible portion of the spectrum, thus increasing the spectrally averaged surface albedo. An additional factor is that clouds change the direct solar radiation beneath the clouds to diffuse radiation, effectively changing the solar zenith angle to about 55°. Deviations of snow albedo under cloudy conditions can be as large as 20% from clear-sky conditions.

Spectrally averaged albedos of polar surfaces are shown in Table 10.1. The albedo of dry snow (snow that is not melting) varies with snow age and with the amount of impurities in the snow. The albedo of melting snow is about 10% less than the albedo of dry snow. Snow-free multi-year ice that is not melting has a surface albedo approximately the same as that of melting snow. Melting multi-year ice has an albedo as low as 0.55. The albedo of multi-year ice is influenced by the presence of air bubbles in the ice and by the presence of a decomposed surface scattering layer. The surface albedo of snow-free young ice exhibits a strong dependence on ice thickness for thickness values less than a meter because of scattering by the underlying seawater for the thinnest ice. The albedo of melt ponds varies primarily with melt pond

Table 10.1 Representative broadband albedos for Arctic Ocean surface types during summer and under cloudy skies. (After Perovich, 1986.)

Surface type	Albedo
New snow	0.87
Dry snow	0.81
Melting old snow	0.77
Melting white ice	0.56–0.68
Bare first-year ice	0.52
Refrozen melt pond	0.40
Old melt pond	0.15–0.35

depth and the optical characterists of the underlying ice, the deepest ponds having albedos as low as 0.15, while shallow ponds typically have albedos of 0.4. The surface albedo of leads is approximately 0.08, roughly the same as that of the open ocean.

A region of sea ice on the scale of a few kilometers may contain bare multi-year ice, leads, snow-covered ice, thin ice, and melt ponds, all having significantly different optical properties. Because of the heterogeneity of the sea ice (Figure 10.7), area-averaged albedo values for a region of order $(100 \text{ km})^2$ in area must be determined using a weighted average reflectance of all the different surface types. Figure 10.8 shows a simulated annual cycle of surface albedo for 80°N for a region of heterogeneous sea ice. During May, in the absence of heavy snowfall, the albedo decreases to the aging snow value. In late May, the snow begins to melt, and the albedo decreases to the melting snow value. In mid-July all snow has melted, leaving bare and ponded ice. The albedo reaches its minimum when the melt pond fraction is at a maximum. In mid-August the melt ponds have refrozen and snow begins to accumulate on the thick ice. The tendency is for melting to decrease the albedo, thereby accelerating the

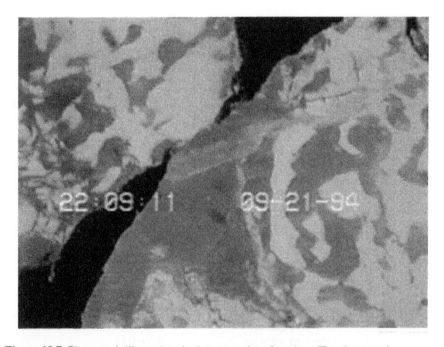

Figure 10.7 Photograph illustrating the heterogeneity of sea ice. The photograph covers an area of approximately 1 km² in the Beaufort Sea. In the photo can be seen a lead (dark wedge), melt ponds (shaded areas with a rounded appearance), snow-covered multi-year ice (white regions), and new ice (darker regions near the lead). (Photograph courtesy of M. Tschudi.)

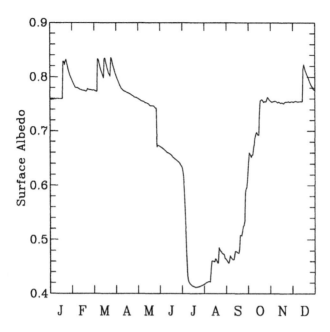

Figure 10.8 Simulated annual cycle of surface albedo for 80°N. In May, the snow begins to melt and the albedo decreases in response to the melting. The albedo reaches a minimum around mid-July, when the melt pond fraction is at a maximum. The albedo begins to increase again in August, as the melt ponds refreeze and snow begins to accumulate. By the end of September, the albedo is again at a maximum, in response to the maximum snow cover. (Courtesy of J. Schramm.)

melt process. The maximum in lead fraction is reached at the end of August, and in early September the open water fraction is decreased drastically by new ice growth. By the end of September the surface is nearly snow-covered, with a small amount of open water (<1%). The asymmetry of the snow albedos in autumn and spring arise primarily from differences in snow age. Autumn snow albedos are effectively new snow values. Springtime snow albedos are lower due to aging of the snow pack and increasing amount of liquid water incorporated into the snow pack. A heavy spring-time snowfall would return the pack to a new snow albedo, but heavy snowfall events over the Arctic Ocean typically occur only a few times per season.

Solar radiation that is not reflected at the surface is transmitted and absorbed in the interior of the ice and snow. Attenuation of solar radiation in the ice or snow can be approximated by the flux form of Beer's law (3.26):

$$F_\lambda(z) = F_\lambda(0)\left(1 - \alpha_{\lambda 0}\right)\exp\left(-k_\lambda z\right) \tag{10.8}$$

where $F_\lambda(0)$ is the shortwave radiation incident at the surface, z is depth below the surface, and k_λ is the volume shortwave extinction coefficient of ice or snow. The extinction coefficient varies both with wavelength (λ) and with depth (z). In the upper 10 cm scattering layer of multi-year ice, k_λ is relatively large, decreasing through the scattering layer to a constant value. The extinction coefficient of snow is much greater than that of ice, and if the snow depth exceeds 30 cm, virtually no radiation reaches the ice. For snow-free ice that is thicker than 1.6 m, virtually no solar radiation reaches the ocean. Most of the solar radiation that reaches the ocean enters through leads or is transmitted through thin ice.

Both surface albedo and the extinction coefficient of ice and snow are influenced strongly by the presence of particulates such as sediment, algae, or soot. In the presence of particulates, the ice and snow albedos decrease while the extinction coefficients increase.

10.6 Surface Energy Balance Over Snow and Sea Ice

The net heat flux at a snow or sea ice surface can be written following (9.1) as

$$F_{Q0}^{net} = F_{Q0}^{rad} + F_{Q0}^{SH} + F_{Q0}^{LH} \tag{10.9}$$

In the absence of surface melting, the net heat flux at the surface F_{Q0}^{net} must balance the conductive flux through the snow or ice at the interface with the atmosphere

$$F_{Q0}^{net} = -\kappa\left(\frac{\partial T}{\partial z}\right) \tag{10.10}$$

For a melting ice surface, the surface temperature is constrained not to exceed the melting point. The resulting flux imbalance causes ice to melt.

Observations of heat flux components at a single point over an individual ice type are not representative of a regional average because of small-scale variations in ice thickness and ice type. An internally consistent high-quality set of observations of the surface energy balance that is representative of a region of order 10–100 km does not exist over a complete annual cycle. In the absence of such a dataset, simulations using an ice thickness distribution model[1] combined with a surface heat flux model are used here to estimate the annual cycle of the surface energy balance over a region that represents the small-scale heterogeneity of the ice-covered ocean.

The simulated annual cycle of surface temperature for several ranges of ice thickness is shown in Figure 10.9, where the ice thickness distribution is represented as a range of ice thicknesses with higher resolution for the lower ice thicknesses because of its more rapid variation. The temperature differences between the various ice thick-

---

[1] The simulation results shown in Figs. 10.8-10.11 and 10.14 are obtained from a single-column ice thickness distribution model (Schramm et al., 1997) for climatological atmospheric forcing corresponding to 80°N.

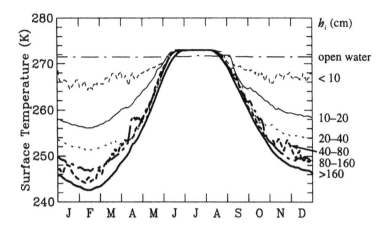

Figure 10.9 Simulated annual cycle of surface temperature for various ice thicknesses (with higher resolution for the lower ice thicknesses). (Courtesy of J. Schramm.)

nesses are the largest in the winter, due to the greater upward conduction of heat through thin ice from the ocean. The ice with thickness between 0.5 and 10 cm is formed by the freezing of leads and its temperature is 3–4°C cooler than that of open water during winter. The largest temperature range over the annual cycle is for the thick multi-year ice that reaches the coldest temperatures during winter.

The simulated annual cycle of sensible heat flux is shown in Figure 10.10. A negative turbulent flux indicates a loss of heat by the water or ice surface to the atmosphere. The open water category has the largest annual cycle, with a maximum loss of -475 W m^{-2} in the winter and a summertime gain of 17 W m^{-2}. Ice with thicknesses between 10 and 20 cm shows the largest variability in the sensible heat flux, related to a similar variability in the surface temperature. The sensible heat flux over ice thinner than 80 cm is negative all year, since the surface temperature of this ice is always warmer than that of the air.

The simulated annual cycle of the surface net heat flux, F_{Q0}^{net} is shown in Figure 10.11. In general, the net flux over open water shows a large net loss of heat in the winter due to sensible and latent heat fluxes. The open water gains heat in the summer due to the incoming shortwave radiation. The timing during spring at which the net surface flux becomes positive varies with ice thickness. Ice thicker than 20 cm begins to gain heat at the surface by late April. In late spring, the sensible heat flux over this thicker ice is relatively small and slowly increasing. During the same period, the thinner ice is losing a large amount of heat due mainly to the turbulent fluxes from the relatively warm surface. This keeps the net heat flux over thin ice negative until late May. The net surface heat flux becomes insensitive to ice thickness once h_i exceeds 80 cm.

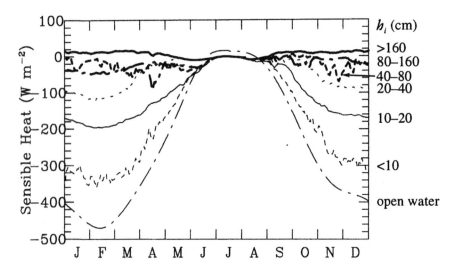

Figure 10.10   Simulated annual cycle of sensible heat flux for various ice thicknesses. (Courtesy of J. Schramm.)

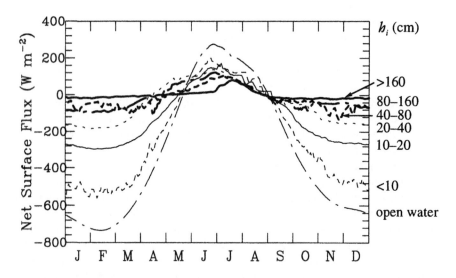

Figure 10.11   Simulated annual cycle of the surface net heat flux for various ice thicknesses. (Courtesy of J. Schramm.)

Leads and other open water are of major importance to the surface energy balance of polar oceans. During winter, when surface air temperatures may be -30 to $-40°C$, an opening in the ice produces a strong temperature gradient between the ocean surface, with a temperature approximately the freezing point $-1.7°C$, and the air in the lowest few meters. Such extreme static instability gives rise to very large wintertimefluxes of sensible and latent heat from leads to the atmosphere. Wintertime sensible heat fluxes from open leads commonly exceed 500 W m^{-2}. During summer, leads have little influence on the surface sensible and latent heat fluxes, since the surface temperature in leads is close to the surface temperature of the melting ice. However, because of the low surface albedo of open water compared to ice, absorption of solar radiation in leads provides a major heat source to the upper ocean. The solar radiation absorbed in leads warms the mixed layer and thus melts the ice from below and also laterally at the edges of leads. In response to the summertime melting, the fractional lead coverage in the Arctic Ocean increases from wintertime values of less than 1% to values exceeding 20% by late summer.

10.7 Growth and Decay of Sea Ice

10.7.1 Young Ice

In the early stages of growth (e.g., grease ice, pancake ice; see Section 5.7), the surface of the ice is wet and is close to the freezing temperature. Once the ice becomes consolidated, free water is no longer available at the surface and the surface temperature cools in response to the surface heat loss. Further ice growth occurs on the underside of the ice (i.e., the ice/ocean interface). The growth rate depends on how rapidly heat can be conducted from the ice/water interface toward the upper surface.

Consider a thin slab of ice with depth h_i and thermal conductivity κ_i (Figure 10.12). The temperature at the bottom of the ice ($z = h_i$) is fixed at the freezing point of the water (T_f) which is determined by the salinity of the ocean mixed layer (see (4.59)). The surface temperature, T_0 varies in response to changes in the surface energy balance. If the ice is thin, the temperature gradient in the ice will be essentially linear and the conductive heat flux in the ice, F_{ci}, will be constant with depth, i.e., $F_{ci}(0) = F_{ci}(h)$. Therefore from (3.37) we have

$$F_{ci} = \frac{\kappa_i}{h_i}\left(T_0 - T_f\right) \tag{10.11}$$

The amount of growth or ablation at the bottom of the ice is determined by the sum of the conductive heat flux F_{ci} and heat flux at the base of the ice from the ocean, F_w. If net heating occurs then the ice will melt, or if net cooling occurs then the ice will grow, exchanging latent heat during the phase change. The growth rate of thin ice is

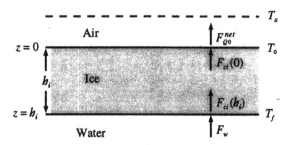

Figure 10.12 Schematic representation of heat fluxes through a thin slab of ice.

therefore determined from

$$-\rho_i L_{il} \frac{dh_i}{dt} = F_{ci} + F_w \tag{10.12}$$

where F_w is the heat flux at the base of the ice from the ocean. If $F_w \ll F_{ci}$, we can use (10.11) to write

$$\rho_i L_{il} \frac{dh_i}{dt} = \frac{\kappa_i}{h_i} \left(T_f - T_0\right) \tag{10.13}$$

Equation (10.13) along with the surface energy balance equation (10.9) to determine T_0 can be solved for any given values of h_i and net heat flux at the surface, F_{Q0}^{net}. Growth rates of young sea ice as a function of ice thickness calculated from (10.13) are given in Figure 10.13. It is seen that growth rate of young ice is very sensitive to h_i, decreasing by almost an order of magnitude between $h_i = 10$ and $h_i = 100$ cm. Effects of differences in surface temperature decrease as h_i increases. The decay of thin ice occurs when F_{Q0}^{net} positive, although if the ocean heat flux, F_w, is significant, it needs to be accounted for.

The presence of a snow cover on thin ice influences substantially the growth of young ice. The principal reason for this is that the thermal conductivity of snow, κ_s, is much smaller than κ_i (see Section 10.8). To determine the growth rate of young ice in the presence of a snow cover, we consider a two-layer system, with a slab of sea ice overlain by a layer of snow. Again we assume that the temperature gradient in the snow and ice are each linear, and thus conductive heat flux is constant with depth. At the snow/ice interface, the conductive flux in the snow must equal the conductive flux in the ice, i.e., $F_{ci} = F_{cs}$. Since the thermal conductivity is much lower in snow

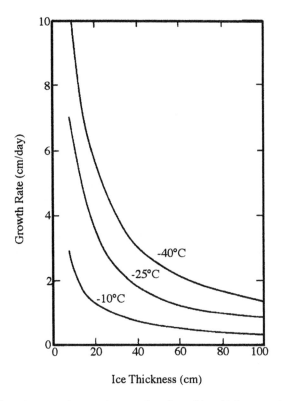

Figure 10.13 Growth rates of young ice as a function of ice thickness and air temperature. (From Maykut, 1986.)

than in ice, the temperature gradient in snow is much steeper than in the ice. The heat conduction F_c for the two-layer system with a snow and ice layer can be written as

$$F_c = \left(\frac{\kappa_i \kappa_s}{\kappa_i h_s + \kappa_s h_i} \right) (T_f + T_0) \tag{10.14}$$

Using the surface energy balance equation (10.9) to determine T_0 allows the modeled new ice growth to respond to time variations in the surface heat fluxes. Because of its low thermal conductivity, the presence of a layer of snow slows the ice growth relative to a snow-free layer of ice.

10.7.2 Multi-year ice

In the young ice model, the temperature gradient in the ice is linear, which causes the ice thickness to respond immediately to variations in the surface heat flux. This assumption is inadequate to describe heat transport through ice that exceeds about 80 cm in thickness (e.g., thick first-year and multi-year ice). In thick ice, there can be a considerable time lag between a change in the surface energy flux and a change in ice thickness. For example, the autumn cooling that begins in September is not felt at the bottom of the ice until mid December. In early summer, the surface of the ice may be melting at the same time that accretion is occurring at the bottom of the sea ice.

Brine pockets contribute to the nonlinear temperature profile in thick ice. Cooling causes some of the brine to freeze, which releases latent heat and slows the cooling. Heating causes the ice surrounding the brine pockets to melt, thus decreasing the rate of warming. Brine pockets thus act as thermal buffers. Melt ponds also act as thermal buffers. As the autumnal freezing commences, some of the surface cooling freezes the melt ponds rather than decreasing the ice temperature, thus retarding the cooling.

To account for nonlinear vertical temperature profiles and vertical variations in the conductive flux, the heat conduction equation (3.41) must be used:

$$\rho_i c_i \frac{\partial T_i}{\partial T} = \kappa_i \frac{\partial^2 T_i}{\partial z^2} + F_\lambda(z) \tag{10.15}$$

where an internal heat source associated with penetrating shortwave radiation, $F_\lambda(z)$, has been included (see (10.7)). A similar equation is used to describe temperatures in the snow:

$$\rho_s c_s \frac{\partial T_s}{\partial t} = \kappa_s \frac{\partial^2 T_i}{\partial z^2} + F_\lambda(z) \tag{10.16}$$

To solve these equations, the surface energy balance equation (10.9) is used as the boundary condition at the air/snow interface. At the snow/ice interface and the ice/ocean interface, heat fluxes at an interface are assumed to be equal in both media.

10.7.3 Underwater frazil production

The heat conduction equation describes the formation of congelation ice that grows at the ice/water interface in response to conductive heat losses along the temperature gradient in the ice. Under certain conditions, underwater frazil production is an important ice formation mechanism in the polar oceans. Ice core data from the Weddell Sea indicates that congelation growth accounts for less than half of the ice thickness, with fine-grained layers of frazil crystals forming the remainder of the ice. Layers of frazil are often sandwiched between layers of congelation ice, suggesting that frazil formation is episodic. Underwater frazil also appears to be important along the Antarctic and Arctic coasts. In the central Arctic Ocean, underwater frazil production occurs in and downwind of leads. The initial frazil ice is "slushy" in texture, but with time the interstitial water between the frazil crystals freezes and produces a solid ice layer.

Several mechanisms have been proposed to explain the underwater production of frazil ice, all of which depend on a slight supercooling or some turbulence in the water column. These mechanisms can be summarized as:

1) Wind- and wave-induced turbulence carries frazil generated at the surface in open leads to lower levels within the mixed layer. Because of their relatively low density, the frazil particles then rise in the water column analogously to an "upside-down" snowfall. The frazil particles may rise in the water beneath the lead to contribute to the ice formation in the lead or the particles may be advected downstream of the lead to form a frazil layer beneath congelation ice.

2) Upward movement of water at its freezing point will cause supercooling as the pressure decrease in the rising water raises the freezing point (4.67). This mechanism is believed to be of importance for frazil production near ice shelves, icebergs, and pressure ridge keels.

3) Contact between two water masses of different salinity but both near the freezing temperature can cause freezing to occur by different rates of diffusion of heat and salt (Section 11.4). This process can occur in coastal regions by the drainage of melting river or glacier water into an ocean whose temperature is colder than the fresh-water freezing point.

4) Freezing of seawater produces descending plumes of brine that cause supercooling in the adjacent water by the differential diffusion rates of heat and salt. This process is of sufficient generality that it can occur over most regions of the polar oceans.

None of these mechanisms has been evaluated carefully using observations. Further observations of frazil ice and its formation mechanisms are needed.

10.7.4 Sea ice mass balance

The physical processes acting to modify the ice mass balance and the ice thickness distribution include net divergence, lateral melting of ice floes, freezing/melting at the ice base, frazil ice production, meltwater runoff into the ocean, and sublimation/evaporation. Ridging acts to rearrange ice mass between different ice thicknesses. Net divergence, lateral melting, and ridging affect ice concentration while the remaining processes affect ice thickness.

A calculation of the Arctic Ocean ice mass balance averaged over the annual cycle is given in Figure 10.14. Sources and sinks represent physical processes that result in the growth or decay of a category of ice thickness. The most striking feature of

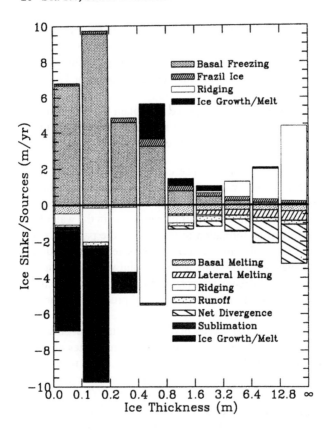

Figure 10.14 Ice mass balance in the Arctic Ocean. Notice that ice thinner than 80 cm gains and loses signficant mass during an annual cycle. (Courtesy of J. Schramm.)

Figure 10.14 is the large amount of ice gained and lost for ice thinner than 80 cm. This is due mainly to ice production by basal freezing and removal by ridging and ice growth. The large conductive flux at the base of this thinner ice is due to strong surface cooling, and results in ice growth via basal accretion. Basal ablation is largest for the thin ice that melts completely in the summer. Ice thinner than 20 cm is affected by sublimation due to warm wintertime surface temperatures and minimal snow cover. The ridging process removes a large amount of the thin ice and creates ice with a wide range of thicknesses. The amount of ice gained and lost by ice thicker than 80 cm is much smaller than that of thinner ice. Lateral melting of the ice occurs during the summer when the water in leads adjacent to the ice is heated, and its magnitude depends on the vertical ice area exposed to the water.

Thus it is the thin ice in an ice thickness distribution that accounts for most of the ice production and the energy exchange with the atmosphere and ocean.

10.8 Metamorphosis of Surface Snow

Snow density and grain size are important determinants of the snow thermal and optical properties. Once snow reaches the surface, wind, surface heating, and the subsequent fall of snow or rain causes the metamorphosis of snow. *Metamorphosis* of snow refers to temporal changes in the snow crystal structure after the snow has reached the ground.

Snow that falls to the ground has an extremely low density relative to solid ice, newly-fallen snow typically having a density in the range 50 to 150 kg m^{-3}. As a result, the topmost layer of newly fallen snow is composed primarily of air that surrounds the crystals. Within a matter of hours, newly fallen snow changes its physical character. Dry snow landing on the surface is frequently redistributed by the wind. The blowing of snow causes abrasion and breakage of the snow crystals. The broken crystals are much more tightly packed and cohesive than new snow. Blown snow is thus more dense than newly fallen snow. The surface of snow that has been compacted by the wind often develops a wavelike appearance, the ridges known as *sastrugi*. Another factor that causes snow crystals to break is the fall of additional snow crystals and the increasing weight of the overlying snow. Snow settles at a rate of about 1% per hour, until the density reaches about 250 kg m^{-3}.

Once snow settles into an even snow pack or into drifts, metamorphic processes alter the crystals in response to their depth in the snow pack and in response to weather conditions. The low density of surface snow implies that the upper snow pack is thoroughly suffused with air. Molecules of water thus sublimate from the surface of the crystals into the entrapped air, and the entrapped air rapidly becomes saturated, with vapor pressures at the ice saturation value.

Destructive metamorphism occurs when water molecules move to new positions on a snow crystal in order to decrease the surface free energy. Pristine snow crystals typically have a very large ratio of surface area to mass, and a low density. Kelvin's equation for snow can be written following (5.14c) as

$$e_{\delta}(r) = e_{\delta} \exp\left(\frac{2\sigma_{\delta v}}{\rho_{\delta} R_v T r_{\delta}}\right) \tag{10.17}$$

where $\sigma_{\delta v}$ is the surface tension between snow and the vapor, ρ_{δ} is the snow density, and r_{δ} is the local radius of the snow element. This equation shows that where the local curvature is high, such as around a small crystal or the branch of a dendritic crystal, the local equilibrium vapor pressure will be higher than on portions of the crystal where there is little curvature. Hence destructive metamorphism acts to lower the surface free energy and thus the saturation vapor pressure by changing small crystals or broken branches into larger, rounded grains. Destructive metamorphism is temperature dependent, and changes occur more rapidly in warmer snow. Destructive metamorphism is important primarily in the period that shortly follows the deposition of the snow, and is of minor importance for snow density in excess of 250 g cm^{-3}.

Constructive metamorphism arises from the vapor transfer within the snow cover due to an overall vertical temperature gradient in the snow. Constructive metamorphism can proceed much more rapidly than destructive metamorphism. Because of conduction of heat upwards through the sea ice during winter from the relatively warm ocean below, a temperature gradient exists in the overlying snow pack. The temperature gradient in the snow causes water vapor diffusion. The water vapor flux within the snow cover can be written following (3.42) as

$$\frac{\partial F_v}{\partial z} = -\frac{\partial}{\partial z}\left(D_v \frac{\partial \rho_v}{\partial z}\right) = -\frac{\partial}{\partial z}\left(D_v \frac{\partial \rho_v}{\partial T} \frac{\partial T}{\partial z}\right) \tag{10.18}$$

where it is assumed that the water vapor content in the interstitial air is saturated with respect to ice, (4.25). The diffusion coefficient for water vapor in a snow pack, D, can be approximated by (Anderson, 1976):

$$D = D_o \frac{p}{p_0}\left(\frac{T}{T_o}\right)^n \tag{10.19}$$

where $p_0 = 1000$ mb, $T_o = 273.15$ K, $D_o = 9 \times 10^{-5}$ m^2 s^{-1}, and $n = 14$. Since air can circulate in the upper portion of the snow pack, water vapor molecules are transferred from the crystals in the warmer lower layers to colder crystals in the upper layers. This results in a change in density at different levels of the snow pack. The crystals in the lower layers become loosely packed, while the crystals in the upper layer become denser and are cemented into a hard crust as bonds form between the ice crystals in response to the diffusion and deposition. The density of the lower layer may become less than half that of the top layer.

Under conditions of prolonged temperature gradients, crystals in the lower layers become needle-like and are called *depth hoar*. Because of its low density and the delicate structure of the ice crystals, depth hoar is structurally weak. The continued formation of depth hoar allows the snow pack to settle and become denser as the lower layer repeatedly collapses. Depth hoar is found when the temperature gradient is large, a situation most often found in relatively shallow, cold snow layers. Formation of depth hoar is a major factor in the initiation of avalanches in mountainous areas.

Melt metamorphism occurs when melting or rainfall results in the presence of liquid water in the snow pack. All or a portion of the meltwater may be retained within the remaining snow cover. If the melting is followed by a freeze, a dense crust is created with increased snow grain size. In liquid-saturated snow, the melting temperature of a snow particle is inferred using (5.19) and (5.6) to be

$$T = T_{\text{o}} \exp\left(\frac{2\sigma_{\jmath} l}{L_{il}\rho_{\jmath} r_{\jmath}}\right) \tag{10.20}$$

where T_{o} is the melting temperature of a flat surface. At equilibrium in saturated snow, small particles have a lower temperature than large particles. This causes the small particles to melt, while the large particles grow. Grain growth occurs faster in saturated snow than in snow with a low liquid water content.

A heavy snowfall on thin sea ice will depress the ice, weighing it down so that seawater can percolate, saturating the lower levels. The saturated snow will freeze solid under continued cold conditions. Ice thus formed is referred to as *white ice*, which is less dense than congelation sea ice and has a granular consistency. The formation of white ice thus limits the depth of snow on sea ice.

Variations in the snow density associated with metamorphism influence the thermal conductivity of snow. The thermal conductivity of snow is a function of the size, shape, and distribution of ice crystals within the snow pack. The thermal conductivity of snow in effect combines heat conduction through the ice particles and the enclosed air pockets, and is thus a function of the density of the snow pack. A quadratic relation between the thermal conductivity and snow density is given by Yen (1969):

$$\kappa_{\jmath} = a\rho_{\jmath}^2 \tag{10.21}$$

where $a = 2.845 \times 10^{-6}$ W m^5 K^{-1} kg^{-1}. Note that expression (10.21) is really an effective conductivity that includes the effects of vapor movement in the snow. From this expression, it can be shown that the thermal conductivity of snow increases from 0.036 to 0.42 W m^{-1} K^{-1} when density increases from 80 to 400 kg m^{-3}. The value of the thermal conductivity for snow is six to eight times smaller than that for ice. As a result of its low thermal conductivity, snow is an excellent insulator. As long as snow remains below freezing, its insulating value is matched by few other naturally occurring substances, accounting for the effectiveness of the snow igloo as a shelter.

Additionally, since the specific heat of seawater is approximately twice as large as that for dry snow, wet snow has an intermediate value of specific heat.

10.9 Glaciers

A *glacier* originates from deepening deposits of snow on land. As the snow deepens and becomes increasingly dense, the ice is subject to alteration and movement owing to its own weight and pressure. A nominal depth of 50 m is required for an ice deposit to be transformed into a glacier.

Firn is snow that has been compacted to a density of 550 kg m^{-3}. This increase in density is caused by pressure compaction and melt–freeze cycles in perennial snow.

In deep firn, bonds between the snow particles are increased through sublimation, and the firn takes on the characteristics of a three-dimensional lattice, a process called *sintering*. As sintering continues, the density increases and the volume of the interstitial air is greatly reduced. When the density of the firn reaches approximately 820 kg m^{-3}, the interstitial air can no longer circulate and the air is contained in closed pockets. At this point the firn has become glacier ice. The trapped air pockets provide useful information on atmospheric composition of the past. The formation of glacial ice is accelerated considerably by the presence of meltwater. Meltwater provides lubrication for the snow grains, allowing increased packing from pressure. Additionally, meltwater may fill in the air pockets, increasing the density of the firn. Hence the presence of meltwater increases the density of the firn and allows the transformation to glacial ice to occur in the upper 10 m. In contrast, the transformation to glacial ice in dry firn may occur at depths as great as 100 m.

Glaciers flow downslope, and in the case of continental glaciers, outward towards the coastline. Glaciers are in constant motion through basal slippage and ice deformation. *Basal slippage* refers to the sliding of the ice mass along the surface of the underlying ground. Because of the pressure of the overlying ice mass, the base of the glacier is commonly at the melting point, a thin layer of liquid water acting as a lubricant between the ice and the ground. *Ice deformation* results from the slippage of planes within the ice crystal in response to stress. The speed of glacier motion ranges from 10 m per year in alpine glaciers up to 1–10 km per year on the coasts of Antarctica and Greenland.

Glaciers flowing towards coastlines transport ice into the sea, where the glacier ice is *calved* off into *icebergs*. Icebergs may become enmeshed into the pack ice or may float in otherwise ice free areas of the ocean. In the Northern Hemisphere, virtually all of the icebergs originate from the Greenland ice sheet. Icebergs are often observed in the shipping lanes of the North Atlantic Ocean, where they present a serious hazard. Icebergs calved from Greenland have been sighted to travel as far south as Bermuda before melting. Icebergs are approximately 90% submerged below the surface of the sea.

As a result of the glacier flow, continental glaciers may extend from the coast in the form of immense *ice shelves* that extend hundreds of kilometers out over the sea and may be several hundred meters thick. Ice shelves occasionally calve very large icebergs, sometimes as large as 100 km. Large icebergs calved from ice shelves along the north shore of Ellesmere Island in the Canadian Archipelago (Figure 10.1b) are called *ice islands*, which may have gravel, soil, and even vegetation on their surfaces.

The bases of continental glaciers often lie below sea level because of isostatic depression. Such glaciers are called *marine ice sheets*, and are believed to be inherently unstable and prone to collapse. Marine ice sheets are grounded on beds well below sea level and float at the grounding line and contribute to large floating ice shelves (Figure 10.15a). The only marine ice sheet remaining since the last glacial period 20,000 years ago is the West Antarctic ice sheet (Figure 10.15b). Fast-flowing

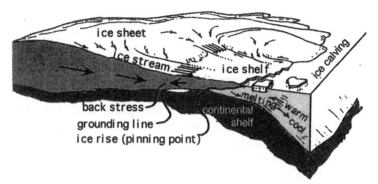

Figure 10.15a Marine ice sheet (Polar Research Board, 1985).

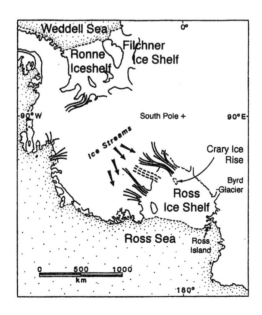

Figure 10.15b The West Antarctic ice sheet, including Ross and Filchner-Ronne Ice Shelves

ice streams drain the West Antarctic ice sheet into the two largest ice shelves in the world (Ross and Filchner-Ronne). The stability of the ice shelves is determined by sea level, ocean temperature, accumulation rate, and ice motion. The marine ice sheet may undergo complete collapse, initiated by the collapse of the ice shelves. The West Antarctic ice sheet is believed to have disappeared at least once during the last 600,000 years, while all other marine ice sheets collapsed completely during the current interglacial period beginning about 18,000 years ago.

The land area of Antarctica covered by glacial ice is about 12 million square kilometers. It is estimated that the average depth of the ice is nearly 2000 m, with a maximum depth of over 4000 m. The total volume if melted would raise sea level by 75 m. Antarctica contains about 90% of the global glacial mass, with Greenland constituting most of the remainder. Alpine glaciers may be several kilometers in width, over 100 km in length, and have a depth of up to 1 km.

Glaciers influence sea level in several ways. The calving of glaciers can raise sea level. Global sea level would rise by 5–6 m if the West Antarctic ice sheet were to collapse completely. On the other hand, accumulation of glacier ice is ultimately derived from water evaporated from the ocean. Therefore, an increase in the volume of glacial ice will decrease the volume of ocean water and therefore lower the sea level. During the last major glaciation of the Northern Hemisphere 18,000 years ago, the volume of glacial ice was about four times greater than the present amount, and it is estimated that sea level was lowered by approximately 100 m. Changes in sea level associated with a change in the volume of seawater are referred to as *eustatic* changes. Since the continents underlying glaciers are not rigid but plastic and float on semi-molten rock, an increase in the weight of a glacier causes the continent to sink deeper into the layer of semi-molten rock. Conversely, the seafloor will be subjected to less weight as glacier ice is accumulated on land. This *isostatic* change in sea level associated with changes in equilibrium of the Earth's crust can partially offset the eustatic changes in sea level. Changes in sea level associated with ice loading can radically alter shorelines. Isostatic adjustments to the Earth's crust can lag some thousands of years behind glacial advances and retreats.

Notes

A comprehensive reference on ice thermodynamics, processes, and characteristics is given by Lock (1990) *The Growth and Decay of Ice.*

A general reference on sea ice is the *Geophysics of Sea Ice*, edited by Untersteiner (1986).

The definitive reference on sea ice thermal properties is Ono (1967).

A thorough description of the growth and decay of sea ice is given by Maykut (1986) in the *Geophysics of Sea Ice.*

An overview of seasonal snow metamorphism is given by Colbeck (1982).

Texts dealing with glaciology include Andrews (1975) *Glacial Systems: An Approach to Glaciers and Their Environment* and Paterson (1981) *The Physics of Glaciers.*

Problems

1. Determine the areally-averaged surface albedo in the visible portion of the spectrum (0.25–0.7 μm) for the following areal distribution of surface types over the Arctic Ocean during the summer melt season:

Ice type	Fractional area	albedo, α_0
Bare first-year ice ($h_i = 0.3$ m)	3%	$0.760 + 0.140 \ln h_i$
Bare multi-year ice	59%	0.778
Open water	16%	0.08
Ponded first-year ice ($h_p = 10$ cm)	7%	$0.075 + 0.5 \exp(-8.1 h_p - 0.47)$
Ponded multi-year ice ($h_p = 20$ cm)	15%	$0.15 + \exp(-8.1 h_p - 0.47)$

where h_p is the depth of the melt pond.

2. Compare the conductive heat flux, F_c, for a slab of snow with thickness 10 cm thick and a temperature gradient of 5°C across this distance for snow of the following characteristics:
a)   new snow, $\rho = 100$ kg m^{-3};
b)   settled snow, $\rho = 250$ kg m^{-3};
c)   wind-packed snow, $\rho = 400$ kg m^{-3};
d)   firn, $\rho = 550$ kg m^{-3};
e)   very wet snow and firn, $\rho = 700$ kg m^{-3};
f)   glacier ice, $\rho = 900$ kg m^{-3}.

3. Consider a two-layer system consisting of a slab of snow 30 cm thick overlying a 3 m thick slab of sea ice. The temperature at the bottom of the sea ice is −1.7°C and the temperature at the top of the snow layer is −40°C. Making reasonable assumptions abaout the thermal conductivities, find the steady-state (diffusive) values of the:
a) temperature at the ice/snow interface;
b) temperature gradient in the ice;
c) temperature gradient in the snow.

4. During wintertime over the Arctic Ocean, the surface temperature over individual ice thicknesses has the following distribution:

h_i (m):	0.31	0.80	1.3	2.9	4.7	10.4
$g(h_i)$ (m^{-1}):	0.07	0.04	0.26	0.30	0.24	0.09
T_0 (°C):	252.2	246.6	244.4	244.2	244.1	244.1

a) Determine the average surface temperature of the ice thickness distribution.
b) Assuming that the temperature at the bottom of the ice is −1.7°C, determine the value of F_c for each ice thickness, assuming a steady-state (diffusive) temperature profile and that there is no snow cover on the ice.
c) Calculate the growth rate for each ice thickness under the conditions in b) assuming that $F_w = 0$.

5. For the problem in **4**, determine the average salt flux, F_s, under the ice thickness distribution, where

$$F_s \approx \frac{\partial h_i}{\partial t}\left(s_w - s_i\right)$$

The salinity of the water $s_w = 30$ psu, and the salinity at the base of ice with different thicknesses can be estimated from Figure 10.4. Compare the contribution of ice thinner than 1 m to F_s with that for ice thicknesses ≥ 1 m.

Chapter 11 | Thermohaline Processes in the Ocean

Heating and freshening at the ocean surface produce a buoyancy flux that can result in density gradients in the upper ocean. Depending on the scale of the temperature and salinity gradients in the ocean, *thermohaline* transfer may occur by molecular diffusion (such as in the millimeter layer right below the air/sea interface), turbulent mixing (in the case of the ocean mixed layer), internal mixing processes, or by large-scale advective transport. Under certain circumstances, the ocean surface buoyancy flux can give rise to deep oceanic convection. The circulation driven principally by the surface buoyancy flux is called the *thermohaline circulation*.

11.1 Radiative Transfer in the Ocean

In Section 9.1.1, we showed that the surface albedo of the ocean is typically about 5%, with values increasing for higher solar zenith angles. Therefore, greater than 90% of the solar radiation incident on the ocean is absorbed.

Extinction of solar radiation in the ocean can be approximated by a flux form of Beer's law (10.8):

$$F_\lambda(z) = F_\lambda(0)\left(1 - \alpha_{\lambda 0}\right)\exp\left(-k_\lambda z\right) \tag{11.1}$$

where $F_\lambda(0)$ is the shortwave radiation incident at the surface, $\alpha_{\lambda 0}$ is the wavelength-dependent albedo of the ocean surface, z is depth below the surface, and k_λ is the volume shortwave extinction coefficient of the ocean which varies with wavelength, λ. The value of the absorption coefficient is increased relative to clear water if the water is turbid. In *turbid* water, selective absorption occurs by chlorophyll and other biotic pigments, suspended sediments, and dissolved matter. Away from coasts, it is principally the chlorophyll content that affects water clarity. Absorption at the near infrared wavelengths is insensitive to turbidity, while absorption in the visible wavelengths depends crucially on the chlorophyll content.

Figure 11.1 shows the spectral extinction of solar radiation with depth in the ocean. At wavelengths greater than about 0.9 μm, nearly all of the radiation is absorbed in the top centimeter of the ocean. In contrast, at wavelengths near 0.5 μm, some of the radiation may penetrate as deep as 100 m.

Solar radiation is the dominant heat source of the ocean. However, the rate of solar heating of the ocean is slow (see (3.33)), because of the large heat capacity of the ocean. In the tropics, where the net surface radiation flux may be as large as 1000 W m^{-2}, the heating rate of the upper 10 m of the ocean is only about 0.036°C hr^{-1}. Although the heating rate is not large, the cumulative effect of solar heating on the ocean is very important. The amount of solar energy absorbed by the oceans is about three times as large as that absorbed by the atmosphere. The heating of the ocean mixed layer by solar radiation is later transmitted to the atmosphere by the surface turbulent heat fluxes and emission of longwave radiation. The portion of the solar radiation that penetrates below the ocean mixed layer can be transported over large horizontal distances by ocean currents.

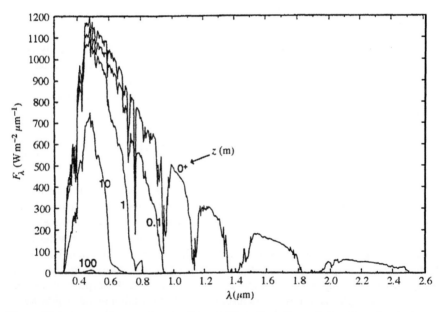

Figure 11.1 Computed spectral composition of the downwelling solar irradiance impinging at the surface (0$^+$) and then propagated to various depths (z = 0.1, 1.0, 10, and 100 m). The propagation is computed assuming a chlorophyll pigment concentration of 0.2 mg m^{-3}. (From Morel and Antoine, 1994.)

11.2 Skin Temperature and the Diffusive Surface Layer

The temperature at the interface between the atmosphere and ocean is called the *skin sea surface temperature*. It is this interfacial temperature that is used in Section 9.1 to calculate the surface sensible and latent heat fluxes and the upwelling longwave flux. It is difficult to measure the skin temperature directly, and remote infrared thermometers must be employed. So-called sea surface temperatures are most commonly measured from ships with thermometers by sampling water at a depth of about 5 m from engine water intake, or from buoys or moorings that measure temperature at a depth of about 0.5 m. These measurements are referred to as *bulk sea surface temperatures* and are typically characteristic of the temperature of the ocean mixed layer some tens of meters deep. Observations show that the skin temperature is invariably a few tenths of a degree cooler than the water a few millimeters below the surface, even during periods of weak winds and strong insolation.

To explain the cool skin, we examine the energy balance of a millimeter-thick layer at the ocean surface. The energy balance for this layer differs from that described in Section 9.1 where we examined the surface energy balance in the context of the heat budget of the entire ocean mixed layer. The difference arises because virtually all of the shortwave radiation is absorbed in the ocean mixed layer, while less than 10% is absorbed in the upper millimeter. Since the surface latent and sensible heat fluxes and the net longwave radiation fluxes are typically negative, there is a net heat loss in this millimeter-thick skin layer, even though the ocean mixed layer may be heating due to solar radiation. The net heat loss in the thin surface layer requires a flux of heat from the upper ocean. On both sides of the interface, the atmosphere and ocean are typically in turbulent motion. However, upon approaching the interface, turbulence is suppressed and the interface is a strong barrier to the turbulent transport between the ocean and atmosphere. Therefore on both sides of the interface the required heat transfer is accomplished by molecular conduction. To balance the large heat loss at the surface by molecular conduction, the temperature gradient just below the surface must be sufficiently large. Since there is a large heat reservoir in the ocean, the surface skin temperature must drop to accommodate the required heat flux from the ocean interior. This results in a cool skin that is a few tenths of a degree cooler than the ocean temperature a millimeter below the surface.

The diffusive sublayer beneath the ocean surface is characterized by a thickness δ, which is given by (e.g., Krauss and Businger, 1994)

$$\delta = \left(\frac{\kappa t^*}{\rho_l c_p} \right)^{1/2} \tag{11.2}$$

where κ is the thermal conductivity. The *surface renewal time scale*, t^*, is the residence time of small eddies which are renewed intermittently after random times of contact with the evaporating surface. The surface renewal time scale is a function of the surface friction velocity u_*, the surface roughness length z_0, and the net surface heat flux in the surface layer, F_{Q0}^{net}.

The temperature drop across the molecular sublayer, ΔT, is the bulk–skin temperature difference, and is given by (Liu and Businger, 1975)

$$\Delta T \propto F_{Q0}^{net} \left(\frac{t^*}{\rho_l c_p \kappa} \right)^{1/2} \tag{11.3}$$

Typical nighttime values of ΔT are 0.3°C, although values may exceed 1°C under some extreme conditions. During the daytime there is significant variability in ΔT that depends on the amount of solar insolation, ocean turbidity, and the magnitude of the wind. While such a small value of ΔT may seem insignificant, use of the bulk temperature instead of the skin temperature to calculate the surface sensible and latent heat fluxes from (9.10) and (9.11) can result in errors in the computed fluxes that exceed 10%. Errors of this magnitude may be large enough to change even the sign of the net surface heat flux and could significantly modify boundary layer and convective processes. Because of the Clausius–Clapeyron relationship, small errors in surface temperature result in larger errors in the latent heat flux, particularly when the surface temperature is high.

When evaporation exceeds precipitation, a "salt skin" forms on the surface, analogously to the cool skin. This tends to occur in the subtropical oceanic regions (see Section 9.2) where precipitation is very light. The cooling at the surface, combined with the higher salinity, causes an increase in density at the surface; depending on the stratification of the ocean below the sublayer, this may promote convective stirring of the ocean mixed layer. A dramatic case of the salt skin occurs over open water in the sea ice pack during winter, when surface sensible and latent heat fluxes may exceed several hundred watts per square meter as the cold air that has been modified over the ice pack streams over open leads or polynyas. Since open water in the ice pack is very near the freezing point, it would seem that any skin cooling would immediately result in surface freezing. However, because of the high evaporation from the surface, surface skin salinity is expected to exceed 80 psu, allowing a cool skin to exist at a temperature below the freezing temperature that corresponds to the mixed layer salinity. This increase in surface salinity will decrease the surface saturation vapor pressure and thus q_{v0}, reducing the surface latent heat and evaporative fluxes.

11.3 Surface Density Changes and the Ocean Mixed Layer

The top few tens of meters of the ocean are usually observed to be well mixed, with uniform temperature and salinity and thus neutral buoyancy (Figure 11.2). This layer is mixed primarily by frictional effects from the wind, although buoyancy effects

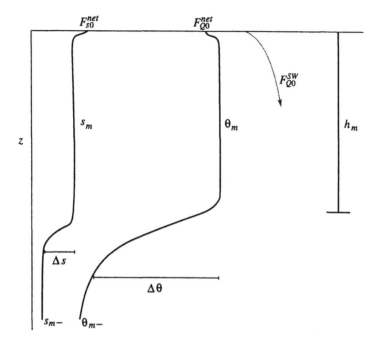

Figure 11.2 Schematic diagram of an ocean mixed layer model. F_{s0}^{net} and F_{Q0}^{net} represent the net surface salinity and heat fluxes, respectively. Δs and $\Delta \theta$ represent the changes in salinity and temperature at the bottom of the mixed layer, h_m is the depth of the mixed layer, and F_{Q0}^{SW} represents the shortwave flux that penetrates into the layer.

(e.g., nighttime surface cooling, enhanced turbulent fluxes) can contribute to the mixing. It is the ocean mixed layer that mediates heat and water exchanges with the atmosphere.

The processes that govern the evolution of the ocean mixed layer are illustrated in the context of a simple bulk mixed layer model, the term "bulk" implying that the temperature and salinity are explicitly specified to be constant over the depth of the mixed layer. Neglecting horizontal advection, the time-dependent equations for temperature and salinity of the ocean (after (3.56) and (3.58)) can be integrated vertically over the mixed layer depth, h_m, to obtain

$$h_m \frac{\partial \theta_m}{\partial t} + u_{ent} \, \Delta\theta = \frac{1}{\rho_m c_p} \left[-\kappa \left(\frac{\partial \theta}{\partial z} \right)_{h_m} + F_{Q0}^{rad} + F_{Q0}^{SH} + F_{Q0}^{LH} + F_{Q0}^{PR} - \left(F_{Q0}^{SW} \right)_{h_m} \right] \quad (11.4)$$

$$h_m \frac{\partial s_m}{\partial t} + u_{ent} \, \Delta s = -D_s \left(\frac{\partial s}{\partial z} \right)_{h_m}$$

$$+ \frac{1}{\rho_m} \left[\left(-\rho_l \dot{P}_r - \rho_s \dot{P}_s + \dot{E} - \dot{R} \right) s_m + \rho_i \frac{dh_i}{dt} \left(s_m - s_i \right) \right] \tag{11.5}$$

where the component terms of the net surface heat and salinity fluxes are described in Sections 7.4 and 7.5. The terms $\Delta\theta = h_{m-} - h_m$ and $\Delta s = s_{m-} - s_m$ represent the jumps in potential temperature and salinity across the base of the mixed layer and u_{ent} is the entrainment velocity at the base of the mixed layer. The subscript h_m on the $\partial\theta/\partial z$ and $\partial s/\partial z$ terms denotes the vertical derivative at the base of the mixed layer, and $(F_{Q0}^{SW})_{h_m}$ denotes the shortwave flux that passes through the base of the mixed layer.

Local changes in the mixed layer depth are determined from[1]

$$\frac{\partial h_m}{\partial t} = u_{ent} \tag{11.6a}$$

The entrainment velocity, u_{ent}, indicates the rate of mixed layer thickening caused by the turbulent entrainment of fluid across h_m and by its incorporation into the mixed layer. The entrainment velocity is determined from the conservation of turbulent kinetic energy in the mixed layer, and responds to the surface momentum and buoyancy fluxes. In effect, it is assumed that a fraction of the energy input by the wind is diffused downward and used to mix fluid up from below the mixed layer, and the remainder of the energy is dissipated. The entrainment velocity can be written (following Chu and Garwood, 1988) as

$$u_{ent} = \frac{c_1 u_*^3 - c_2 \left(F_{B0}/\rho \right) h_m}{h_m \left(\alpha g \, \Delta T - \beta g \, \Delta s \right)} \tag{11.6b}$$

where u_* is the surface friction velocity (9.15), $c_1 \approx 2$ and $c_2 \approx 0.2$. If the surface is cooling, then the resulting convective instability can provide mixing energy; if it is being heated, some of the mechanical mixing energy is used to overcome this stability effect. Therefore, the depth of the mixed layer will increase with increasing surface wind speed and decreasing surface buoyancy flux.

The ocean mixed layer characteristics vary in response to varying surface buoyancy and momentum forcing. In the following, we examine the characteristics of the ocean mixed layer at the two thermodynamic extremes of the world ocean: the tropical western Pacific "warm pool" and the ocean mixed layer beneath Arctic sea ice.

---

[1] See Kantha and Clayson (1999) for a detailed description of determination of the mixed layer depth.

11.3.1 Tropical Ocean Warm Pool

The warmest sea surface temperatures in the global ocean occur in the tropical western Pacific Ocean and the equatorial Indian Ocean, which is termed the *warm pool* (Figure 11.3). In the western Pacific warm pool, skin temperatures can reach as high as 34°C. The surface winds are typically light (averaging about 5 m s^{-1}), precipitation is very heavy ($\dot{P}_r - \dot{E} \sim 2$ m yr^{-1}), and there is net surface heating (see Figure 9.4). As a result, the mixed layer is relatively shallow, with a mean depth of about 30 m. Figure 11.4 shows a representative vertical structure of the western Pacific warm pool mixed layer. The large amount of precipitation generates a distinct, stable mixed layer near the surface, about 40 m deep (indicated by the depth of the constant density layer). The nearly isothermal layer extends to a depth of about 100 m.

The low surface wind speeds and the presence of a relatively fresh isohaline layer just below the surface implies that the heat content of the ocean mixed layer and the skin temperature are very sensitive to heating from above. Processes occurring in the warm pool are illustrated by the time series shown in Figure 11.5. Between days 318–326, surface wind speed averaged 4 m s^{-1} and a diurnal cycle is seen in skin temperature with an amplitude of 1–2°C. The depth of the mixed layer is very shallow during this period, deepening to about 20 m with the nocturnal cooling but shallowing to about 4 m during daytime. During the subsequent period 327–331, wind speed averaged 7 m s^{-1} and a progressive increase in mixed layer depth is seen. Between days 332–337, wind speed diminishes to about 3 m s^{-1} and steady warming resumes. The period 350–370 is characterized by much stronger winds, with peak values reaching 15 m s^{-1}. As a result, the fresh layer is breached and the ocean mixed layer depth increases markedly to 75 m as a result of entrainment. During this period, the amplitude of the diurnal cycle in skin temperature and mixed layer depth is virtually zero.

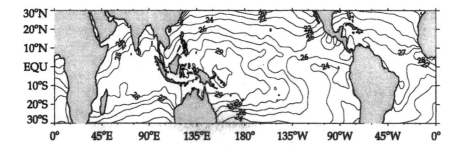

Figure 11.3 The geographic distribution of mean winter (DJF) bulk sea surface temperatures in the tropical oceans.

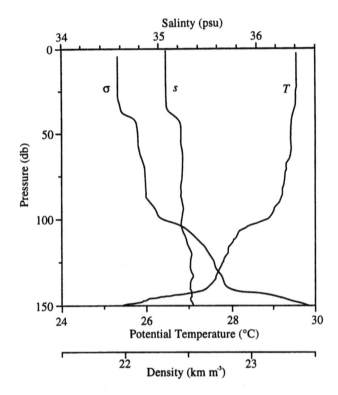

Figure 11.4 Typical vertical structure of the western Pacific warm pool. In this region, surface temperatures are relatively high, the isothermal layer extends to a depth of about 100 m, the mixed layer extends to about 40 m, and the isohaline layer is relatively fresh, due to the large amount of precipitation in the region. (Data from the Western Equatorial Pacific Ocean Circulation Study (WEPOCS) at 1°S, 155°E, Webster and Lukas, 1992.)

11.3.2 Ice/Ocean Interactions

The exchange of heat, fresh water, and momentum between the atmosphere and ocean is modulated strongly by the presence of a sea ice cover. Ice growth and melting causes buoyancy fluxes that affect the ocean mixed layer below the ice (see Section 9.3). Since the momentum flux from the atmosphere to the ocean is damped by the presence of ice, buoyancy fluxes play a more important role in maintaining the mixed layer than does wind.

  Figure 11.6 shows a time series of monthly averaged salinity and temperature profiles in the ocean beneath the ice in the Arctic Ocean. The salinity of the mixed layer is relatively low, between 30 and 31 psu. This low value arises from the large amount of river runoff ($\dot{R}$) into the Arctic basin. Mixed layer salinity reaches a maximum in early spring and a minimum in late summer, because the annual cycle of salinity is dominated

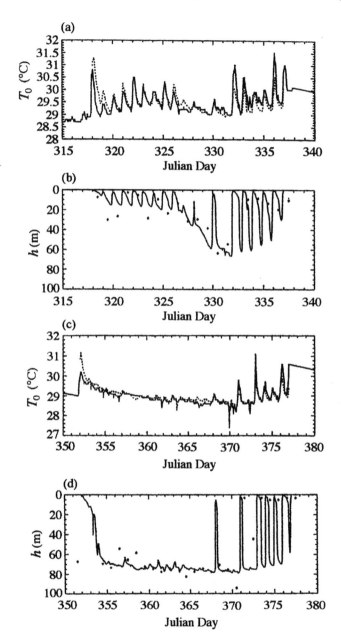

Figure 11.5 Modeled and observed SST (T_0) and thermocline depths (h_m) during the TOGA-COARE experiment in the tropical Pacific Ocean. Day 315 corresponds to November 12, 1992, and day 380 corresponds to January 16, 1993. (From Clayson, 1995.)

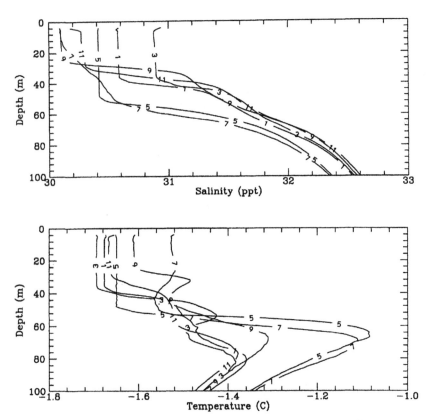

Figure 11.6 Time series of monthly averaged salinity and temperatures profiles in the water beneath the Arctic Ocean ice obtained during the Arctic Ice Dynamics Joint Experiment (AIDJEX) in the Beaufort Sea during 1995. Numbers correspond to the month (e.g., 1=January, 3=March, etc.). (Data provided by M. McPhee. Figure courtesy of J. Schramm.)

by the freeze–melt cycle of the sea ice. The mixed layer salinity exhibits its most rapid changes during the summer melt period, from late June to early August. During the remainder of the year, changes in the mixed layer salinity are more gradual and act to salinize the mixed layer. This salinization is caused both by the salinity flux at the mixed layer base and by the brine rejection and drainage that occurs during ice growth.

Variations in the freezing temperature occur over the annual cycle occur in response to the salinity variations. The mixed layer temperature does not depart significantly from the freezing temperature, although small departures occur during winter as a result of constitutional supercooling (see Section 5.7) and during summer, owing largely to absorption of solar radiation. Only about 4% of the incident shortwave radiation is transmitted to the ocean, and this is transmitted through the thinnest ice and open water areas.

The mixed layer depth is modulated by mechanical stirring (arising from the relative velocities of the ice and the ocean) and buoyancy effects. Variations in density occur primarily as a result of variations in salinity and hence the mixed layer depth is determined by the salinity profile. During autumn, winter, and spring, the mixed layer becomes more saline as a result of brine rejection and drainage due to ice growth. This increases the density of the ocean just beneath the ice, which produces static instability and enhances the turbulent activity in the mixed layer, causing it to deepen to a maximum depth of 60 m in mid-May. During the melt season, fresh water from the melting snow and ice enters the mixed layer and the layer warms due to the increased flux of solar radiation. The warming and freshening stabilize the water column and cause the mixed layer to retreat to a minimum depth of approximately 15 m in mid-July.

The Arctic Ocean mixed layer overlies a very stable density structure down to a depth of about 200 m, arising from a strong increase of salinity with depth. The density structure is a significant factor for sea ice growth because the stable stratification implies that the convective depth to which cooling occurs before freezing commences at the surface is generally not large (see Section 5.6). Therefore, heat and salt from the water underneath the top 200 m generally do not upwell to the surface. Hence the surface water, influenced also by the fresh-water river inflows, remains relatively fresh and cold.

11.4 Instability and Mixing in the Ocean Interior

Upper water masses are modified by heat and fresh-water exchanges with the atmosphere and from solar radiation transmitted through the atmosphere. By contrast, intermediate and deep water masses are modified only by advective or diffusive exchanges with other water masses; there are no internal sources of heat or salinity. Instability can be generated in the ocean interior by mixing processes, which can lead to rapid water mass modification and generation of deep oceanic convection and associated circulations.

In considering mixing and instability in the ocean interior, it is useful to employ isopycnal (constant density) surfaces, rather than the usual cartesian x–y surfaces. An isopycnal surface can also be thought of as a neutral surface (Figure 11.7), such that if a water parcel moves a small distance isentropically along an isopycnal surface (i.e., at constant θ and s), then the parcel will not experience any buoyant restoring forces. Denoting the two-dimensional gradient operator along an isopycnal surface by ∇_i, we can write (after McDougall, 1984):

$$\rho^{-1} \nabla_i \rho = \beta' \nabla_i s - \alpha' \nabla_i \theta + \gamma' \nabla_i p \tag{11.7}$$

Values of the thermal expansion coefficient α', saline contraction coefficient β', and

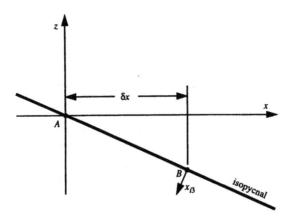

Figure 11.7 Isopycnal surfaces. A water parcel moving isentropically from A to B along an isopycnal surface will experience no restoring buoyancy forces. In the diagram, x_{i3} is perpendicular to the isopycnal surface, and z is the vertical coordinate.

compressibility coefficient $\gamma'^{,2}$ are defined as (following Gill, 1982)

$$\alpha' = -\frac{1}{\rho}\left(\frac{\partial \rho}{\partial \theta}\right) = \alpha\left(\frac{\partial \theta}{\partial T}\right)^{-1} \tag{11.8a}$$

$$\beta' = \frac{1}{\rho}\left(\frac{\partial \rho}{\partial s}\right) = \beta + \alpha'\left(\frac{\partial \theta}{\partial s}\right) \tag{11.8b}$$

$$\gamma' = \frac{1}{\rho}\left(\frac{\partial \rho}{\partial p}\right) = \gamma + \alpha'\left(\frac{\partial \theta}{\partial p}\right) \tag{11.8c}$$

where we have from (1.31a), (1.31b), and (1.31d) expressions for the usual coefficients

$$\alpha = -\frac{1}{\rho}\left(\frac{\partial \rho}{\partial T}\right) \quad \beta = \frac{1}{\rho}\left(\frac{\partial \rho}{\partial s}\right) \quad \gamma = \frac{1}{\rho}\left(\frac{\partial \rho}{\partial p}\right)$$

If a water parcel moves isentropically along an isopycnal surface, implying conservation of θ and s, we can write from (11.7)

$$\rho^{-1}\,\nabla_i\,\rho = \gamma'\,\nabla_i\,p \tag{11.9}$$

---

[2] The speed of sound, c_s, is defined in terms of the adiabatic compressibility coefficient γ' as $c_s^2 = 1/(\rho\gamma')$.

which implies

$$\alpha' \nabla_i \theta = \beta' \nabla_i s \tag{11.10}$$

Equation (11.10) can be used to define uniquely the isopycnal surface. By analogous arguments, we can also show that

$$\beta' \frac{\partial s}{\partial t} = \alpha' \frac{\partial \theta}{\partial t} \tag{11.11}$$

for the time derivatives on an isopycnal surface.

The isopycnal surface is the natural reference frame from which to interpret convective instabilities. Velocity of the fluid perpendicular to the isopycnal surfaces is associated with instability.   As described in Section 7.1, the stability of a parcel of water is determined by the buoyancy frequency, N^2

$$g^{-1} N^2 = \alpha \left(\frac{dT}{dz} + \Gamma'_{ad} \right) - \beta \frac{ds}{dz}$$

Using (11.8a) and (11.8b), we can write the buoyancy frequency as

$$g^{-1} N^2 = \alpha' \frac{d\theta}{dz} - \beta' \frac{ds}{dz} \tag{11.12}$$

where the density is regarded as a function of θ, s, and p (rather than T, s, and p). Neutral stability $N^2 = 0$ therefore corresponds to $\alpha' d\theta/dz = \beta' ds/dz$.

To interpret further the criterion for instability, we can expand the expression (11.12) by using the conservation equations for heat (3.6) and salt (3.11). The conservation equations for heat and salt including only the source terms associated with molecular diffusion are written as

$$\frac{\partial \theta}{\partial t} + u_i \frac{\partial \theta}{\partial x_i} = \frac{\partial}{\partial x_i} K_i \frac{\partial \theta}{\partial x_i} + \frac{\kappa}{\rho c_p} \frac{\partial^2 \theta}{\partial x_3^2}$$

$$\frac{\partial s}{\partial t} + u_i \frac{\partial s}{\partial x_i} = \frac{\partial}{\partial x_i} K_i \frac{\partial s}{\partial x_i} + D_s \frac{\partial^2 s}{\partial x_3^2}$$

where we have written the turbulent flux terms as eddy diffusivity terms, following (3.50) and (3.51). We assume that the eddy diffusivity coefficient, K, is identical for heat and salt, although the value of the eddy diffusivity coefficient in the vertical direction, K_3, is different from the values in the horizontal directions, $K_1 = K_2$. The conservation equations can be written in the reference frame of an isopycnal surface by formulating an orthogonal coordinate system consisting of the isopycnal surface and a coordinate that is perpendicular, called the *diapycnal coordinate*. Since the typical slope of an isopycnal surface in the ocean is approximately 10^{-3}, we can consider the isopycnal coordinate system to consist simply of a small rotation of the cartesian coordinate system. Thus the scale factors are unity and the conservation equations have exactly the same form in the isopycnal reference frame as in the cartesian reference frame. We can therefore write the conservations of heat and salt in the isopycnal reference frame as

$$\frac{\partial \theta}{\partial t} + u_i \nabla_i \theta + u_3 \frac{\partial \theta}{\partial x_3} = \nabla_i K_i \nabla_i \theta + \frac{\partial}{\partial x_3} K_3 \frac{\partial \theta}{\partial x_3} + \frac{\kappa}{\rho c_p} \frac{\partial^2 \theta}{\partial x_3^2} \qquad (11.13)$$

$$\frac{\partial s}{\partial t} + u_i \nabla_i s + u_3 \frac{\partial s}{\partial x_3} = \nabla_i K_i \nabla_i s + \frac{\partial}{\partial x_3} K_3 \frac{\partial s}{\partial x_3} + D_s \frac{\partial^2 s}{\partial x_3^2} \qquad (11.14)$$

where the notation ∇_i represents the horizontal gradient along the isopycnal surface, u_i represents the horizontal velocity components along the isopycnal surface, u_3 is the vertical velocity in the diapycnal direction, and $\partial/\partial x_3$ represents the derivative in the diapycnal direction.

Multiplying (11.14) by β' and subtracting α' times (11.13) gives

$$\left(u_3 - \frac{\partial K_3}{\partial x_3}\right) \frac{N^2}{g} = K_i \left(\alpha' \nabla_i^2 \theta - \beta' \nabla_i^2 s\right) + K_3 \left(\alpha' \frac{\partial^2 \theta}{\partial x_3^2} - \beta' \frac{\partial^2 s}{\partial x_3^2}\right) \qquad (11.15)$$

$$+ \left(\frac{\alpha' \kappa}{\rho c_p} \frac{\partial^2 \theta}{\partial x_3^2} - \beta' D_s \frac{\partial^2 s}{\partial x_3^2}\right)$$

where

$$g^{-1} N^2 = \alpha' \frac{d\theta}{du_3} - \beta' \frac{ds}{du_3} \qquad (11.16)$$

and we have ignored horizontal variations in the eddy diffusivity coefficient. Note that use of the isopycnal coordinate system results in elimination of the time derivative and advection terms, providing a principal motivation for using this coordinate system. The term $(u_3 - \partial K_3/\partial x_3)$ is the diapycnal velocity, and gives the speed at which the isopycnal surface moves through the fluid. Note that spatial variations of diffusivities are referred to as *pseudovelocities*, since they appear in conservation diapycnal velocity is a direct result of the various mixing processes represented on the right-hand side of (11.15). The first term on the right-hand side describes the instabilities that arise from horizontal mixing on an isopycnal surface. The second term represents an instability associated with diapycnal turbulent mixing. The last term in (11.15) is associated with double-diffusive convection.

More specifically, the first term of the right-hand side of (11.15) represents the isopycnal mixing of parcels of water with different values of θ and s. A typical value of K_i in the ocean is $1000 \text{ m}^2 \text{ s}^{-1}$. The instabilities associated with this term can be elucited as follows. Taking the isopycnal divergence of the identiy $\alpha' \nabla_i \theta = \beta' \nabla_i s$ (11.10), we obtain (following McDougall, 1987)

$$
\begin{aligned}
K_i \left(\alpha' \nabla_i^2 \theta - \beta' \nabla_i^2 s \right) &= -K_i \left| \nabla_i \theta \right|^2 \left(\frac{\partial \alpha'}{\partial \theta} + 2 \frac{\alpha' \partial \alpha'}{\beta' \partial s} - \frac{\alpha'^2 \partial \beta'}{\beta'^2 \partial s} \right) \\
&\quad - K_i \nabla_i \theta \cdot \nabla_i p \left(\frac{\partial \alpha'}{\partial p} - \frac{\alpha' \partial \beta'}{\beta' \partial p} \right)
\end{aligned}
\tag{11.17}
$$

The first term on the right-hand side of (11.17) is called the *cabbeling instability*. When two parcels of water are mixed along an isopycnal surface, the density of the mixture increases because of the nonlinearity of the equation of state of seawater. The cabbeling parameter is given by

$$
\frac{\partial \alpha'}{\partial \theta} + 2 \frac{\alpha'}{\beta'} \frac{\partial \alpha'}{\partial s} - \frac{\alpha'^2}{\beta'^2} \frac{\partial \beta'}{\partial s}
$$

A typical value of the cabbeling parameter is 10^{-5} K^{-2}. Figure 11.8 shows isopycnals near $0°C$ and 34.6 psu. Two different water types, A and B, are identified with different values of temperature and salinity, but having the same density ($\sigma = 27.8$). If the two water types mix, their final temperature and salinity will be a mass-weighted average of the two different water types (analogous to Section 6.4). It is easily shown that the properties of the mixture lie on the straight line AB, and hence the mixture is more dense than either of the original water types. Consequently, any such mixture will tend to sink. Thus the cabbeling instability arises from the nonlinear relation between density, temperature, and salinity, particularly at low temperatures.

The second term on the right-hand side of (11.17) represents the *thermobaric instability*, which arises from the variation of the thermal expansion and saline contraction coefficients with pressure. The thermobaric parameter is given by

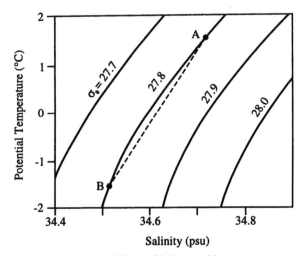

Figure 11.8 Isopycnals near 0°C and 34.6 psu. If two water masses having different temperatures and salinities, but the same density (e.g., σ = 27.8) mix together, the final water mass will have mass-weighted average values of temperature and salinity. The properties of the final water mass will lie along the dotted line that joins the original water masses, A and B. Hence, the density of the mixture is greater than either of the original masses, and will therefore tend to sink.

$$\frac{\partial \alpha'}{\partial p} - \frac{\alpha'}{\beta'} \frac{\partial \beta'}{\partial p}$$

A typical value of the thermobaric parameter is 2.6×10^{-12} K^{-1} Pa^{-1}. Table 11.1 shows the variation of α and β with pressure at $T = 2$°C and $s = 35$ psu. The compressibility of cold water is generally greater than that of warm water, so a cold parcel displaced downward could in principle become heavier than its surroundings. The variation of the saline contraction coefficient with pressure is small, and hence its contribution to generating thermobaric instability is much less. The thermobaric instability is illustrated in Figure 11.9. Along a neutral surface, p and θ both vary so that $\nabla_i \theta \cdot \nabla_i p \neq 0$. Consider two parcels of fluid with different temperatures on an isopycnal surface. If $\nabla_i \theta \cdot \nabla_i p > 0$, the parcels move below the neutral surface, and if $\nabla_i \theta \cdot \nabla_i p < 0$ the parcels rise above it. If the parcels are moved back to their original positions, no irreversible effects will have occurred. However, the mixing of the two parcels together off the isopycnal surface (cabbeling) is irreversible and hence consolidates the vertical movement of the parcels away from the neutral surface by thermobaricity.

Table 11.1 Variation of α and β with pressure, for a temperature of 2°C and salinity of 35 psu. Note that β differs from β' by less than 0.3% and that $\alpha' = \alpha(T/\theta)\,[c_p(p_r,\theta)/c_p(p,T)]$ where p_r is a reference pressure of 1 bar. (Data from UNESCO, 1981.)

p (bar)	$\alpha\,(10^{-7}\ \mathrm{K}^{-1})$	$\beta\,(10^{-6}\ \mathrm{psu}^{-1})$
0	781	779
100	1031	767
200	1269	758
300	1494	747
400	1707	737
500	1907	728
600	2094	719

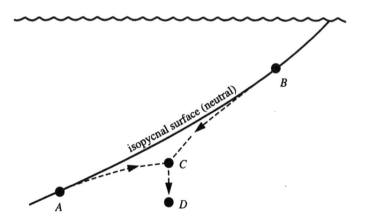

Figure 11.9 Thermobaric instability. Along the neutral surface, p and θ vary, so that two water parcels, A and B, both of which lie along the isopycnal, have different values of temperature and pressure and therefore different thermal expansion and saline contraction coefficients. The pressure difference along the neutral surface causes the water parcels to move off the isopycnal. If the two parcels are forced back to their original positions, no irreversible effects will have occurred. However, if the parcels mix (as at point C), the mixing process is irreversible. Once the parcels have mixed at C, further vertical movement of the mixed parcel from C to D may occur as a result of cabbeling, since the mixed water parcel has a higher density than the two original parcels. (After McDougall, 1987.)

Thermobaricity and cabbeling are processes caused by lateral mixing along isopycnal surfaces that result in vertical advection because the equation of state for seawater is nonlinear. Thermobaricity arises from the variation of α'/β' with pressure on an isopycnal surface, while cabbeling arises from the dependence of α'/β' on θ and s along an isopycnal surface. While cabbeling and thermobaricity are effective at producing diapycnal downwelling (or upwelling) of water, they are ineffective at producing vertical diffusion of tracers. Over most of the global ocean, thermobaricity and cabbeling are weak. However, where the slope of the isopycnal surfaces is significant, the downwelling due to cabelling is quite large. Cabbeling and thermobaricity are largest in the North Atlantic and oceans surrounding Antarctica, causing contributions to the diapycnal downwelling velocity of the order of -1×10^{-7} m s^{-1}. Thermobaricity is generally smaller than cabbeling except in the Antarctic Circumpolar Current where it is equivalent in magnitude and also of the same sign.

The last term in (11.15) is associated with *double-diffusive instability*. When a one-component fluid is stably stratified, no vertical velocities are generated by buoyancy forces. In a multi-component fluid such as seawater, in which density is determined by concentration of the solute in addition to the temperature of the solution, buoyant instabilities can arise simply from the different values of the diffusivity of heat and the diffusivity of the scalar concentration. The ensuing instability in the ocean is referred to by convention as double-diffusive instability. The thermal diffusivity of seawater, $\kappa/\rho c_p \approx 1.4 \times 10^{-7}$ m s^{-1} is two orders of magnitude larger than the diffusivity of salt in water, $D_s \sim 1.1 \times 10^{-9}$ m^2 s^{-1}. Therefore, heat is transferred by diffusion more rapidly than salt. Even if the mean density distribution is hydrostatically stable, the difference in molecular diffusivities allows instability to be initiated. For double–diffusive instability to occur in the ocean, the temperature and salinity gradients across the isopycnal surface must have the same sign, since temperature and salinity act in opposite directions to influence density. There are two kinds of double–diffusive convection: diffusive staircases, in which relatively fresh, cold water overlies warmer, saltier water; and salt finger staircases, in which warm, salty water overlies cooler, fresher water.

Consider a layer of warm, salty water overlying a cooler, fresher layer of water, both with equal densities, so that $T_1 > T_2$, $s_1 > s_2$, and $\rho_1 = \rho_2$. The warm salty layer loses heat to the cooler, fresher layer faster than it loses salt. By this double diffusion process, the density of the upper layer increases and the density in the lower layer decreases. Buoyancy forces cause the upper, denser layer to sink. This results in salt fingering, with the instability arising from the lower diffusive flux of salinity. In this case, long narrow convection cells or salt fingers rapidly form (Figure 11.10). Individual salt fingers are a few millimeters across and up to 25 cm long. After the salt fingers form, lateral diffusion occurs between the fingers and produces a uniform layer. The process may start again at the two new interfaces so that eventually a number of individually homogeneous layers develop with sharply defined interfaces in terms of temperature and salinity. These layers may be meters to tens of meters

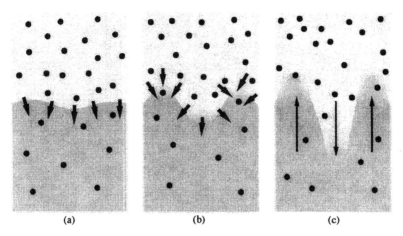

(a)	(b)	(c)

Figure 11.10 Double diffusion of heat and salt in the ocean occurs when warm, saline water overlies cooler, less saline water (a). The upper layer loses heat (arrows) to the lower layer faster than it loses salt (dots), and hence the density of the upper layer becomes greater than that of the lower layer. The upper layer sinks, creating "salt fingers" as shown in (b) and (c). (After The Open University, 1989.)

thick, separated by thinner interface zones of sharp gradients of temperature and salinity, giving the vertical temperature and salinity profiles a step-like appearance. Double diffusion occurring by this mechanism was first observed below the outflow of relatively warm, salty Mediterranean water into the Atlantic.

Now consider the situation of cold, less salty water overlying warmer, saltier water, with a stable stratification. The instability is driven by the larger vertical flux of heat relative to salt through the "diffusive" interfaces, which are in turn kept sharpened by the convection in the layers. This gives rise to staircases that exhibit regularity in the steps and in which both s and T increase systematically with depth. Such layers have been observed in the Arctic Ocean, Weddell Sea, and the Red Sea (Figure 11.11).

It is now recognized that salt fingering and related processes can make a significant contribution to vertical mixing within the oceans. It seems that double-diffusive staircases could in principle be observed at any interface between water masses and in the vicinity of all thermoclines. The buoyancy flux ratio for salt fingers, R_ρ, is given by

$$R_\rho = \frac{\alpha' \Delta\theta}{\beta' \Delta s} \tag{11.18}$$

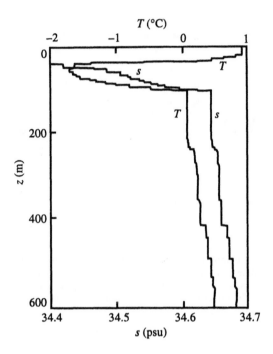

Figure 11.11 Staircases can form in regions where cold, less salty water overlies warmer, saltier water. (From Foster and Carmack, 1976.)

It appears that salt fingers do not occur if $R_\rho > 0.6$. Salt fingers are not always observed even if conditions are favorable if more vigorous processes are operative, such as turbulent mixing.

In regions where there are significant horizontal density gradients (sloping isopycnal surfaces), thermobaricity, cabbeling, and double diffusion can produce significant diapycnal velocities in the ocean. These processes influence water mass transformation in the ocean interior, particularly at the boundaries of water masses. However, in regions of the ocean where horizontal density gradients are small, processes such as thermobaricity, cabbeling, and double diffusion are often negligible relative to other processes.

11.5 Oceanic Convection and Deep Water Formation

Thermodynamic mechanisms driving oceanic deep convection are caused by the convective instabilities described in the preceding section. Any mechanism that causes a denser water mass to lie over a lighter water mass will give rise to such an instability.

In addition to the thermodynamic mechanisms, there are dynamical instabilities that contribute to deep convection. Dynamical instabilities in the atmosphere and ocean are the dynamic response on a rotating earth to thermodynamic imbalances. These dynamical instabilities include Kelvin–Helmholtz instability, baroclinic instability, centrifugal instability, and symmetric instability.[3] Dynamical instabilities exchange water mass in the vertical direction, which reduces the stratification, making the ocean more susceptible to thermodynamic instabilities.

Deep convection in the open ocean is observed to occur in the Labrador, Greenland, Weddell, and Mediterranean Seas. These regions are the principal source regions for deep and intermediate waters of the world's oceans. No deep water is formed in the North Pacific under the present climate regime, primarily because the static stability is too large and surface waters too fresh even at the freezing point to sink below a few hundred meters.

Deep convection in the open ocean occurs through the following sequence of events.

i)   A background cyclonic (counterclockwise) gyre circulation (Figure 11.12a) develops in the ocean in response to the surface winds. The cyclonic gyre is associated with rising motion in the ocean (upward vertical velocity). The gyre forms an elongated "dome" of isotherms, isohalines and isopycnals in the center of the cyclonic gyre, on a horizontal scale of about 100 km. This reduces the vertical stability of water columns within the gyre.

ii) Formation of a preconditioning pool (Figure 11.12b) occurs by a combination of density increase in the upper ocean due to surface cooling and salinization plus strong surface winds, leading to mixed layer penetration into the intermediate-water dome. This creates a background of low static stability within the gyre. Formation of an oceanic frontal system by hydrodynamical instability also supports preconditioning.

iii) Mixed-layer penetration into the intermediate water dome may result in a mixed water mass with an increased density via cabbeling. A more efficient mode of mixing occurs via mixing of two different water masses across a front (associated with baroclinic instability). This can result in a dramatic increase in the depth of the mixing.

iv) Narrow plumes with width on the order of 1 km develop, with strong downward motion (Figure 11.12c) reaching as high as 10 cm s^{-1}. The plumes broaden as they descend through lateral entrainment. The mode of convection seems to be nonpenetrative rather than penetrative. In other words, mixing occurs in such a way as to keep the density structure a continuous function of depth.

v)   The plumes rapidly mix properties over the preconditioned pool, forming a vertical "chimney" of homogeneous fluid. When the surface forcing ceases,

---

[3] A discussion of dynamical instabilities in the ocean is beyond the scope of this book; for further information, the reader is referred to *Deep Convection and Deep Water Formation in the Oceans* (1991) by Chu and Gascard.

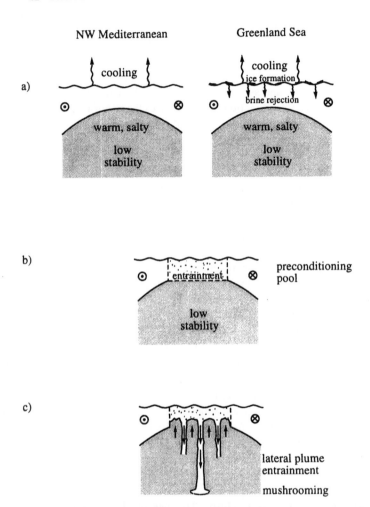

Figure 11.12 Preconditioning events and deep convection in the open ocean. a) Cyclonic circulation (represented by ⊙ and ⊗, the tip and tail of an arrow, respectively) during early winter leads to an uplifting of the isotherms, isohalines, and isopycnals into an elongated "dome" in the center of the gyre. The surface mixed layer on top of the dome cools, and in regions where ice forms, brine rejection increases the salinity of the mixed layer. b) Continued cooling and strong winds during mid-winter increase the depth of the mixed layer, so that it can penetrate into the relatively warm and salty intermediate-water dome. This entrainment brings higher salinity water out of the dome and into the layer above it. Further mixed layer deepening and cooling both act to decrease the stability within the gyre. c) In response to the unstable stratification, narrow plumes develop in late winter, with downward motions of 5–10 cm s^{-1}. The vertical "chimneys" formed in this manner eventually break up due to bottom topography, shear in the mean circulation, dynamical instabilities, and mixing associated with internal waves. (After Malanotte-Rizzoli, 1994.)

convective overturning declines sharply and the predominantly vertical heat transfer gives way to horizontal transfer.

vi) The chimneys eventually break up due to mushrooming around bottom topographical features, vertical shear of the mean circulation, dynamical instabilities, and mixing associated with internal waves. The horizontal spreading of the deep water is an efficient way for anomalous wintertime atmospheric features to extend their effect far below the surface layer of the ocean and far from their location of origin.

The best documented open-ocean convection occurs during most winters in the Gulf of Lions of the western Mediterranean Sea. During February, strong wind outbursts of cold continental air blow from the north-northwest over the cyclonic gyre in the western Mediterranean, cooling the surface by evaporation. In this region, the surface buoyancy flux results in a gradual erosion of the thermocline by a deepening surface mixed layer. In the center of the preconditioned dome, intense convection can occur, although it does not usually reach the ocean bottom.

North Atlantic Deep Water is formed in the Greenland Sea. The circulation of the Greenland Sea is characterized by cyclonic upper-layer flow (Figure 11.13). Relatively warm and saline water is carried northward by the Norwegian-Atlantic Current. One branch of this current curls cyclonically around the Greenland Sea, thus transporting salt into the region. During early winter, the formation of ice and the associated brine rejection increase mixed layer salinity at freezing point temperatures and reduce the stability of the underlying weakly stratified waters. In the later part of winter, strong surface winds generate ice export and cause deepening of the exposed mixed layer. Thus a pool of dense water is generated in the central Greenland Sea that is several 100 m deep and 100 km wide. Within this pool, convection begins to develop in late winter in response to favorable atmospheric heat and fresh-water fluxes.

Deep ocean convection also occurs sporadically in the Southern Ocean. Strong cooling at the surface of polynyas drives the convection throughout the winter period, leading to substantial deep water formation. The marginal stability of the Southern Ocean water column leaves it susceptible to ocean overturning, but the strong ocean heat flux acts to limit sea ice growth and minimizes the salinization associated with ice formation. Surface cooling, cyclonic circulation in the ocean, decrease in pycnocline strength, or decreasing the oceanic heat flux all tend to destabilize the stratification during winter, thus increasing the possibility of deep convection.

In addition to sinking in the open ocean, deep convection can also occur near an ocean boundary, such as a continental shelf (Figure 11.14). Near-boundary convection in the high latitudes occurs in the following sequence of events:

i) Dense water forms over the continental shelf, due to surface cooling and salt rejection that occurs during ice formation. The width of the shelf seems to be positively correlated with the amount of deep water formed. Providing that the

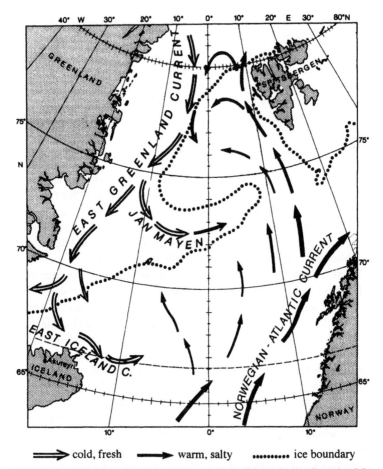

   ➾ cold, fresh   ➝ warm, salty   •••••••• ice boundary

Figure 11.13 Preconditioning of the North Atlantic Deep Water. In the Greenland Sea, upper-layer flow is cyclonic, bringing warm, salty water northward into the region via the Norwegian-Atlantic Current, and cold, dense water southward via the East Greenland Current. A combination of increased salinity and surface wind strength leads to the formation of a dense pool of water in the center of the Greenland Sea. Under the favorable conditions that occur in late winter, this pool becomes a source of deep water convection and the North Atlantic Deep Water mass.

     depth of the reservoir is sufficiently shallow (500 m in the Antarctic, 50–100 m in the Arctic), the resulting vertical convection yields cold, salty water at depth on the shelf.

ii) A horizontal density gradient parallel to the coast produces local circulations that are favorable for the reservoir to empty. The horizontal density gradient may be induced by variations in shelf geometry.

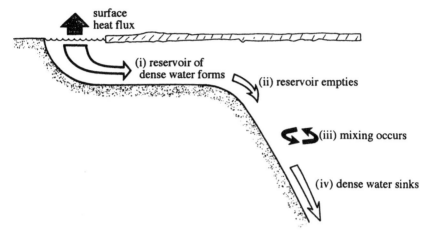

Figure 11.14 Idealized diagram representing deep ocean convection near a continental shelf. (i) A reservoir of dense water forms over the continental shelf. (ii) The reservoir empties and (iii) mixes with an off-shore water mass, (iv) producing a high-density mass of water that descends the slope of the shelf. (From Carmack, 1990.)

iii) Mixing with an off-shore water mass increases the density. While mixing does not appear to be a necessary element of the near-boundary convection, it appears consistently in the observations.

iv) Dense, salty water on the shelf descends the slope under a balance of Coriolis, gravity, and frictional forces. The thermobaric effect may also contribute to the sinking.

Near-boundary convection has been observed along coastal regions surrounding Antarctica, the coastal regions of the Arctic Ocean, and in the coastal regions surrounding the Greenland and Norwegian Seas. Outside the polar regions, boundary deep convection has been observed in the Red Sea and the Northern Adriatic Sea.

The lower 2000 m of the global ocean is dominated by Antarctic Bottom Water (AABW), with temperature below 2°C. AABW forms both by open ocean convection and along coasts. Polynyas (Section 10.1) play a major role in the formation of AABW. Along the coasts, large amounts of sea ice form within the coastal polynyas which is continually swept away by coastal winds, resulting in further ice formation and salt rejection. The AABW formation arises from the mixing of shelf and slope water across the shelf break. The dense water formed at this front sinks along the continental slope to the deep and bottom depths to a thin plume of concentrated AABW. During the mid-1970s, a large polynya in the Weddell Sea formed each winter, associated with vigorous oceanic convection that eliminated the sea ice cover. It has been suggested (Gordon, 1982) that there may be a tradeoff between open ocean and shelf production of deep water in the Southern Ocean. Advection of sea ice northward acts to destabilize the shelf region (where more ice forms) while simultaneously stabilizing the open ocean region (where the ice melts). Hence, conditions favoring open ocean convection coincide with unfavorable conditions on the shelves and vice versa.

The North Atlantic Deep Water (NADW), formed by deep convection in the Greenland Sea, is warmer and saltier (2.5°C, 35 psu) than Antarctic Deep Water that forms in the Weddell Sea (−1.0°C, 34.6 psu). More than half the ocean's volume is filled with cold water from these two sources. The rates of production and flow of deep water has been inferred from the distributions of tritium (^{3}H) in the decades after atmospheric nuclear weapons tests deposited large amounts in the northern North Atlantic and Arctic Oceans. It has been hypothesized that the formation of North Atlantic Deep Water is especially vulnerable to changes in the high-latitude surface salinity flux (e.g., runoff, precipitation, evaporation, sea ice transport).

11.6 Global Thermohaline Circulations

The thermohaline circulation is basically an overturning of the ocean in the meridional-vertical plane. In contrast to the quasi-horizontal wind-driven gyres, which are constrained to limited ranges of latitude and depth, the thermohaline-driven overturning cells are global in scale. The thermohaline circulation is that part of the large-scale ocean flow driven by diapycnal fluxes, rather than directly by the wind. The term thermohaline circulation is used increasingly to apply not only to the direct response to surface buoyancy fluxes but also to flows whose characteristics are altered significantly by upwelling or mixing. The interhemispheric and interbasin water mass exchanges associated with the thermohaline circulation have an important influence on the distribution of properties in the deep ocean and on global-scale heat and fresh-water transports. Although thermohaline processes in the ocean are commonly considered separately from wind-driven processes, many aspects of the ocean thermohaline and wind-driven circulations are inextricably linked by nonlinearities in the system and cannot be easily separated.

Surface heating at the equator and cooling at high latitudes implies a thermohaline circulation associated with the meridional (latitudinal) density differences. However the circulation associated with this overturning remains weak since the heating near the equator occurs only in the shallow mixed layer. For an efficient thermohaline "engine," expansion must take place at higher pressure, and compression must take place at lower pressure (Figure 11.15a). Inclusion of salinity effects shows overall that freshening in high latitudes counteracts the cooling effect. Figure 11.15b shows a schematic of the upper ocean thermohaline circulation, including both thermal and salinity effects. An efficient thermohaline circulation seems to be restricted to the upper layers in tropical and subtropical regions. This upper thermohaline circulation is of minor importance compared to the wind-driven circulation.

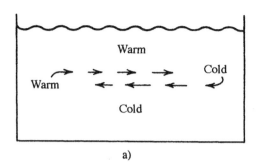

a)

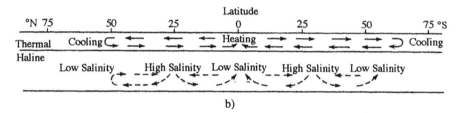

b)

Figure 11.15 Schematic of upper-level thermohaline circulation in the ocean. In a), the hot and cold sources are separated by only a slight difference in depth (or pressure), leading to only a very shallow layer of overturning. b) Poleward of about 25°, the haline effect counteracts the thermal effect. Between the Equator and about 25°, the salinity-driven circulation reinforces the thermal-driven circulation. (Adapted from Neumann and Pierson, 1966.)

In the deep ocean (below the thermocline), where the ocean is not influenced directly by winds, thermohaline circulations are very important, if not very efficient. Local thermohaline circulations occur in certain marginal seas, associated with warm waters rendered dense by their high salinity resulting from evaporation. Water from the Mediterranean, made dense by evaporation in the eastern basin, flows westward below the surface along that sea out through the Strait of Gibraltar into the Atlantic. There it spreads westward across the ocean, being identifiable as a high salinity tongue between about 500 and 3000 m. In the Red Sea, evaporation increases the salinity in the summer and then winter cooling of the saline upper waters causes it to sink to mid-depth and flow out and southward in the Indian Ocean as a high salinity tongue between 500 and 2500 m.

The Atlantic Ocean thermohaline circulation is illustrated in Figure 11.16. The abbreviations for the water masses shown in Figure 11.16 are given in Table 11.2. In the high latitudes of the North Atlantic, surface water is advected northward and cools as it enters the polar seas. Upper North Atlantic Deep Water (UNADW) is formed

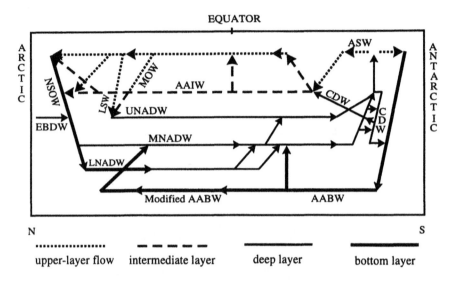

Figure 11.16 Thermohaline circulation in the Atlantic Ocean. Abbreviations are detailed in Table 11.2. Circulation and water mass formation are described in the text. (Adapted from Schmitz, 1996.)

from a mixture of Labrador Sea Water (LSW) and Mediterranean Overflow Water (MOW), along with entrained Antarctic Intermediate Water (AAIW). AAIW forms as the relatively cool, fresh water of the Southern Ocean flows northward and sinks, spreading at a level near 1000 m below the main thermocline. Deep water is formed on the shelves of the Arctic Ocean. The Eurasian Basin Deep Water (EBDW) is advected into the North Atlantic through the Fram Strait. Lower North Atlantic Deep Water (LNADW) is formed from Nordic Seas Overflow Water (NSOW), Denmark Strait Overflow Water (DSOW), and EBDW that mixes with intermediate water plus modified AABW. The LNADW comprises the bottom water north of 45°N, while AABW comprises the bottom water in the rest of the global ocean. The deep water that moves southward across the equator hence originated as AABW, EBDW, NSOW, DSOW, LSW, AAIW, and MOW. The Circumpolar Deep Water (CDW) forms initially from NADW, but becomes modified by circulation in the Antarctic Circumpolar Current System and transits in the Pacific and Indian Oceans, undergoing mixing along the way. The circulation is completed by weak but widespread upwelling in the low- and mid-latitudes, and also by the wind-driven overturning that drives the Antarctic circumpolar current.

Table 11.2 Abbreviations of water masses used in Figure 11.16 and elsewhere in the text. (After Schmitz, 1996.)

Abbreviation	Full name
AABW	Antarctic Bottom Water
AAIW	Antarctic Intermediate Water
ASW	Antarctic Surface Water
CDW	Circumpolar Deep Water
DSOW	Denmark Strait Overflow Water
EBDW	Eurasian Basin Deep Water
LNADW	Lower North Atlantic Deep Water
LSW	Labrador Sea Water
MNADW	Middle North Atlantic Deep Water
MOW	Mediterranean Outflow Water
NACW	North Atlantic Central Water
NADW	North Atlantic Deep Water
NSOW	Nordic Seas Overflow Water
UNADW	Upper North Atlantic Deep Water

A schematic of the global ocean thermohaline circulation, or *conveyor belt,* is shown in Figure 11.17. The Antarctic Circumpolar Current System is seen to be the conduit for transporting NADW to the Indian and Pacific Oceans, either as pristine NADW or modified CDW. The source of NADW has multiple paths, with upwelling and water mass conversion occurring in numerous locations. Radioactive tracer measurements show that the ocean thermohaline circulation brings deep ocean water into contact with the atmosphere every 600 years or so. The overturning is driven by both buoyancy forces generated in high-latitude oceans and the mechanical process driven by the wind stress in the region of the Antarctic Circumpolar Current System. While the Atlantic branch of the global thermohaline circulation is fairly well known, substantial uncertainties remain, particularly regarding the thermohaline flow patterns near the equator and the deep circulation in the North Pacific.

Since the density differences driving the global ocean thermohaline circulation are small, slight changes in surface salinity at high latitudes caused by precipitation, evaporation, ice melt, or river runoff can have a large effect on the circulation strength. Paleoceanographers hypothesize that the shutdown of the North Atlantic thermohaline circulation was a key feature of past Northern Hemisphere glaciation. Meltwater discharge at the end of the glaciation provided enough fresh water to insulate the deep ocean from the atmosphere on a large scale, turning off the North Atlantic thermohaline circulation and returning Europe to ice age conditions. A modern instance of fresh-water capping in the North Atlantic has been documented, called the *great*

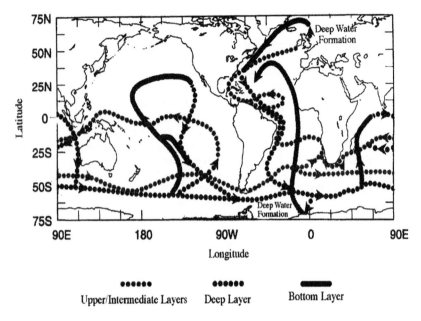

Figure 11.17 A schematic view of the ocean thermohaline conveyor belt. North Atlantic Deep Water (NADW) flows southward along the eastern coast of South America, and eventually reaches the Indian and Pacific Oceans. It is generally believed that the Antarctic Circumpolar Current System (ACCS) is responsible for transporting NADW to the Indian and Pacific Oceans. The return flow to the Atlantic occurs via two methods. Some of the water rises in the Pacific and flows as a surface current through the Indonesian Archipelago and the South Indian Ocean, around the southern tip of Africa, and north again into the North Atlantic. A portion of the NADW sinks in the Weddell Sea and flows northward with the Antarctic Bottom Water (AABW) into the North Atlantic. Primary deep water sources are found in the North Atlantic and in the Weddell Sea. (After Schmitz, 1995.)

salinity anomaly. Salinity records suggest that a large discharge of Arctic sea ice led to patch of fresh upper ocean water, which circulated around the subpolar gyre during the 1960s and 1970s, which caused deep convection locally to cease. The insulating effect of such low salinity patches causes reduced heat and moisture flux to the atmosphere and cooler conditions in Europe. Buoyancy-driven overturning circulations are vulnerable to shutdown by relatively small increases in high-latitude fresh-water fluxes. An overturning that includes a significant degree of mechanical forcing by Southern Hemisphere winds over the Drake Passage may be less vulnerable to such changes.

Notes

Surface renewal theory is described in *Atmosphere-Ocean Interaction* (1994, Chapter 5) by Kraus and Businger and in *Small-scale Processes in Geophysical Flows* (1999, Chapter 4) by Kantha and Clayson.

Characteristics of the ocean mixed layer and the associated physical processes are described in *Small-scale Processes in Geophysical Flows* (1999, Chapter 2) by Kantha and Clayson. An overview of the Arctic Ocean mixed layer and ice/ocean interactions is given by Holland *et al.* (1997).

Mixing in the ocean is described by Turner (1981). The roles of thermobaricity and cabbeling in water mass conversion are described by McDougall (1987). Kantha and Clayson (1999, Chapter 7) give a thorough description of double-diffusive convection.

An extensive overview of oceanic deep convection is given in *Deep Convection and Deep Water Formation in the Oceans* (1991) by Chu and Gascard. An additional review is given by Killworth (1983).

A discussion of the relative roles of thermohaline and wind-driven processes in the general circulation of the ocean is given in *Atmosphere-Ocean Dynamics* (1982) by Gill.

Warren (1981) provides an overview of the deep ocean thermohaline circulations and Schmitz (1995) gives a recent review of the interbasin-scale thermohaline circulation.

Problems

1. Using the data given in problem **9.2** for noon, determine the error in the surface upwelling longwave flux and the sensible and latent heat fluxes if the bulk sea surface temperature (ocean mixed layer temperature) of 29°C was used instead of the skin temperature.

2. Assume that the incident solar flux at the surface is 500 W m^{-2}. Use the following spectral extinction coefficients for the ocean:

Spectral interval (μm)	k_λ (m^{-1})	Spectral weight (%)
0.25–0.69	0.126	46
0.69–1.19	17	32.6
1.19–2.38	1,000	18.1
2.38–4.00	10,000	3.3

a) Calculate the radiative heating rate in the ocean at depths 1, 5, 10, 20, 50, and 100 m that is associated with solar radiation (using Beer's law).

b) What is the buoyancy frequency, N^2, after 6 hours of solar heating, assuming that no further mixing takes place? Assume that the net heating of the ocean mixed layer is determined only by the solar radiation flux, and that the temperature of the mixed layer is 30°C and its depth is 100 m.

3. In the marginal ice zone of the North Atlantic, compare the contributions of mechanical mixing energy and the buoyant mixing to the ocean mixed layer entrainment heat flux, F_{Q0}^{ent}. What percent error in the ocean mixed layer entrainment heat flux would be made by incuding only the contribution from mechanical mixing energy? Assume the following values: $u_* = 0.017$ m s^{-1}; $\rho = 1028$ kg m^{-3}, $h_m = 75$ m; $dh_i/dt = 0.1$ m day^{-1}; $(\dot{P} - \dot{E}/\rho) = -0.005$ m day^{-1}; $F_{Q0} = -500$ W m^{-2}; $s_m = 34.6$ psu; $T_m = -1.89$°C; $T_{m-} = 0$°C; $s_m = 34.6$ psu; $s_{m-} = 34.95$ psu; $\alpha = 0.025 \times 10^{-3}$ K^{-1}, $\beta = 0.79 \times 10^{-3}$ psu^{-1}; $c_p = 4186$ J kg^{-1} K^{-1}. A value $c_1 = 2$ is appropriate for a high-latitude deep mixed layer.

Chapter 12 | Global Energy and Entropy Balances

In earlier chapters, we have considered thermodynamic processes that are for the most part small scale, such as those associated with individual cloud drops, the local interface of the atmosphere with the ocean, small-scale mixing events in the ocean, and brine pockets in sea ice. This chapter examines planetary-scale energetics and entropy. Consideration of the overall energy and entropy of the planet provides insight into the maintenance and stability of the Earth's climate and is a starting point for understanding climate change. Such a consideration requires examination not only of internal processes in the Earth–atmosphere thermodynamic system, but also interactions with the system's environment, particulary with respect to the intensity of solar radiation and the geometrical relationship between the Earth and sun.

Latitudinal variation in the net radiation flux at the top of the atmosphere results in an overall heat transport from equatorial to polar regions. In effect, the atmosphere operates as a heat engine, whereby a portion of the absorbed radiation (heat source) is converted into kinetic energy (work). The efficiency of the atmospheric heat engine is low, because of strong irreversibilities in the system arising primarily from a highly irreversible heat transfer of solar radiation to the Earth. Finally, the global hydrological cycle modulates the Earth's energy and entropy budgets through radiative and latent heating.

12.1 Planetary Radiation Balance

When the Earth is in radiative equilibrium, the amount of solar radiation absorbed by the planet is equal to its emitted terrestrial radiation when averaged globally over an annual cycle. Because the atmosphere emits and absorbs radiation, the surface of the Earth is much warmer than it would be in the absence of an atmosphere.

The *luminosity* of the sun, $L_0 \approx 3.9 \times 10^{26}$ W, is the total rate at which energy is released by the sun. From this value, it can be estimated that the sun emits radiation at an equivalent black-body temperature of about 6000 K. Since space is essentially a vacuum and energy is conserved, the amount of radiation which reaches a planet is inversely proportional to the square of the distance between the planet and the sun (*inverse square law*). The distance between a planet and the sun is determined from the mean planet–sun distance and the eccentricity and obliquity of the orbital plane. The axial tilt, which is the angle between the axis of rotation and the normal to the plane of orbit, influences the seasonal and latitudinal variation of insolation. The solar constant, S, is defined as the amount of solar radiation received per unit time and per unit area, perpendicular to the sun's rays at the top of the atmosphere, at the mean Earth-sun distance. The solar constant can be evaluated from the inverse square law to be

$$S = \frac{L_0}{4\pi \overline{d}^{\,2}}$$

where $\overline{d}$ is the mean Earth–sun distance. The solar constant has been monitored by satellite and is found to be about 1370 ± 4 W m^{-2}.[1]

The amount of solar radiation received by the Earth and its atmosphere is equal to the solar constant minus the amount of shortwave radiation reflected to space, times the cross-sectional area of the Earth that is perpendicular to the beam of parallel solar radiation. Assuming that the terrestrial emission is equivalent to the black-body flux at temperature T_e^*, the amount of longwave radiation emitted by the Earth is the equivalent black-body flux times the surface area of the Earth. Hence we can write the following expression for the Earth's radiative energy balance under conditions of radiative equilibrium (Figure 12.1):

$$S\left(1 - \alpha_P\right)\pi r_P^2 = \sigma T_e^{*4} \, 4\pi r_P \tag{12.1a}$$

or

$$\frac{S}{4}\left(1 - \alpha_P\right) = \sigma T_e^{*4} \tag{12.1b}$$

where r_P is the radius of the solid Earth and α_P is the planetary albedo. The factor 4 in the denominator arises from the ratio of the surface area of a sphere to its cross-sectional area. The amount of solar radiation intercepted by the Earth is $S\pi r_P^2$. The globally averaged insolation at the top of the atmosphere is $S/4 = 342$ W m^{-2}.

Solving (12.1b) for the equivalent black-body temperature, T_e^*, we obtain

$$T_e^* = \sqrt[4]{\frac{S\left(1 - \alpha_P\right)}{4\sigma}} \tag{12.1c}$$

---

[1] Since the formation of the solar system, the solar luminosity (and hence S) has increased by 20–30%.

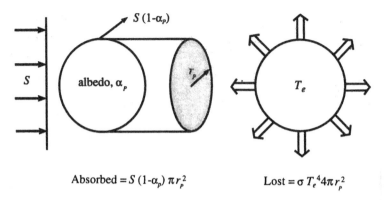

$$\text{Absorbed} = S\,(1\text{-}\alpha_p)\,\pi r_p^2 \qquad\qquad \text{Lost} = \sigma\,T_e^4 4\pi r_p^2$$

Figure 12.1 Radiative energy balance of the Earth. The Earth absorbs energy from the sun and loses energy through longwave emission.

The temperature T_e^* is not necessarily the actual surface or atmospheric temperature of the planet; it is simply the equivalent black-body emission temperature a planet requires to balance the solar radiation it absorbs. Using a value of $\alpha_P = 0.31$, we obtain from (12.1c) a value of $T_e^* = 254$ K. Note that this temperature is much less than the observed global mean surface temperature $T_0 = 288$ K. The difference between T_e^* and T_0 arises from the emission of thermal radiation by atmospheric gases and clouds at temperatures colder than T_0.

Figure 12.2 depicts the Earth's globally and annually averaged radiation and energy balance. The net incoming solar radiation of 342 W m^{-2} is partially reflected by the surface, clouds, aerosols, and atmospheric gases (107 W m^{-2}), with 67 W m^{-2} (20%) absorbed in the atmosphere and 168 W m^{-2} (49%) absorbed by the surface. The thermal radiation leaving the atmosphere originates from both the Earth's surface and from the atmosphere. The internal energy exchanges of longwave radiation between the surface and atmosphere have a larger magnitude even than the insolation at the top of the atmosphere. These large magnitudes indicate the significance of the so-called *greenhouse effect* of the Earth's atmosphere. The term greenhouse is used as a loose analogy since most of the solar radiation passes through the atmosphere unimpeded, while thermal radiation is emitted by the atmosphere at temperatures colder than the Earth's surface. Water vapor and clouds provide about 80% of the greenhouse effect, with minor contributions coming from carbon dioxide, ozone, and several other trace gases. Figure 12.2 shows that the atmosphere is subject to a net cooling of about 90 W m^{-2}; this is balanced by latent and sensible heat transfer from the surface into the atmosphere.

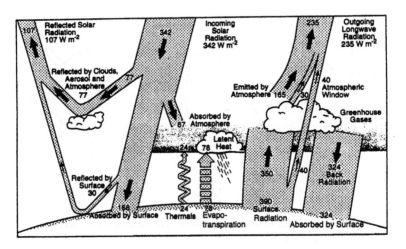

Figure 12.2 Estimated annual mean global energy balance for the Earth. Units are W m⁻²
(Kiehl and Trenberth, 1997).

The solar flux at the top of the atmosphere, F_{TOA}^{SW}, is then given by

$$F_{TOA}^{SW} = S\left(\frac{\bar{d}}{d}\right)^2 \cos Z \qquad (12.2)$$

where Z is the solar zenith angle, d is the Earth–sun distance, and $\bar{d} = 150 \times 10^{11}$ m is
its average value.

Expansion of a Fourier series for the squared ratio of the mean Earth–sun distance
to the actual distance is given by

$$\left(\frac{\bar{d}}{d}\right)^2 = \sum_{n=0}^{2} a_n \cos\left(n\varphi_d\right) + b_n \sin\left(n\varphi_d\right) \qquad (12.3)$$

where

$$\varphi_d = \frac{2\pi d_n}{365}$$

and d_n is the day number (January 1 $\equiv$ 0; December 31 $\equiv$ 364). The Fourier coefficients in (12.3) are given by (Spencer, 1971):

n	a_n	b_n
0	0.006918	0
1	-0.399912	0.070257
2	-0.006758	0.000907
3	-0.002697	0.001480

The solar zenith angle, Z, is defined as the angle between the vertical direction and the direction of the incoming solar beam (see Section 3.3) and is given by

$$\cos Z = \sin \phi \sin \delta + \cos \phi \cos \delta \cos \psi \qquad (12.4)$$

where ϕ is the latitude (ϕ is negative in the southern hemisphere), δ is the solar declination angle, and ψ is the hour angle. The *hour angle* is zero at solar noon and increases by 15° for every hour before or after solar noon. The *solar declination angle* is a function only of the day of the year and is independent of location. It varies from 23°45' on June 21 to –23°45' on December 21, and is zero on the equinoxes. The declination angle can be approximated by

$$\delta = \sum_{n=0}^{3} a_n \cos\left(n\varphi_d\right) + b_n \sin\left(n\varphi_d\right)$$

where the coefficients are given by (Spencer, 1971):

n	a_n	b_n
0	0.006918	0
1	-0.399912	0.070257
2	-0.006758	0.000907
3	-0.002697	0.001480

The daily average insolation calculated using (12.1) is shown in Figure 12.3. The annual average radiation recieved at the poles is approximately half that received at the equator. A very small annual cycle is seen at the equator, with a slight semi-annual oscillation having maxima at the equinoxes and minima at the solstices. At the pole, direct sunlight is absent for exactly half the year. However, near the summer solstice the daily amount of radiation received at the pole exceeds that received at the equator because the sun is above the horizon for 24 hours per day at the pole. A slight asymmetry in insolation between the hemispheres is seen, arising from the fact that the Earth is closer to the sun during the northern hemisphere winter because of the eccentricity of its orbit.

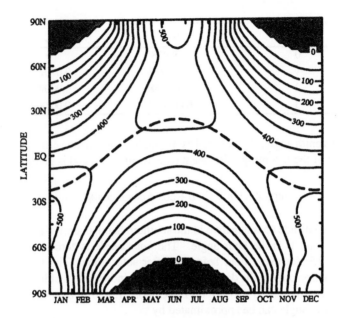

Figure 12.3 Daily average insolation at the top of the atmosphere as a function of latitude and season. Units are W m⁻². Dashed line indicates the latitude at which the sun is directly overhead. (From Hartmann, 1994.)

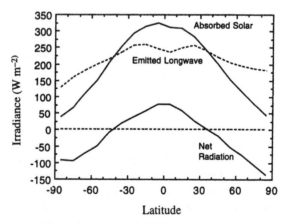

Figure 12.4 Annual mean absorbed shortwave, outgoing longwave, and net radiation averaged around latitude circles. (From Hartmann, 1994.)

The latitudinal variation of net shortwave and longwave radiation at the top of the atmosphere is shown in Figure 12.4. In spite of significant hemispheric differences in surface albedos that arise from the larger land masses in the northern hemisphere, the TOA radiation in both hemispheres are nearly identical. The hemispheric symmetry of the net TOA radiation arises since hemispheric differences in cloud properties cancel the differences in surface albedos.

12.2 Global Heat Engine

Figure 12.4 shows that the annual mean net radiation is positive equatorward of 40° latitude and negative at higher latitudes. Since polar temperatures are not observed to cool and tropical temperatures are not observed to warm on average, a transport of heat from equatorial to polar regions must occur. This transport occurs via fluid motions in the atmosphere and ocean that are driven by horizontal pressure gradients generated by the uneven heating.

An estimate of the total annual mean meridional energy transport required to equalize the pole–equator radiative imbalance is given in Figure 12.5. While the total energy transport required is easily determined by the energy balance at the top of the atmosphere, partitioning the transport between the atmosphere and ocean is more difficult. One method is to evaluate the transport from meteorological analyses of temperature, humidity and wind velocity, and then determine the oceanic transport as a residual between the total and atmospheric energy transports. An alternative method to

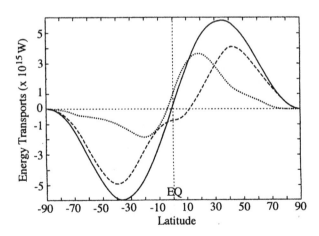

Figure 12.5 Annual mean northward energy transports required to equalize the pole–equator radiative imbalance. The solid line represents the top-of-the-atmosphere radiation budget, the dashed line represents the atmosphere, and the dotted line represents the ocean (From Zhang and Rossow, 1997).

infer the partition of energy transport is from the difference in the energy balance at the top of the atmosphere and at the atmosphere–ocean interface (9.1).

Since the nearly symmetric hemispheric TOA radiation balances are produced by offsetting asymmetries of surface and cloud properties, the partitioning between oceanic and atmospheric energy transport must differ between the two hemispheres. Poleward atmospheric transports are larger in the Southern Hemisphere than in the Northern Hemisphere, but oceanic transport is larger in the Northern Hemisphere than in the Southern Hemisphere. Because of near hemispheric balance in TOA net radiative flux, there is little net cross-equatorial transport of heat. However, because of hemispheric asymmetries in partitioning between oceanic and atmospheric transport, there is a northward cross-equatorial transport in the ocean and a southward cross-equatorial transport in the atmosphere. Figure 12.5 shows that oceanic transport peaks near 20° in both hemispheres, and atmospheric transport dominates at higher latitudes.

To understand the partitioning of energy transport in the atmosphere and ocean, we must consider interconversions among kinetic energy, potential energy, and internal energy. When radiation is absorbed at the Earth's surface or in the atmosphere, it appears as internal energy. The internal energy per unit horizontal area of a layer of air with unit mass is $u = c_v T$. The internal energy for a unit area column of the atmosphere, which extends from the Earth's surface to the top of the atmosphere, is given by $E_{\mathscr{I}}$:

$$E_{\mathscr{I}} = \int_0^\infty \rho u \, dz = c_v \int_0^\infty \rho T \, dz = \frac{c_v}{g} \int_0^{p_0} T \, dp \qquad (12.5)$$

Within the Earth–atmosphere system, internal energy may be transferred from one location to another and in particular between the atmosphere and the underlying surface through radiation and conduction, but the net heating of the system through these processes is zero. Latent heat is part of the internal energy (see (6.5)), but it is useful here to consider latent heat separately from the internal energy.

One of the basic problems of atmospheric science is to determine how the internal energy generated by the sun is converted into potential and kinetic energy. The *potential energy*, gz, per unit mass for a unit area column of the atmosphere is given by $E_{\mathscr{P}}$:

$$E_{\mathscr{P}} = \int_0^\infty \rho g z \, dz = \int_0^{p_0} z \, dp = R_d \int_0^\infty \rho T \, dz = \frac{R_d}{c_v} E_{\mathscr{I}} \qquad (12.6)$$

which shows that the potential and internal energy in the atmosphere are not independent forms of energy. Hence it is convenient to consider the *total potential energy*, $E_T = E_{\mathscr{I}} + E_{\mathscr{P}}$. The conversion of internal energy into kinetic energy occurs reversibly via the pressure gradient force. The corresponding kinetic energy, E_K, is given by

$$E_K = \frac{1}{2} \int_0^\infty \rho u^2 \, dz = \frac{1}{g} \int_0^{p_0} u^2 \, dp \qquad (12.7)$$

where u is the wind velocity. Only those processes involving a force can produce or destroy kinetic energy. Upward or downward motion of the atmosphere (or the ocean) with or against the force of gravity converts potential energy into kinetic energy, or vice versa, reversibly and adiabatically. Motion of the atmosphere against the force of friction, and the frictional heating which accompanies it, converts kinetic energy into internal energy.

In effect, the atmosphere operates as a heat engine, whereby a portion of the absorbed radiation (heat source) is converted into kinetic energy (work). In the atmospheric heat engine, heat flows on the whole from the warm sources to the cold sinks. The work performed by the atmospheric heat engine maintains the kinetic energy of the circulation against a continuous drain of energy by frictional dissipation. The frictional heating produces the necessary increase in entropy. The momentum flux associated with the surface frictional dissipation acts to drive oceanic motions.

The strength of the thermal circulation depends on the efficiency of the heat engine. For a reversible Carnot engine, we have from (2.30b)

$$\mathscr{E} = 1 - \frac{T_2}{T_1} = \frac{T_1 - T_2}{T_1}$$

If we identify $T_1 = 300$ K with the tropical surface heat source and $T_2 = 200$ K with the high-latitude upper atmosphere cold sink, we obtain $\mathscr{E} = 33\%$. Since the Earth's climate system is irreversible, the actual efficiency of the Earth as a heat engine is much smaller. A more meaningful estimate of the efficiency can be determined from (2.30a):

$$\mathscr{E} = \frac{w}{q_1}$$

The heating term is the mean incoming solar radiation, $q_1 = (1 - \alpha_p)S/4 = 238$ W m^{-2}. To estimate the work term, w, it is assumed that the production of kinetic energy is balanced by frictional dissipation, maintaining the average kinetic energy of the atmosphere. This term has been estimated by Oort and Peixoto (1983) to be $w = 2$ W m^{-2}, yielding an efficiency of $\mathscr{E} \sim 0.8\%$.

To understand the reason for the low efficiency of the atmospheric heat engine, we must examine why only a fraction of the total potential energy is available to be converted into kinetic energy, whereas most of the total potential energy is unusable.

Consider an atmosphere that is hydrostatic with no horizontal temperature gradients on isobaric surfaces (a *barotropic atmosphere*). The total potential energy of such an atmosphere can be calculated by summing (12.5) and (12.6). Although the total potential energy of a barotropic atmosphere can be very large, there are no mechanisms for generating kinetic energy and the atmosphere will remain at rest.

Suppose an initially barotropic atmosphere is heated at low latitudes and cooled at high latitudes in a manner such that there is no net heating over the globe. In a hydrostatic atmosphere, the thickness of a layer between isobaric surfaces (1.45) increases at low latitudes and decreases at high latitudes, tilting the isobaric surfaces. This produces a nonuniform distribution of density and temperature on isobaric surfaces (a *baroclinic atmosphere*), and hence a horizontal pressure gradient that results in potential energy being available for conversion to kinetic energy.

This process is illustrated in Figure 12.6 by two immiscible fluids of different densities that are adjacent to each other. Assuming that both fluids are in hydrostatic equilibrium, a pressure gradient force is directed from the heavier fluid to the lighter one, causing the heavier fluid to accelerate towards the lighter one. The ensuing motion will result in the heavier fluid lying beneath the lighter one. Through the sinking of denser fluid and the rising of the lighter fluid, the center of gravity of the system is lowered and potential energy is converted into kinetic energy of fluid motions.

The *available potential energy* is defined relative to an ideal reference state of the atmosphere with minimum total potential energy and maximum entropy. The reference state is barotropic, hydrostatic, and statically stable. If mass is conserved and redistribution of the mass is allowed to occur isentropically, then the sum of internal, potential, and kinetic energies is invariant. The available potential energy, Λ, is therefore defined by

$$\Lambda = \int \left(E_g + E_p\right) dm - \int \left(E_g + E_p\right)_r dm \tag{12.8}$$

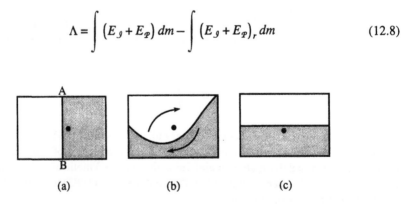

(a)   (b)   (c)

Figure 12.6 (a) Heavier (shaded) and lighter fluids separated by a movable partition, AB. The dot represents the center of gravity. (b) Fluids in motion following the removal of the partition. (c) Equilibrium configuration of the fluids after the motion has dissipated.

where the subscript r indicates the reference state and m is the mass. This expression represents the maximum possible amount of total potential energy that can be converted into kinetic energy. In the long term, the available potential energy removed by conversion to kinetic energy must be replaced by heating if the climate system is to remain in equilibrium.

If the Earth were not rotating, the atmospheric transport of heat from pole to equator would occur as a direct thermal circulation: heating at the surface in the equatorial regions causes rising motion → heat is transported polewards at upper levels → sinking occurs over the polar regions → the circulation is completed by a low-level return flow of cold air from high to low latitudes. The actual mean equator-to-pole transport of heat in the atmosphere is complicated considerably by the Earth's rotation, angular momentum considerations and subsequent hydrodynamical instabilities, especially poleward of the subtropics. The large-scale eddies (e.g., storms) produced in midlatitudes rapidly transfer heat poleward to satisfy the global energy balance.

In addition to the global meridional transfer of heat from low to high latitudes, heat transfer occurs on large horizontal scales, primarily in response to turbulent heat fluxes into the atmosphere arising from surface temperature gradients arising from the geographical distribution of continents. The *Walker Circulation* (Figure 12.7) is generally symmetric about the equator with ascending motion in the warm pool regions of the Indian and Pacific Oceans and the Indonesian rchipelago, and descent in the western Indian Ocean and the eastern Pacific Ocean. Weakening or reversal of the Walker circulation, where there is rising motion in the eastern Pacific and sinking motion in the western Pacific, occurs several times in a decade and is referred to as *El Niño*. The Asian–Australian *monsoon* (Figure 12.8) is a global circulation pattern which is asymmetric about the equator and has its focus and basic forcing in the land/ocean distribution of the Eastern Hemisphere. If there were no tropical continents and the ocean continued completely around the equator, a broad maximum sea surface temperature would span the equator at low latitudes with the location of the warmest sea

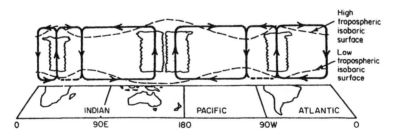

Figure 12.7 Schematic view of the east–west Walker circulation along the equator, indicating low-level convergence in regions of convection where mean upward motion occurs. (From Webster, 1987. © John Wiley & Sons, Inc. Reprinted with permission.)

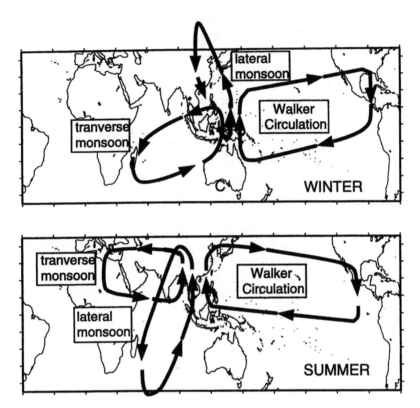

Figure 12.8 The Asian–Australian monsoon circulation pattern. The somewhat complicated wind circulation patterns and the seasonal differences in the patterns result from the variations in the heating of the continents. (From Webster *et al.*, 1998.)

surface temperature following the location of maximum solar insolation. The atmospheric cross-equatorial flow would follow the annual migration of the maximum sea surface temperature about the equator. However, because of continents and the alterations of atmospheric heating associated with them, the general wind circulations are complicated considerably and seasonal monsoon circulations develop.

12.3 Entropy and Climate

The Earth–atmosphere system is able to perform mechanical work (e.g., when generating wind) by virtue of zonal differences in heat input and output at the top of the atmosphere, which was used in the previous section to estimate the efficiency of the atmospheric heat engine. However, the observed temperature distribution in the Earth and its atmosphere is not determined solely by the astronomical constellation of Earth and sun, but rather from the extremely complex response of a geophysical fluid system to the external heating. The magnitude and zonal distribution of the heating at the top of the atmosphere is not sufficient to explain the temperature distribution of the Earth-atmosphere system. The low efficiency of the atmosphere as a heat engine implies the action of strong irreversible processes. Since the temperature of the Earth–atmosphere system is considerably lower than the sun, which is the source of the solar radiation, there is a highly irreversible heat transfer from the sun to the Earth. Scattering of radiation is an additional source of irreversibility.

For a thermodynamic system that is not isolated (i.e., a system that is allowed to exchange energy with its environment), the entropy production defines the variation of entropy $d\eta$ that results from entropy flowing into the system ($d_{ext}\eta$) and entropy that is produced by irreversible processes inside the system ($d_{int}\eta$),

$$\frac{d\eta}{dt} = \frac{d_{ext}\eta}{dt} + \frac{d_{int}\eta}{dt} \tag{12.9}$$

where

$$\frac{d_{ext}\eta}{dt} = -\int_A \boldsymbol{J} \cdot \boldsymbol{n} \, dA \tag{12.10}$$

$\boldsymbol{J}$ is the total flow of entropy per unit area and unit time through the top of the atmosphere and $\boldsymbol{n}$ is the unit normal to the top of the atmosphere, defined as positive going out of the system. According to the second law of thermodynamics, $d_{int}\eta > 0$ for irreversible internal processes. If the variation of η is small over some time period, we can write

$$\frac{d_{int}\eta}{dt} = \int_V \boldsymbol{J} \cdot \boldsymbol{n} \, dA \tag{12.11}$$

Hence, an estimate of the entropy flow through the top of the atmosphere that is associated with radiation provides a constraint on the production of entropy by irreversible processes inside the system.

The total flow of entropy at the top of the atmosphere is determined by the radiative transfer (Figure 12.9). Although the net incoming and outgoing radiation at the top of the atmosphere are equal when averaged globally and over an annual cycle (12.1), the net incoming and outgoing radiation entropies are never equal to each other. The solar radiation brings in a small amount of entropy in comparison with the entropy that longwave radiation removes from the system.

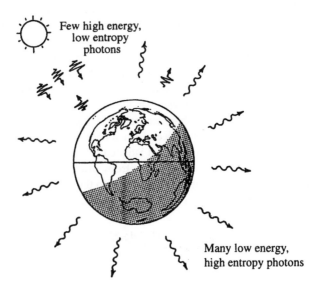

Figure 12.9 Radiative transfer determines the flow of entropy at the top of the atmosphere. Outgoing longwave radiation removes entropy from the system, while incoming solar radiation adds entropy. (After Stephens and O'Brien, 1993.)

To examine the entropy of radiation, consider Planck's radiation law (3.19)

$$I_\lambda^* = \frac{2hc^2}{\lambda^5 \left[\exp\left(\dfrac{hc}{k\lambda T} \right) - 1 \right]} \tag{12.12}$$

written in terms of the radiance, I_λ^*, by using (3.13). *Frequency*, ν, is defined as

$$\nu = \frac{c}{\lambda} \tag{12.13}$$

where c is the speed of light. The Planck radiation law may be written in terms of frequency in such a way that the energy integrated over the same spectral domain is equivalent. Thus

$$I_\lambda^* d\lambda = -I_\nu^* d\nu$$

and it follows that

$$I_\nu^* = \frac{2h\nu^3}{c^2 \left[\exp\left(\frac{h\nu}{kT}\right) - 1\right]}$$

(12.14)

Associated with each monochromatic beam of radiation is the monochromatic brightness temperature, T_ν^*, which is determined from (12.14) to be

$$T_\nu^* = \frac{h\nu}{k} \left[\ln\left(\frac{h\nu^3}{c^2 I_\nu} + 1\right)\right]^{-1}$$

(12.15)

Since T_ν^* is determined in the context of the radiance, this temperature is also a function of the direction of the flow or radiation. For black-body radiation, T_ν^* is independent of both frequency and direction, and T_ν^* is equal to the thermal temperature T of the system.

From the definition of entropy (2.25a), we can write an expression for the monochromatic radiant entropy, η_ν^*, as

$$d\eta_\nu^* = \frac{1}{T_\nu^*} dI_\nu$$

(12.16)

In a certain frequency interval, η_ν^* represents the amount radiant entropy transported across an element of area in directions confined to an element of solid angle during a certain time interval. From (12.16) and (12.14), we can write

$$\eta_\nu^* = \frac{2k\nu^2}{c^2}\left[(1+y)\ln(1+y) - y\ln y\right] = \frac{I_\nu^*}{T_\nu^*} + \frac{2k\nu^2}{c^2}\ln(1+y)$$

(12.17)

where $y = c^2 I_\nu^*/2h\nu^3$. Hence, η_ν^* can be calculated given I_ν^*. For a given amount of energy, high-frequency radiation is associated with a lower amount of radiant entropy than is low-frequency radiation. This can be interpreted in the context of the Planck equation for the energy of a photon, E_o:

$$E_o = h\nu$$

The same amount of energy contains fewer photons in the form of solar radiation than in the form of terrestrial radiation, and hence the solar radiation is associated with a lower radiant entropy than is the terrestrial radiation.

The entropy flux density, J, from (12.11) and the spectral entropy flux density, J_ν, are defined as

$$J = \int J_\nu \, dv \qquad (12.18)$$

and

$$J_\nu = \int_{2\pi} \eta_\nu(\theta) \, \theta \cdot n \, d\Omega(\theta) \qquad (12.19)$$

where $d\Omega$ is an element of solid angle centered on the direction of the radiation, θ, and n is the normal to the surface in question.

Substitution of (12.17) into (12.19) leads to the definition of the black-body entropy flux density

$$J = \frac{4}{3} \sigma T^3 \qquad (12.20)$$

The broad-band entropy flux density received on a horizontal surface from a distant black-body sun at temperature T_{sun} therefore follows as

$$J_0 = \frac{4}{3} \sigma T_{sun}^3 \cos Z \frac{\Omega_0}{\pi} \qquad (12.21)$$

where Z is the solar zenith angle and Ω_0 is the the solid angle subtended by the sun at a point on the planetary surface ($\Omega_0 = 67.7 \times 10^{-6}$ sr for the Earth). Diffuse reflection of solar radiation by the Earth and its atmosphere is written (following Stephens and O'Brien, 1993) as

$$J = \frac{4}{3} \sigma T_{sun}^3 \chi \qquad (12.22)$$

where

$$\chi = \alpha_p \cos Z \frac{\Omega_0}{\pi} \left[-0.2777 \log \left(\alpha_p \cos Z \frac{\Omega_0}{\pi} \right) + 0.9652 \right]$$

The processes of scattering and diffuse reflection increases entropy.

The planetary entropy balance, η_p, can therefore be written as

$$\eta_P = \frac{4}{3} \sigma T_{sun}^3 \chi - J_o + \frac{4}{3} \sigma T_e^{*3} \qquad (12.23)$$

where the planetary temperature T_e^* is determined from the emitted longwave flux at the top of the atmosphere. Evaluations of the terms in (12.23) by Stephens and O'Brien (1993) show that the reflected shortwave entropy flux density exceeds the incoming shortwave entropy flux density, for a net shortwave entropy flux density of 0.02 W m^{-2} K^{-1}. The entropy flux density for the terrestrial radiation is evaluated to be 1.23 W m^{-2} K^{-1}, for a total radiation entropy flux density of 1.25 W m^{-2} K^{-1}.

Radiation from a hot emission temperature (e.g., the sun) and subsequent scattering is associated with small entropy relative to radiation from a cool emission temperature (e.g., the Earth). In fact, the entropy exported by the outlong longwave radiation is found to be about two orders of magnitude larger than the entropy associated with the net solar radiation. This negative entropy stream at the top of the atmosphere allows internal production of entropy by the Earth while at the same time maintaining order in the atmosphere. If the overall entropy of the atmosphere were increasing, the atmosphere would approach a state of maximum entropy, leading to a uniformity of the climate.

12.4 Global Hydrological Cycle

Transport of water in the Earth–atmosphere system and changes of phase modulate substantially the global energy and entropy balances. Figure 12.10 shows the main reservoirs of water in the Earth system and the fluxes of water between these reservoirs. The ocean contains about 96% by mass of the Earth's total water content. The atmosphere contains only about 0.001% of the total water on Earth, mainly in vapor form. The water on the continents is distributed among several reservoirs, including glaciers, ground water, lakes and rivers, and living matter.

The continual movement of water among the reservoirs of ocean, atmosphere, and land is called the *hydrological cycle*. The total amount of water on Earth remains effectively constant on time scales of thousands of years, but it changes state between its liquid, solid, and gaseous forms as it moves through the hydrological cycle. The sun provides the energy necessary to evaporate water from the surface, which condenses in ascending moist air. Water vapor and condensed water can be transported horizontally in the atmosphere for great distances by the winds, where it may eventually precipitate out of the atmosphere and onto the ocean or land. This horizontal movement of water vapor is critical to the water balance of land areas, since about one-third of the precipitation that falls on the land areas of Earth is water that was evaporated from ocean areas and then transported to the land in the atmosphere (Figure 12.10). The precipitated water then returns partly to the atmosphere through evaporation, infiltrates into the ground, or runs off to the rivers which return the water to the ocean.

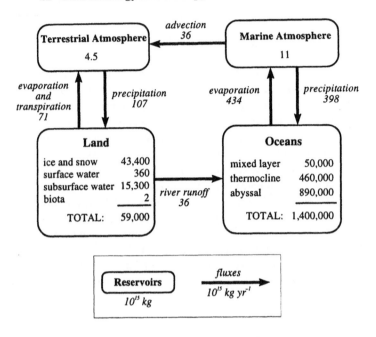

Figure 12.10 The global water cycle and the reservoirs of water. Arrows indicate fluxes of water from one reservoir to another. Note that water in the atmosphere over the ocean and the land accounts for only 0.0001% of the total water in the system. Units are in 10^{15} kg. (Data from Chahine, 1992.)

   The data in Figure 12.10 allow the residence times of water in the various reservoirs to be calculated from the ratio of the amount of water in a particular reservoir to the accumulation or depletion rate. The atmospheric residence time (mass of water vapor divided by total precipitation) is about 10 days; that is, the atmosphere recycles its water over 30 times per year. Surface water over land has a residence time of about 5 years, although the residence time for soil moisture is about 1 year and the total residence time for glaciers is 6000 years. The residence time of water in the oceans is about 3000 years, although not all parts of the ocean recycle at the same rate. In the ocean surface layers the time scales may be on the order of days to weeks, while deep bottom water may take thousands of years to recycle.

   An additional recycling rate can be calculated that refers to how much moisture that precipitates out comes from horizontal tranpsort into a domain versus local evaporation. It is seen in Table 12.1 that the results depend greatly on the scale of the domain under consideration. However even for 1000 km scales, less than 20% of the annual precipitation typically originates within the domain. This highlights the mismatches in the local rates of rainfall versus evaporation, which imply that precipitating systems of all kinds feed mostly on atmospheric water vapor that is advected.

Table 12.1 Evaluation of the recycling rate (%), which is the amount of precipitation that comes from local evaporation versus horizontal transport, evaluated for scales of 500 and 1000 km. (Data from Trenberth, 1998.)

	scale (km)	
	500	1000
Global	9.6	16.8
Land	8.9	15.4
Ocean	9.9	17.3

The net excess of precipitation compared to evaporation over the continents must be maintained by a net flux of atmospheric water from the large oceanic source regions. In addition, global meridional transport of atmospheric water vapor also occurs. Examination of Figure 9.5 shows that while precipitation is nearly equal in the Northern and Southern Hemispheres, there is significantly more evaporation in the Southern Hemisphere because of of the smaller land mass. As a result, the Southern Hemisphere supplies a considerable amount of water vapor to the Northern Hemisphere. While the atmospheric transport of water substance balances the local inequalities that arise in precipitation and evaporation, local water imbalances do not drive the advection; it is the thermal imbalances (Section 12.2) that drive the advection. The transport of latent energy in the atmosphere is for practical puposes proportional to the transport of water vapor, which balances the excess of evaporation over precipitation.

Notes

General references for this chapter are *Global Physical Climatology* (1994; chapters 2,5, and 6) by Hartmann and *Energy and Water Cycles in the Climate System* (1993) by Raschke and Jacob.

The entropy of radiation is described by Callies and Herbert (1988) and Stephens and O'Brien (1993).

Problems

1. Determine simple expressions for the solar zenith angle, Z, that do not involve trigonometric functions for the following conditions:
a) $\phi = 90°N$;

b) solar noon, where $\cos\psi = 1$.

Show that:

c) the half day length, H, can be written as $\cos H = -\tan\phi \tan\delta$;

d) the latitude of the polar night is equal to $90° - |\delta|$.

2. If all sources of energy were instantly turned off and the atmosphere were allowed to radiate to space as a black body with an effective temperature of 250 K, compute the rate at which its average temperature would begin to drop.

3. If the potential temperature θ in a column of the atmosphere is constant, evaluate the total potential energy in the column.

4. Consider a planet that is identical to Earth in all respects except for its albedo.

a) Determine the equivalent black-body emission temperature of this planet if $\alpha_P = 0$.

b) Determine the net entropy production of the planet with $\alpha_P = 0$, which is the maximum possible entropy production of the planet.

c) Determine the net entropy production for the planet with $\alpha_P = 1$, which is the lower limit to the entropy production of the planet.

d) Is the entropy of the actual planet Earth closer to the maximum or minimum possible entropy production?

Chapter 13 | Thermodynamic Feedbacks in the Climate System

The Earth's climate varies over many time scales, ranging from interannual and interdecadal variations to changes on geological time scales associated with ice ages and continental drift. The climate can vary as a result of alterations in the internal dynamics and exchanges of energy within the climate system, or from external forcing. Examples of external climate forcing include variations in the amount and latitudinal distribution of solar radiation at the top of the atmosphere, varying amounts of greenhouse gases (e.g., CO_2) caused by human activity, and variations in volcanic activity.

To understand and simulate climate and climate change, it is necessary to interpret the role of various physical processes in determining the magnitude of the climate response to a specific forcing. The large number of interrelated physical processes acting at different rates within and between the components of the climate system makes this interpretation a difficult task. An anomaly in one part of the system may set off a series of adjustments throughout the rest of the climate system, depending on the nature, location, and size of the initial disturbance.

The relationship between the magnitude of the climate forcing and the magnitude of the climate change response defines the climate sensitivity. A process that changes the sensitivity of the climate response is called a *feedback mechanism*. A feedback is positive if the process increases the magnitude of the response and negative if the feedback reduces the magnitude of the response. The concepts behind feedbacks as applied to climate change are derived from concepts in control theory that were first developed for electronics. By examining separate feedback loops, one can gain a sense of the direction of the influence of the feedback on a change in the state of the system, whether it is reinforcing or damping, and the relative importance of a given feedback when compared with other feedbacks. Climate change can therefore be viewed as the result of adjustment among compensating feedback processes, each of which behaves in a characteristically nonlinear fashion. The fact that the climate of the Earth has varied in the past between rather narrow limits despite large variations in external forcing is evidence for the efficiency and robustness of these feedbacks.

A variety of climate feedback mechanisms have been identified, including radiation feedbacks that involve water vapor and clouds, ocean feedbacks that involve the hydrological cycle, and biospheric feedbacks that involve the carbon cycle. In our

351

consideration of thermodynamic feedbacks in the climate system, we will concentrate on the radiative and ocean thermohaline feedbacks. The radiation feedback processes that are of interest in the context of the climate system include the snow/ice–albedo feedback, water vapor feedback, and cloud–radiation feedbacks.

13.1 Introduction to Feedback and Control Systems

A *control system* is an arrangement of physical components that are connected in such a manner as to regulate itself or another system. The *input* to a control system is the stimulus from an external energy source that produces a specified response from the control system. The *output* is the actual response obtained from the control system.

An *open-loop* control system is one in which the control action is independent of the output. A *closed-loop* control system is one in which the control action is somehow dependent on the output. Closed-loop control systems are more commonly called *feedback control systems*. Feedback is said to exist in a system when a closed sequence of cause-and-effect relations exists between system variables.

In order to solve a control systems problem, the specifications or description of the system configuration and its components must be put into a form amenable to analysis and evaluation. Three basic representations (models) are employed in the study of control systems:

1) block diagrams;
2) signal flow graphs;
3) differential equations and other mathematical relations.

A *block diagram* is a shorthand, graphical representation of cause-and-effect relationships between the input and output of a control system. It provides a convenient and useful method for characterizing the functional relationships among the various components of a control system. System components are also called *elements* of the system. The simplest form of the block diagram is the single block, with one input and one output:

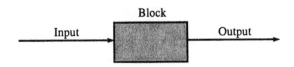

The *input* is the stimulus applied to a control system from an external energy source. The *output* is the actual response obtained from a control system, which may or may not be equal to the specified response implied by the input. The basic configuration of a simple closed-loop (feedback) control system is illustrated in Figure 13.1. In the

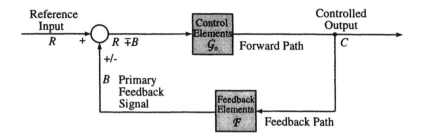

Figure 13.1 Basic block diagram including feedback elements. Negative feedback means that the summing point is a subtractor, i.e., $R - B$. Positive feedback means that the summing point is an adder, i.e., $R + B$.

absence of feedback, the term G_o represents the *gain* of the system, which is the ratio of output to input. The *feedback* $\mathcal{F}$ is the component required to establish the functional relationship between the primary feedback signal and the controlled output. It is emphasized that the arrows of the closed loop, connecting one block with another, represent the flow of control energy or information, and not the main source of energy for the system.

A *signal flow graph* displays pictorially the transmission of signals through the system, as does the block diagram; however, a signal flow graph is easier to construct than the block diagram. A signal flow graph is shown in Figure 13.2, corresponding to the block diagram shown in Figure 13.1. A *feedback loop* is a path which originates and terminates on the same node. For example, in Figure 13.2, C to E and back to C is a feedback loop.

Referring to Figures 13.1 and 13.2, we define the following parameters. The *feedback factor, f*, is defined as:

$$f = G_o \, \mathcal{F} \tag{13.1a}$$

The gain of the system in the presence of feedbacks is given by G_f, which is also referred to as the *control ratio*:

$$G_f = \frac{G_o}{1 - f} \tag{13.1b}$$

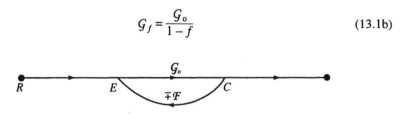

Figure 13.2 Signal flow graph for the system described in Figure 13.1

We define the *feedback gain ratio* as

$$\mathcal{R}_f = \frac{G_f}{G_o} = \frac{1}{1-f} \qquad (13.1c)$$

Mathematical models, in the form of system equations, are employed when detailed relationships are required. Every control system may be characterized theoretically by mathematical equations. Often, however, a solution is difficult if not impossible to find. In these cases, certain simplifying assumptions must be made in the mathematical description, typically leading to systems that can be described by linear ordinary differential equations.

Evaluations of feedbacks in the climate system are useful in the following contexts:

- conceptual understanding of how the climate system operates;
- evaluation of climate model performance;
- quantification of climate system response to different forcing and the role played by various physical processes.

Applications of the theory of feedback control systems to the Earth's climate are considered in the following sections.

13.2 Radiation Climate Sensitivity and Feedbacks

We may express a climate system in the following manner:

$$\mathcal{L} = \mathcal{L}(X) = \mathcal{L}(X_1, X_2, \dots)$$

where

$$X_1 = X_1(\mathcal{L}, X) = X_1(X_2, X_3, \dots)$$

In this nonlinear system, $\mathcal{L}$ describes a family of climate variables which depend on X_i. In a fully interactive (closed-loop) system, X_1 may also be a function of $\mathcal{L}$ and X_i. In an open-loop system, X_1 is a constant and therefore changes to the climate system, which are produced by a change in specification of X_1, are not allowed to feedback onto X_1. The *sensitivity*, δ, of the climate variable X_2 to a change in X_1 is defined as

$$\delta = \frac{X_1}{X_2} \frac{\partial X_2}{\partial X_1} \qquad (13.2)$$

An example of a tractable closed-loop climate system is a simple planetary energy balance model (as described in Section 12.1) that is used to examine radiative feedback processes in the Earth's climate. A planetary energy balance climate model predicts the change in temperature at the Earth's surface, ΔT_0, from the requirement that $\Delta F_{TOA}^{rad} = 0$, where F_{TOA}^{rad} is the net radiative flux at the top of the atmosphere from (12.1b):

$$F_{TOA}^{rad} = \frac{S}{4}\left(1 - \alpha_P\right) - F_{TOA}^{LW}$$

F_{TOA}^{LW} is the upwelling longwave radiation at the top of the atmosphere, frequently referred to as *outgoing longwave radiation* (OLR), and α_P is the planetary albedo. To assess changes in the net radiative flux at the top of the atmosphere, F_{TOA}^{rad} can be expressed as a symbolic function $\mathcal{L}$:

$$F_{TOA}^{rad} = \mathcal{L}\left(\mathcal{E}_i, T_0, I_j\right) \tag{13.3}$$

The term $\mathcal{E}_i$ represents quantities that can be regarded as external to the climate system, which are quantities whose change can lead to change in climate but are independent of the climate (e.g., volcanic eruptions, change in solar output). The term I_j represents quantities that are internal to the climate system, which are quantities that can change as the climate changes, and in so doing feed back to modify the climate change. The internal quantities include all of the variables of the climate system other than T_0. Because T_0 is the only dependent variable in the energy balance climate model, the internal quantities must be represented as a function of T_0.

A small change in the energy flux at the top of the atmosphere ΔF_{TOA}^{rad} can be expressed in terms of $\mathcal{E}_i$, T_0, and I_j as

$$\Delta F_{TOA}^{rad} = \sum_i \frac{\partial F_{TOA}^{rad}}{\partial \mathcal{E}_i} \Delta \mathcal{E}_i + \left(\frac{\partial F_{TOA}^{rad}}{\partial T_0} + \sum_j \frac{\partial F_{TOA}^{rad}}{\partial I_j} \frac{dI_j}{dT_0}\right) \Delta T_0 \tag{13.4}$$

By examining Figure 13.1 and (13.4), we can identify the following relationships:

$$\text{input:} \quad \frac{\partial F_{TOA}^{rad}}{\partial \mathcal{E}_i} \Delta \mathcal{E}_i \equiv \Delta Q \quad \frac{\partial F_{TOA}^{rad}}{\partial \mathcal{E}_i} \Delta \mathcal{E}_i \equiv \Delta Q \tag{13.5}$$

$$\text{output:} \quad \Delta T_0$$

$$G_0^{-1} = \frac{\partial F_{TOA}^{rad}}{\partial T_0} = 4\sigma T_0^3 \tag{13.6}$$

$$\mathcal{F} = \sum_j \frac{\partial F_{TOA}^{rad}}{\partial I_j} \frac{dI_j}{dT_0} \tag{13.7}$$

$$f = \mathcal{G}_0 \sum_j \frac{\partial F_{TOA}^{rad}}{\partial I_j} \frac{dI_j}{dT_0} \tag{13.8}$$

where we have defined a new variable Q in (13.5), which is referred to as the *external climate input*.

Incorporating the above relationships into (13.5), we obtain

$$\Delta F_{TOA}^{rad} = \Delta Q + \left(\mathcal{G}_0^{-1} - \mathcal{F} \right) \Delta T_0 = \Delta Q - \mathcal{G}_f^{-1} \Delta T_0 \tag{13.9}$$

Using the energy balance requirement that $\Delta F_{TOA}^{rad} = 0$, we obtain

$$\Delta T_0 = \mathcal{G}_f \Delta Q \tag{13.10a}$$

If $\mathcal{F} \neq 0$, the response of the surface temperature, ΔT_0, to the forcing, ΔQ, is modulated by feedback. If $\mathcal{F} = 0$, we have

$$\Delta T_0 = \left(\Delta T_0 \right)^{\wedge} = \mathcal{G}_0 \Delta Q \tag{13.10b}$$

where $\left(\Delta T_0 \right)^{\wedge}$ represents the zero-feedback temperature change.

From the definition of the feedback gain ratio, we can write

$$\mathcal{R}_f = \frac{\mathcal{G}_f}{\mathcal{G}_0} = \frac{1}{1-f} = \frac{\Delta T_0}{\left(\Delta T_0 \right)^{\wedge}} \tag{13.11}$$

Figure 13.3 plots the feedback gain ratio, $\mathcal{R}_f$, against the feedback factor, f. Values of $f < 0$ and $0 < \mathcal{R}_f < 1$ represents negative feedback. Values of $\mathcal{R}_f > 1$ for $0 < f < 1$ represents positive feedback.

This simple application of control theory has considered a linear climate system. The expression for the feedback parameter,

$$f = \mathcal{G}_0 \sum_j \frac{\partial F_{TOA}^{rad}}{\partial I_j} \frac{dI_j}{dT_0}$$

derived from this analysis assumes that the feedbacks are additive and thus independent.

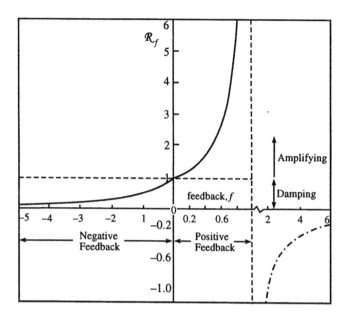

Figure 13.3 Feedback gain ratio versus feedback. (From Schlesinger, 1986.)

In principle, the contribution of each mechanism to the total feedback could be individually determined and ranked. In a nonlinear system, however, the feedbacks are not independent and addition of the individual terms will not give the true feedback of the nonlinear climate system. Applications of this type of linear feedback analysis have been made to the climate system, justified by considering only small perturbations to the radiative flux and surface temperature. Because of the difficulty of nonlinear control analysis, particularly for a system as complex as the climate system, no attempt has been made to apply control theory to the climate system beyond the type of linear analysis described above.

   In spite of its simplicity, the linear analysis described here can be used to provide useful insights about the climate system. Direct evaluation of the feedback gain ratio, $\mathcal{R}_f$, from (13.11) (and then f from (13.8)) using a numerical model can be done in the following way. Three different model simulations are required:

1. a baseline simulation, representing the current unperturbed climate conditions;
2. a simulation in which the climate is subject to an external perturbation and all feedbacks are operative;
3. a run in which the climate is subject to an external perturbation, and selected feedbacks are "turned off" (for example, the snow/ice–albedo feedback mechanism can be switched off by keeping the surface albedo fixed in the perturbed run to the same values used in the baseline simulation).

The feedback gain ratio is then evaluated from

$$\mathcal{R}_f = \frac{\Delta T_0}{\left(\Delta T_0\right)^\wedge} = \frac{\left(T_0\right)_2 - \left(T_0\right)_1}{\left(T_0\right)_3 - \left(T_0\right)_1} \tag{13.12}$$

where the numerical subscripts refer to the enumerated model simulations above and $(\Delta T_0)^\wedge$ represents the temperature change without feedback. This method of determining the feedbacks has the advantage that the feedbacks are not assumed *ab initio* to be additive.

An alternative application of the linear analysis is the evaluation of sensitivity. Examination of the feedback factor

$$f = \mathcal{G}_0 \sum_j \frac{\partial F_{TOA}^{rad}}{\partial I_i} \frac{dI_i}{dT_0}$$

shows that the following term can be identified with the sensitivity, δ (13.2)

$$\delta = \mathcal{G}_0 \frac{\partial F_{TOA}^{rad}}{\partial I_i} \tag{13.13}$$

The sensitivity of the radiative flux to changes in climate variables such as water vapor amount and cloud characteristics can then be determined from (13.13).

The individual feedback terms and net feedback determined from expressions like (13.12) must be interpreted with caution, because of the nonlinearity of the system and the associated feedbacks. However, the signs of the individual feedback terms determined from (13.12) will probably be correct in response to a small perturbation to the climate system. Additionally, the sensitivity terms themselves provide useful diagnostics of the climate system and numerical simulations. To simulate feedbacks correctly, it is necessary (but not sufficient) for a climate model to reproduce the observed derivatives.

The general approach here outlined for a planetary energy balance model can also be applied easily to determine feedbacks in other simple models such as a surface energy balance model or radiative–convective model.

13.3 Water Vapor Feedback

The feedback between surface temperature, water vapor, and the Earth's radiation balance is referred to as the *water vapor feedback*. The water vapor feedback may be written following (13.12) as

$$f = \mathcal{G}_0 \frac{\partial F^{rad}}{\partial \mathcal{W}_v} \frac{d\mathcal{W}_v}{dT_0} \tag{13.14}$$

where $\mathcal{W}_v$ is the vertically-integrated amount of water vapor (precipitable water; (4.41)).

Since water vapor emits strongly in the thermal (infrared) part of the spectrum, the net radiative flux at both the top of the atmosphere and at the surface increases as the amount of water vapor increases (positive $\partial F^{rad}/\partial \mathcal{W}_v$). The concentration of water vapor decreases approximately exponentially with height (see Figure 1.1) and is very small in the stratosphere. The outgoing longwave radiation flux (OLR) at the top of the atmosphere, F_{TOA}^{LW}, decreases with increasing water vapor amount, since water vapor in the atmosphere emits at a colder temperature than the surface. The downwelling longwave radiation flux at the surface increases with increasing water vapor amount.

Climate modeling results have shown that the water vapor path increases with increasing surface temperature ($d\mathcal{W}_v/dT_0 > 0$). This increase arises from increased evaporation from a warmer ocean surface, providing additional water vapor to the atmosphere. A consistent result of climate models has been that atmospheric relative humidity remains approximately constant in a perturbed climate and that the water vapor feedback is among the chief mechanisms amplifying the global climate response to a perturbation. Since condensation and precipitation are associated with important sources and sinks of water vapor (Section 8.6), the water vapor feedback simulated by climate models depends on the model's parameterizations of cloud, precipitation, and convective processes. Given the current deficiencies in climate model parameterization of these processes, the water vapor feedback determined by these models must be questioned.

A more thorough understanding of $d\mathcal{W}_v/dT_0$ and the water vapor feedback can be obtained by using (4.41) to expand the derivative $d\mathcal{W}_v/dT_0$ as[1]

$$\frac{d\mathcal{W}_v}{dT_0} = \frac{d}{dT_0} \frac{1}{g} \int_0^{p_0} w_v \, dp = \frac{1}{g} \int_0^{p_0} \frac{dw_v}{dT_0} \, dp \tag{13.15}$$

---

[1] Let

$$\phi(\alpha) = \int_{u_1}^{u_2} f(x,\alpha) \, dx \text{ where } u_1 \text{ and } u_2 \text{ may depend on the parameter } \alpha. \text{ Then}$$

$$\frac{d\phi}{d\alpha} = \int_{u_1}^{u_1} \frac{\partial f}{\partial \alpha} \, dx + f(u_2,\alpha)\frac{du_2}{d\alpha} - f(u_1,\alpha)\frac{du_1}{d\alpha}$$

Since u_1 and u_2 in the above are constants, the last two derivatives are zero.

Using the definition of the water vapor mixing ratio (4.36), the derivative dw_v/dT_0 can be written as

$$\frac{dw_v}{dT_0} = \frac{\varepsilon}{p}\left(e_s(p)\frac{\partial\mathcal{H}(p)}{\partial T_0} + \mathcal{H}(p)\frac{\partial e_s(p)}{\partial T_0}\right) \tag{13.16}$$

where $\mathcal{H}$ is the relative humidity. We can use the chain rule to expand further the derivatives in (13.16) to obtain

$$\frac{dw_v}{dT_0} = \frac{\varepsilon}{p}\frac{\partial T(p)}{\partial \Gamma}\frac{d\Gamma}{dT_0}\left(e_s(p)\frac{\partial\mathcal{H}(p)}{\partial T(p)} + \mathcal{H}(p)\frac{\partial e_s(p)}{\partial T(p)}\right) \tag{3.17}$$

where Γ is the atmospheric lapse rate (1.2). Using the Clausius–Clapeyron equation (4.19), we can write

$$\frac{dw_v}{dT_0} = w_s\frac{\partial T(p)}{\partial \Gamma}\frac{\partial \Gamma}{\partial T_0}\left(\frac{\partial\mathcal{H}(p)}{\partial T(p)} + \frac{\mathcal{H}(p)L_{lv}}{R_v T^2(p)}\right) \tag{13.18}$$

where w_s is the saturation mixing ratio (4.37). Incorporating (13.18) into (13.15), we obtain finally

$$f = \mathcal{G}_0\frac{\partial F^{rad}}{\partial W_v}\frac{1}{g}\int_0^{p_0}\left[w_s\frac{\partial T}{\partial \Gamma}\frac{d\Gamma}{dT_0}\left(\frac{\partial\mathcal{H}}{\partial T} + \frac{\mathcal{H}L_{lv}}{R_v T^2}\right)\right]dp \tag{13.19}$$

To illuminate the physics behind the water vapor feedback, consider the following simple example. If $\partial\mathcal{H}/\partial T_0 = 0$, we can simplify (13.19) to be

$$f = \mathcal{G}_0\frac{\partial F^{rad}}{\partial W_v}\frac{L_{lv}}{R_v g}\int_0^{p_0}\frac{\partial T}{\partial \Gamma}\frac{d\Gamma}{dT_0}\frac{w_v}{T^2}dp \tag{13.20}$$

where we have used $w_v = \mathcal{H}w_s$. If we assume a simple functional relationship for w_v, such as

$$w_v = w_{s0}\,\mathcal{H}_0\left(\frac{p}{p_0}\right) = \mathcal{H}_0\,\varepsilon\frac{e_s(T_0)}{p_0}\,\mathcal{H}_0\left(\frac{p}{p_0}\right) \approx \frac{p}{p_0^2}\,a\,\exp\left(-b/T_0\right)\mathcal{H}_0^2$$

we obtain

$$W_v = \frac{1}{2g} \mathcal{H}_0^2 \, a \, \exp\left(-b/T_0\right)$$

where the constants a and b are easily determined from (4.31) and $\mathcal{H}_0$ is the surface air humidity. We can then write (13.20) as

$$f = \mathcal{G}_0 \frac{\partial F^{rad}}{\partial W_v} \frac{L_{lv}}{R_v} \frac{2}{p_0^2} W_v \int_0^{p_0} \frac{\partial T}{\partial \Gamma} \frac{d\Gamma}{dT_0} \frac{p}{T^2} \, dp \qquad (13.21)$$

If we further assume that the lapse rate is constant with height in the atmosphere, we can write from (1.48)

$$T = T_0 \left(\frac{p}{p_0}\right)^{R_v \Gamma/g}$$

and

$$\frac{dT}{d\Gamma} = \frac{dT_0}{d\Gamma} \left(\frac{p}{p_0}\right)^{R_v \Gamma/g} + T_0 \left(\frac{p}{p_0}\right)^{R_v \Gamma/g} \frac{R_v}{g} \ln \frac{p}{p_0}$$

We can then evaluate the integral in (13.21) to obtain

$$f = \mathcal{G}_0 \frac{\partial F^{rad}}{\partial W_v} \frac{L_{lv}}{R_v} 2 W_v \left(\frac{1}{T_0^2 \left(2 - \frac{R_v \Gamma}{g}\right)} + \frac{R_v}{gT_0} \frac{1}{\left(2 - \frac{R_v \Gamma}{g}\right)^2} \frac{d\Gamma}{dT_0} \right) \qquad (13.22)$$

If the lapse rate remains constant, $d\Gamma/dT_0 = 0$ and we can write

$$f = \mathcal{G}_0 \frac{\partial F^{rad}}{\partial W_v} \frac{L_{lv}}{R_v} \frac{2 W_v}{T_0^2 \left(2 - \frac{R_v \Gamma}{g}\right)} \qquad (13.23)$$

which is a positive quantity since $\partial F^{rad}/\partial W_v > 0$ and $(R_v \Gamma/g) < 2$ for all values of Γ. From (13.23), we can estimate that

$$\frac{1}{\mathcal{W}_v}\frac{\partial \mathcal{W}_v}{\partial T_0} = \frac{L_{lv}}{R_v}\frac{2}{T_0^2\left(2 - \frac{R_v \Gamma}{g}\right)} \approx 7.7\%$$

The water vapor path increases with temperature by a fractional rate of about 7.7% K^{-1} under standard atmospheric conditions. The rationale behind the simple expression (13.23) has dominated the thinking on water vapor feedback, whereby it is determined primarily by increased evaporation from the ocean surface according to the Clausius–Clapeyron relationship. However, some recent research has brought into question this simplified view of the water vapor feedback.

Variations of the lapse rate associated with a change in surface temperature, $d\Gamma/dT_0$, can alter the sign of the water vapor feedback. In the highly convective tropics, $d\mathcal{W}_v/dT_0$ is dominated by $d\Gamma/dT_0$, which is negative. Hence the lapse rate variation diminishes the water vapor feedback in the tropics. At higher latitudes, the mid- and upper-troposphere has been simulated in climate models to warm less rapidly than the surface so that the lapse rate increases, and hence $d\Gamma/dT_0 > 0$. Therefore, the lapse rate variation acts to enhance the water vapor feedback in higher latitudes.

The net radiative flux is sensitive to vertical variations in the distribution of water vapor, even if there is no net change to $\mathcal{W}_v$. The net radiative flux is more sensitive to changes in water vapor amount in regions of the atmosphere where the water vapor mixing ratio is low, such as the upper troposphere (globally) and the lower troposphere in the polar regions. The relatively high sensitivity arises primarily from the infrared radiative transfer in the water vapor rotation band, at around 20 μm. Since low values of water vapor mixing ratio occur typically at low temperatures, the wavelength of maximum emission (3.21) occurs at longer wavelengths, and radiative transfer in the water vapor rotation band becomes increasingly important. Therefore the mechanisms controlling the humidity in these dry zones, particularly those that occur at cold temperatures, are of particular importance for understanding the water vapor feedback.

Two of these dry zones have been studied in detail, the upper tropical troposphere and the lower polar troposphere. In applying the type of feedback analysis described above, it must be remembered that this model represents the global energy budget, and care must be taken in using such a model to characterize regional responses. In applying expressions such as (13.20) and (13.23) to the examination of regional feedback processes, variations in large-scale dynamical transport may dominate the local thermodynamic processes in determining the sign and magnitude of the feedback.

13.3.1 Tropical upper troposphere

In the tropics, outgoing longwave radiation (OLR) is very sensitive to changes in upper tropospheric water vapor content. The lower tropospheric moisture content is very high, with a corresponding high water vapor emissivity. Therefore, small changes in lower tropospheric moisture content have little influence on either the local tropical OLR or the surface downwelling longwave flux, F_{Q0}^{LW}. The sign of the tropical water vapor feedback therefore depends on the sign of dW_v/dT_0 in the upper troposphere.

The simple relationship in (13.16) associates a warmer surface with higher water vapor contents of the air above that surface. Even in the convectively active tropics, a direct link between surface temperature and water vapor amount occurs only in the turbulent boundary layer, below about 700 hPa. Above the boundary layer, in the free troposphere, other less well-understood processes control the humidity of the air.

Figure 13.4 shows a histogram of relative humidity values as a function of atmospheric pressure, obtained from radiosonde observations in the tropical western Pacific. The levels below 700 hPa are moist, with $\mathcal{H} = 70$–90%, as would be expected for the turbulent boundary layer with the tropical ocean as a moisture source. In the layer between 500 and 250 hPa, much drier conditions exist, with a peak in the frequency distribution at $\mathcal{H} = 15\%$. At higher levels, the average humidity moistens to form a peak near $\mathcal{H} = 35\%$ just below the tropopause, which is typically near 100 hPa in the tropics.

The sharp decrease in $\mathcal{H}$ above 700 hPa rules out the possibility of large-scale upward transport of water vapor from the boundary layer. The source of upper tropospheric moisture in the convectively-active regions of the tropics is most likely to be

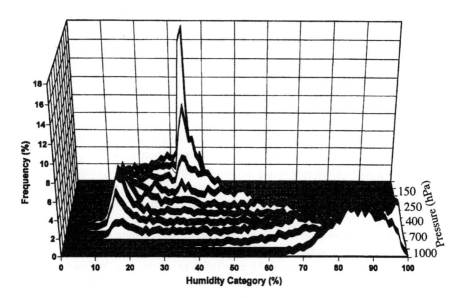

Figure 13.4 Relative humidity values as a function of atmospheric pressure obtained during 1,600 radiosonde ascents in the tropical western Pacific during January–May of 1994 and 1995. (From Spencer and Braswell, 1997.)

the vertical tranpsort of condensed water and the subsequent convective detrainment. *Detrainment* is a process whereby air and cloud particles are transferred from the organized convective updraft to the surrounding atmosphere (the opposite of entrainment). Deep convection in the tropics detrains at a level just below the tropopause, accounting for the maxima in relative humidity at this level. Deep convective clouds are not the only convective process of importance in determining the tropical water vapor profile. The influence of the full range of convective clouds, particularly mesoscale convective complexes, are important in determining the vertical moisture distribution in convectively-active regions. Mesoscale convective systems contain convective cells of various depths and stages of development with detraining tops supplying moisture over the entire depth of the troposphere.

Deep convection in the tropics also generates widespread and often deep upper-level anvil clouds which generate precipitation. Some of the anvil precipitation evaporates into the subsaturated air below the cloud, often occurring at some distance away from the convective core. The outflow region is also a region of widespread compensating subsidence. Subsidence advects the water vapor downward, decreasing the upper tropospheric water vapor mixing ratio. Subsidence also decreases the relative humidity through compressional heating. Hence we have detrainment and evaporation of anvil precipitation acting to moisten the tropical environment, while compensating subsidence dries the tropical environment.

Lindzen (1990) has hypothesized that the tropical water vapor feedback is negative, whereby a warmer T_0 results locally in decreased upper tropospheric water vapor content. Such a negative relationship between T_0 and $\mathcal{W}_v$ might arise from:

1) increased surface temperature producing deeper convection with higher, colder cloud tops, with relatively dry air being detrained in the upper troposphere;
2) a warmer atmosphere resulting in increased precipitation efficiency (warm rain process), resulting in the water vapor cycling through the atmosphere at relatively low altitudes, with less water transported to the upper troposphere.

The processes controlling the vertical mass flux in deep convection and precipitation efficiency are not well known for tropical mesoscale convective systems. It is not clear whether deep convection acts to dry or moisten the upper troposphere. Even more uncertain is how this will change in a warmer climate. The magnitude and sign of the local tropical water vapor feedback remains controversial. An additional difficulty is that an analysis of the local tropical thermodynamic water vapor feedback can be misleading, since advection is an important part of the moisture budget.

When determining water vapor feedback using a numerical climate model, the parameterizations of convection and precipitation processes (see Section 8.6) are crucial in determining the vertical distribution of water vapor, especially in the tropics. In view of the uncertainties in parameterization of these processes, the ubiquity of the positive water vapor feedback in the tropics found by these models should be questioned.

13.3.2 Polar troposphere

The second regional example of water vapor feedback considered here occurs in the wintertime polar regions. During winter, the polar troposphere is very stable (see Figure 8.17), due to strong radiative cooling of the surface. Because of this great stability, there is a lack of convective coupling between the surface and troposphere. Associated with the vertical temperature inversions are humidity inversions (see Section 8.4). Radiative cooling of the lower troposphere results in the formation of low-level clouds that are crystalline (diamond dust) at temperatures below about −15°C. Subsequent fallout of these ice crystals dehydrates the lower atmosphere, constraining the relative humidity so that it does not exceed the ice saturation value.

The process of diamond dust formation results in a positive value of $\partial \mathcal{H}/\partial T$ in (13.19), since the relative humidity is constrained not to exceed $\mathcal{H}_i$, the ice saturation value. Observations (Figure 13.5) show that the mean monthly relative humidity with respect to ice in the wintertime lower Arctic troposphere is $\mathcal{H}_i = 93\%$, for all air temperatures colder than about −10°C, so $\partial \mathcal{H}_i/\partial T = 0$. However, (4.35) and Table 4.4 imply that $\partial \mathcal{H}/\partial T > 0$ when $\partial \mathcal{H}_i/\partial T = 0$. Thus the additional positive term $\partial \mathcal{H}/\partial T$ in (13.19) contributes to a larger positive water vapor feedback in the polar regions than would be expected from the simple expression (13.21).

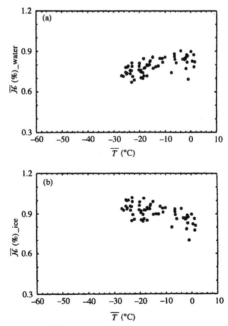

Figure 13.5 Monthly averaged values of relative humidity over the Arctic Ocean (a) with respect to liquid water and (b) with respect to ice. Humidity values are averaged over the 500- to 300-hPa layer and plotted against mean monthly temperature. (From Curry *et al.*, 1995.)

13.4 Cloud-radiation Feedback

Changes in cloud characteristics induced by a climate change would modify the radiative fluxes, thus altering the surface and atmospheric temperatures and further modify cloud characteristics. The feedback between surface temperature, clouds, and the Earth's radiation balance is referred to as the *cloud-radiation feedback*. The cloud-radiation feedback may be described using (13.8) as

$$f = \mathcal{G}_0 \left(\frac{\partial F^{rad}}{\partial A_c} \frac{dA_c}{dT_0} + \frac{\partial F^{rad}}{\partial T_c} \frac{dT_c}{dT_0} + \frac{\partial F^{rad}}{d\tau_c} \frac{d\tau_c}{dT_0} \right) \qquad (13.24)$$

where F^{rad} is the net radiative flux, A_c is the cloud fraction, T_c is the cloud temperature, and τ_c is the cloud optical depth. The first two terms on the right-hand side of (13.24) constitute the cloud-distribution feedback, while the third term represents the cloud-optical depth feedback.

The cloud-radiation feedback is illustrated using a signal flow graph in Figure 13.6. A perturbation to the Earth's radiation balance modifies surface temperature and possibly surface albedo. Changes in surface temperature will modify fluxes of radiation and surface sensible and latent heat, which will modify the atmospheric temperature, humidity and dynamics. Modifications to the atmospheric thermodynamic and dynamic structure will modify cloud properties (e.g., cloud distribution, cloud optical depth), which in turn modify the radiative fluxes. An understanding and correct simulation of the cloud-radiation feedback mechanism requires understanding of changes in: cloud fractional coverage and vertical distribution as the dynamics and vertical temperature and humidity profiles change; and changes in cloud water content, phase and particle size as atmospheric temperature and composition changes. The cloud-radiative feedback is generally regarded as one of the most uncertain aspects of global climate simulations.

In the following subsections, the individual derivatives in (13.24) are discussed.

13.4.1 Cloud-radiative effect

To the extent that the net radiative flux at the top of the atmosphere is linearly related to cloud fraction, the sensitivity term $\partial F^{rad}/\partial A_c$ in (13.24) can be related to a parameter called the *cloud-radiative effect*,[2] CF^{net}

$$\frac{\partial F^{rad}}{\partial A_c} dA_c \approx CF^{net} = F^{rad}(A_c) - F^{rad}(0) \qquad (13.25)$$

[2] The term "cloud forcing" is typically used to refer to the cloud-radiative effect. We believe that the word "force" is a misnomer for this effect.

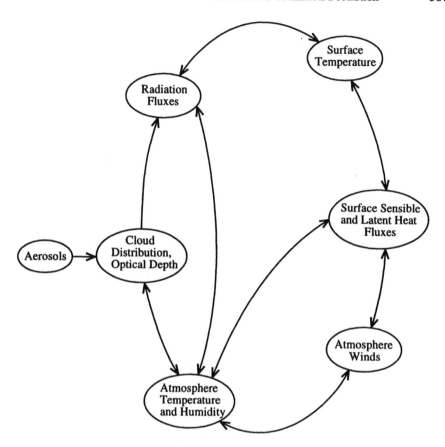

Figure 13.6 Signal flow graph illustrating the cloud-radiation feedback. Feedback loops may be positive or negative, depending on the atmospheric state and the horizontal scale under consideration. (After Curry *et al.*, 1996.)

(note that the radiative fluxes are defined to be positive downwards). The cloud-radiative effect is defined to be the actual radiative flux (which depends on cloud amount) minus the radiative flux for cloud-free conditions, all other characteristics of the atmosphere and surface remaining the same. The values of the cloud-radiative effect are negative for cooling and positive for warming. The cloud-radiative effect is most often defined in the context of the net radiative flux at the top of the atmosphere, F_{TOA}^{rad}, although the cloud-radiative effect can also be defined in the context of the surface radiative flux, F_{Q0}^{rad}. We can also separate the cloud-radiative effect into longwave and shortwave components:

$$CF^{LW} = -\left[F^{LW}\left(A_c \right) - F^{LW}\left(0 \right) \right] \qquad\qquad (13.26\text{a})$$

$$CF^{SW} = F^{SW}\left(A_c \right) - F^{SW}\left(0 \right) \qquad\qquad (13.26\text{b})$$

where

$$CF^{net} = CF^{SW} + CF^{LW} \qquad\qquad (13.26\text{c})$$

The cloud-radiative effect provides information on the overall effect of clouds on radiative fluxes, relative to a cloud-free Earth. The cloud-radiative effect can be evaluated exactly using a radiative transfer model, where fluxes obtained from a calculation for a cloud-free but otherwise exactly similar atmosphere is subtracted from a calculation for the actual cloudy atmosphere. Determination of the cloud-radiative effect from satellite is accomplished by separating the clear from the cloudy observations. In spite of the simplicity of evaluating the cloud-radiative effect at the top of the atmosphere using satellite data, such evaluations are somewhat ambiguous. Ambiguities arise since the distinction between clear and cloudy regions is not always simple (particularly in polar regions) and because other characteristics of the atmosphere (e.g., water vapor amount, atmospheric and surface temperature) change in cloudy versus clear conditions, even in the same location.

Table 13.1 provides some estimates of the mean annual global cloud-radiative effect at the top of the atmosphere. Clouds reduce the longwave emission at the top of the atmosphere since they are emitting at a colder temperature than the Earth's surface. At the same time, clouds decrease the net shortwave radiation at the top of the atmosphere because clouds overlying the Earth reflect more shortwave radiation than does the cloud-free Earth-atmosphere. Because of the partial cancellation of these effects, the net cloud-radiative effect has a smaller magnitude than either the individual longwave or shortwave terms. Both satellite and model estimates agree that

Table 13.1. Estimates of the mean annual, globally averaged cloud-radiative effect (W m^{-2}) at the top of the atmosphere derived from satellite observations and general circulation models.

Basis	Investigation	CF^{LW}	CF^{SW}	CF^{net}
Satellite	Ramanathan et al. (1989)	31	−48	−17
Satellite,	Ardanuy et al. (1991)	24	−51	−27
Models	Cess and Potter (1987)	23 to 55	−45 to −75	−2 to −34

the net cloud-radiative effect at the top of the atmosphere is negative and that short-wave effect dominates, i.e. that clouds reduce the global net radiative energy flux into the planet by about 20 W m^{-2}.

While clouds appear to have a net cooling effect on the global planetary radiation balance, there are regional and seasonal variations in the sign and magnitude of the cloud-radiative effect. The effect of an individual cloud on the local radiation balance depends on the cloud temperature and optical depth, as well as the insolation and characteristics of the underlying surface. For example, a low cloud over the ocean reduces substantially the net radiation at the top of the atmosphere, because it increases the planetary albedo while having little influence on the longwave flux at the top of the atmosphere. On the other hand, a high thin cloud can greatly decrease the outgoing longwave fluxes while having little influence on the solar radiation, therefore increasing the net radiation at the top of the atmosphere.

13.4.2 Cloud-distribution feedback

Changes in cloud fraction and in the vertical distribution of clouds induced by a climate change could modify the radiative fluxes and further modify cloud properties. From (13.24), the *cloud-distribution feedback* can be written as

$$f = G_0 \left(\frac{\partial F^{rad}}{\partial A_c} \frac{dA_c}{dT_0} + \frac{\partial F^{rad}}{\partial T_c} \frac{dT_c}{dT_0} \right) \qquad (13.27)$$

The general behavior of $\partial F^{rad}/\partial A_c$ and $\partial F^{rad}/\partial T_c$ has been described in subsection 13.4.1, in the context of the cloud-radiative effect. Determination of the terms dA_c/dT_0 and dT_c/dT_0 are far more difficult, since it requires understanding how cloud amount and its vertical distribution will vary in response to an altered value of T_0.

Simulations using global climate models of a doubled carbon dioxide (greenhouse warming) scenario generally show:

- cloud cover overall reduced at middle and low latitudes;
- increased cloudiness near the tropopause at middle and high latitudes;
- increased cloudiness near the surface at middle and high latitudes.

Most climate models produce a positive cloud-distribution feedback, resulting from a decrease of low-level cloudiness (with a large albedo but warm temperature) and an increase in high clouds (with low temperature and a smaller albedo). However, simulation of clouds by climate models remains uncertain.

In addition to changes in T_c arising from changes in the altitude of the cloud, changes in T_c arise from changing atmospheric temperatures. Climate models generally predict a warming of the troposphere in a scenario with increased greenhouse gases.

An additional aspect of the cloud-distribution feedback is the temporal distribution of clouds over the diurnal cycle. Shifting 10% of the nighttime cloud cover to daytime produces an effect that is large enough to offset the effects of doubling atmospheric CO_2. This is a consequence of the delicate balance between shortwave effects (confined to daytime) and longwave effects.

13.4.2 Cloud optical depth feedback

The *cloud optical depth feedback* is written from (13.8) as

$$f = G_0 \frac{\partial F^{rad}}{\partial \tau_c} \frac{d\tau_c}{dT_0} \tag{13.28}$$

The partial derivatives $\partial F^{rad}/\partial \tau_c$ have the same sign as the derivative $\partial F^{rad}/\partial A_c$: the partial derivative is negative for F^{SW} and positive for F^{LW}, with similar variations for changes in cloud height. Assuming that the cloud distribution remains the same, if $d\tau_c/dT_0 > 0$, then the cloud optical depth feedback will be negative, since $\partial F^{rad}/\partial \tau_c < 0$.

Since the cloud optical depth is a function of the amount of condensed water, its phase, and the size of the particles, we can use (8.12) to expand $d\tau_c/dT_0$ as

$$\frac{d\tau_c}{dT_0} = \frac{3}{2}\left(\frac{1}{r_{el}\rho} \frac{dW_l}{dT_0} + \frac{1}{r_{ei}\rho_i} \frac{dW_i}{dT_0} + \frac{W_l}{\rho_l} \frac{dr_{el}}{dT_0} + \frac{W_i}{\rho_i} \frac{dr_{ei}}{dT_0} \right) \tag{13.29}$$

where r_{el} and r_{ei} are the effective radii for liquid and ice particles. To assess the cloud optical depth variation with surface temperature, we next consider the individual terms in (13.29).

The derivative dW_l/dT_0 has been hypothesized to be positive, whereby a warmer (and presumably moister) lower atmosphere would increase the liquid water content of clouds. To assess this hypothesis, we consider a single layer liquid water cloud and expand the derivative dW_l/dT_0 using the definition of the liquid water path (8.6):

$$\frac{dW_l}{dT_0} = \frac{d}{dT_0} \frac{1}{g} \int_{p_t}^{p_b} w_l \, dp = \frac{1}{g} \int_{p_t}^{p_b} \frac{dw_l}{dT_0} \, dp + \frac{1}{g}\left(w_{lt} \frac{dp_t}{dT_0} - w_{lb} \frac{dp_b}{dT_0} \right) \tag{13.30}$$

where w_l is the liquid water mixing ratio and p_t and p_b are, respectively, the cloud top and base pressures. The second term on the right-hand side of (13.30) arises from the variation of p_t and p_b with T_0 (see footnote 1 in Section 13.3). If we assume that w_l is constant within the cloud, we have $w_l = gW_l/(p_b - p_t)$ and therefore can write (13.30) as

$$\frac{dW_l}{dT_0} = \frac{d}{dT_0} \frac{1}{g} \int_{p_t}^{p_b} w_l \, dp = \frac{1}{g} \int_{p_t}^{p_b} \frac{dw_l}{dT_0} \, dp + W_l \frac{d \ln (p_b - p_t)}{dT_0} \tag{13.31}$$

If we assume that the cloud forms in saturated adiabatic ascent, the integral on the right-hand side of (13.31) can be expanded as

$$\frac{1}{g} \int_{p_t}^{p_b} \frac{dw_l}{dT_0} \, dp = \mathcal{D} \frac{\partial W_l^{ad}}{\partial T_0} + W_l^{ad} \frac{\partial \mathcal{D}}{\partial T_0} \tag{13.32}$$

where the term W_l^{ad} is the adiabatic liquid water path defined in Section 8.2. The term $\mathcal{D}$ represents the fractional cloud water dilution relative to the adiabatic liquid water path, which arises from entrainment, precipitation, and conversion to the ice phase. We can therefore write (13.31) as

$$\frac{dW_l}{dT_0} = \mathcal{D} \frac{\partial W_l^{ad}}{\partial T_0} + W_l^{ad} \frac{\partial \mathcal{D}}{\partial T_0} + W_l^{ad} \frac{d \ln (p_t - p_b)}{dT_0} \tag{13.33}$$

The term $\partial W_l^{ad} / \partial T_0$ is positive as long as the lower atmospheric temperature increases with increasing surface temperature, which is a reasonable assumption. The sign of the term $\partial \mathcal{D}/\partial T_0$ depends on issues such as whether the precipitation efficiency in a warmer climate will increase and whether the clouds will be more convective or stratiform in a warmer climate and thus modify dilution by entrainment. The term $\partial \mathcal{D}/\partial T_0$ has a negative component if the phase of mid-level clouds is more likely to be liquid (rather than crystalline) in a warmer environment. The last term on the right-hand side of (13.33) arises from a possible variation of cloud depth with changing surface temperature; the sign of this term is unknown. Although the adiabatic liquid water path in the lower atmosphere is expected to increase in a warmer climate, the sign of $\partial W_l^{ad} / \partial T_0$ remains unknown because of uncertainties in processes that control cloud depth, entrainment, and precipitation efficiency, which vary with cloud type.

A similar analysis can be done for the term dW_i/dT_0. To interpret this term, we consider a single-layer ice cloud in the upper troposphere and expand the derivative dW_i/dT_0:

$$\frac{dW_i}{dT_0} = \frac{1}{g} \int_{p_t}^{p_b} \frac{dw_i}{dT_0} \, dp + W_i \frac{d \ln (p_t - p_b)}{dT_0} \tag{13.34}$$

where w_i is the ice water mixing ratio. The integral on the right-hand side of (13.34) can be expanded as

$$\frac{1}{g} \int_{p_t}^{p_b} \frac{dw_i}{dT_0} \, dp = \frac{1}{g} \int_{p_t}^{p_b} \frac{\partial w_i}{\partial T_c} \frac{dT_c}{dT_0} \, dp \tag{13.35}$$

where T_c is the cloud temperature. We can therefore write (13.34) as

$$\frac{d\mathcal{W}_i}{dT_0} = \frac{1}{g} \int_{p_t}^{p_b} \frac{\partial w_i}{\partial T_c} \frac{dT_c}{dT_0} \, dp + \mathcal{W}_i \frac{d \ln (p_t - p_b)}{dT_0} \tag{13.36}$$

Observations such as those shown in Figure 13.7 show that the derivative $dw_i/dT_c > 0$. Simulations from climate models indicate that upper tropospheric temperatures increase with a surface warming, so that $dT_c/dT_0 > 0$ and the first term on the right-hand side of (13.36) is positive. The sign of the second term on the right-hand side of (13.36) is unknown. Variations in cirrus cloud depth with a surface warming depend, in complex ways, on changes to deep convection and the strength of mid-latitude frontal systems.

Several mechanisms have been suggested for changes in cloud particle size in a global warming scenario. Recall the definition of r_e for a distribution of spherical particles (8.13):

$$r_e = \frac{\displaystyle\int_0^\infty r^3 n(r) \, dr}{\displaystyle\int_0^\infty r^2 n(r) \, dr}$$

The effective radius can vary with the amount of condensed water and the number of condensed particles, N. We can therefore write

$$\frac{dr_{el}}{dT_0} = \frac{\partial r_{el}}{\partial w_l} \frac{\partial w_l}{\partial T_{cl}} \frac{dT_{cl}}{dT_0} + \frac{\partial r_{el}}{\partial N_l} \frac{dN_l}{dT_0} \tag{13.37a}$$

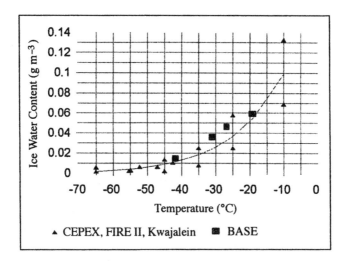

Figure 13.7 Observed relationship between cirrus ice water content and cloud temperature, obtained from research aircraft flights during CEPEX (tropics), FIRE II (mid-latitudes), and BASE (Arctic). (Courtesy of J. Intrieri.)

$$\frac{dr_{ei}}{dT_0} = \frac{\partial r_{ei}}{\partial w_i} \frac{\partial w_i}{\partial T_{ci}} \frac{dT_{ci}}{dT_0} + \frac{\partial r_{ei}}{\partial N_i} \frac{dN_i}{dT_0} \tag{13.37b}$$

where T_{cl} is the temperature of the liquid cloud, T_{ci} is the temperature of the ice cloud, N_l is the number concentration of water drops, and N_i is the number concentration of ice particles.

The partial derivatives $\partial r_e/\partial w$ in (13.37a) and (13.37b) are positive; assuming that N remains constant, an increase in the amount of condensed water implies an increase in particle size. The derivatives $\partial w/\partial T_c$ are also positive, as discussed previously in this section. Furthermore, as described in the discussion regarding the dW/dT_0 terms, it was shown that $dT_c/dT_0 > 0$. Hence the first term is positive on the right-hand sides of (13.37a) and (13.37b).

If the number concentration increases, and all other things remain constant (such as liquid and ice water mixing ratio), then the effective radius will decrease, so that $\partial r_e/\partial N < 0$. Changes in droplet and ice particle concentrations, N_l and N_i, could arise from:

- changes in the concentrations of cloud condensation nuclei (CCN) and ice-forming nuclei (IFN);
- changes in the rate of entrainment that evaporate or sublimate cloud particles;

- changes in the efficiency of precipitation which would alter the number of cloud particles that fall out of the cloud; and
- changes in the phase of precipitation due to freezing or melting, where $dN_l = -dN_i$.

For there to be a feedback associated with N, there must be some relation between N and T_0. An increase in the number of CCN can arise from anthropogenic pollution, which is an external climate forcing. An internal source of CCN has been hypothesized to occur via the oxidation of dimethylsulfide (DMS), which is emitted by phytoplankton in seawater (Section 5.2). Therefore, DMS from the oceans may determine the concentrations and size spectra of cloud droplets. If it is assumed that the DMS emissions increase with increasing ocean temperature, there would be an increase in atmospheric aerosol particles and $dN_l/dT_0 > 0$. However, the increase of DMS emissions with increasing ocean temperature has not been verified from observations. Additionally, the factors which most enhance biological productivity are sunlight and nutrients; incoming sunlight would be depleted by additional aerosols, which might reduce the production of DMS.

Additional relationships between N and T_0 might arise from changes in cloud type in an altered climate, whereby an increase in convective clouds would increase both entrainment and precipitation; both processes would decrease N. An increase in droplet concentration may in itself reduce precipitation efficiency and hence increase the lifetime (cloud fraction) and optical depth of the cloud. Warmer air temperature at heights where atmospheric temperature ranges between about 0 and $-15°C$ would result in an increasing amount of liquid relative to ice phase clouds. Since clouds with ice in them are more likely to form precipitation-sized particles, the cloud water content would increase as the atmosphere warms.

In summary, the cloud–optical depth feedback is very complex. Climate models that include at least some of the cloud microphysical processes involved in the cloud–optical depth feedback generally find that this is a negative feedback. However, the sign and magnitude of the feedback depends on the cloud parameterizations that are used in the model, introducing substantial uncertainty into the estimation.

13.5 Snow/Ice-albedo Feedback

The possible importance of high-latitude snow and ice for climate change has been recognized since the 19th century. It has been hypothesized that when climate warms, snow and ice cover will decrease, leading to a decrease in surface albedo and an increase in the absorption of solar radiation at the Earth's surface, which would favor further warming. The same mechanism works in reverse as climate cools. This climate feedback mechanism is generally referred to as the *snow/ice-albedo feedback*, which is a positive feedback mechanism. The ice-albedo feedback has proven to be quite important in simulations of global warming in response to increased greenhouse gas concentrations.

In the context of a surface energy balance model, we can write the following expression for the snow/ice-albedo feedback mechanism over the ocean:

$$f = \mathcal{G}_0 \frac{\partial F_{Q0}^{rad}}{\partial \alpha_0} \frac{d\alpha_0}{dT_0} \tag{13.38}$$

where by definition of the surface albedo, α_0, we have $\partial F_{Q0}^{rad}/\partial \alpha_0 < 0$. To date, most of the research on ice-albedo feedback has focused on the terms

$$\frac{d\alpha_0}{dT_0} = \frac{dA_i}{dT_0} \alpha_i + \frac{dA_l}{dT_0} \alpha_l$$

where the subscript i denotes ice, the subscript l denotes open water, and A is the fractional area coverage. In a simple model where the surface is either ice-covered or open water, then $dA_i = -dA_l$ so we can write

$$\frac{d\alpha_0}{dT_0} = \frac{dA_i}{dT_0} (\alpha_i - \alpha_l)$$

Since the area coverage of sea ice will decrease in a warmer climate ($dA_i/dT_0 < 0$) and $\alpha_i > \alpha_l$ the term $d\alpha_0/dT_0 < 0$, and from (13.38), we have $f > 0$.

As discussed in Section 10.5, the surface albedo of an ice-covered ocean is quite complex, including contributions from melt ponds, snow-covered ice, open water in leads, and bare ice. To assess the contribution of each of these different surface types to the albedo feedback (Figure 13.8), we can write the surface snow/ice albedo as the fractional-area-weighted sum of the albedos of the individual surface types that characterize ice-covered oceans:

$$\alpha_0 = A_i \alpha_i + A_l \alpha_l + A_p \alpha_p + A_s \alpha_s \tag{13.39}$$

where A denotes the fractional area/time coverage of the individual surface types and the subscripts i, l, p, and s denote, respectively, bare ice, open water, melt ponds, and snow. Differentiating (13.39) with respect to T_0 yields

$$\frac{d\alpha_0}{dT_0} = A_i \frac{d\alpha_i}{dT_0} + \frac{dA_i}{dT_0} \alpha_i + A_l \frac{d\alpha_l}{dT_0} + \frac{dA_l}{dT_0} \alpha_l$$
$$+ A_p \frac{d\alpha_p}{dT_0} + \frac{dA_p}{dT_0} \alpha_p + A_s \frac{d\alpha_s}{dT_0} + \frac{dA_s}{dT_0} \alpha_s \tag{13.40}$$

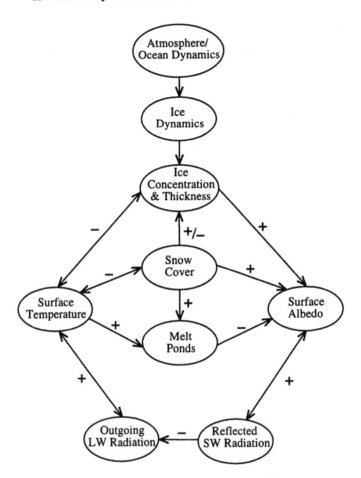

Figure 13.8 Signal flow graph illustrating the sea-ice-albedo feedback mechanism locally over the ice pack. In this context, large-scale atmospheric and oceanic circulations are regarding as external forcing. Note that ice concentration includes both large-scale sea ice extent and small-scale features such as leads. (After Curry *et al.*, 1996.)

Analogously to the cloud-radiation feedback, we can divide the ice-albedo feedback into an ice-area-distribution feedback and a surface-optical properties feedback.

Consider first the terms in (13.40) that include dA/dT_0, which constitute the ice-area-distribution feedback (outer loop in Figure 13.8)

$$\frac{dA_j}{dT_0}\,\alpha_j \equiv \frac{dA_i}{dT_0}\,\alpha_i + \frac{dA_l}{dT_0}\,\alpha_l + \frac{dA_p}{dT_0}\,\alpha_p + \frac{dA_s}{dT_0}\,\alpha_s \tag{13.41}$$

Table 13.2 Magnitude and/or sign of the terms in (13.41).

Ice type	Summertime albedo	dA/dT_0
Bare ice	0.56	< 0
Open water	0.10	> 0
Melt ponds	0.25	< 0
Snow	0.77	< 0

where the subscript j on the left-hand side of (13.41) is simply an index. Table 13.2 summarizes mid-summer values of α_j (Table 10.1) along with the expected signs of dA_j/dT_0. It seems reasonable to expect that $dA_s/dT_0 < 0$, $dA_l/dT_0 > 0$, and $dA_i/dT_0 < 0$, since a warmer climate would likely be associated with more open water and less snow and bare ice. In a warmer climate as the ice thins, melt ponds may become "melt holes" as the ponds melt through the ice, decreasing A_p and increasing A_l. In a warmer climate, the area of the high-albedo surfaces will decrease and the area of the low-albedo surfaces will increase; hence, $\alpha \, dA/dT_0 < 0$.

Next consider the terms in (13.40) that include $A \, d\alpha/dT_0$, which comprise the surface-optical properties feedback (inner loop in Figure 13.8):

$$A_j \frac{d\alpha_j}{dT_0} \equiv A_i \frac{d\alpha_i}{dT_0} + A_l \frac{d\alpha_l}{dT_0} + A_p \frac{d\alpha_p}{dT_0} + A_s \frac{d\alpha_s}{dT_0} \tag{13.42}$$

where again, j is an index.

The albedo of bare ice depends on ice thickness and the age of the sea ice. We can therefore write

$$\frac{d\alpha_i}{dT_0} = \frac{dh_i}{dT_0} \left(\frac{\partial \alpha_i}{\partial h_i} + \frac{\partial \alpha_i}{\partial y_i} \frac{\partial y_i}{\partial h_i} \right) \tag{13.43}$$

where the term y_i denotes the age of the ice and h_i is ice thickness. Since thick ice is associated with colder surface temperatures (Figure 10.8), $dh_i/dT_0 < 0$. If the ice thickness is less than 2 m, the albedo is influenced by the underlying ocean, and thus $\partial \alpha_i/\partial h_i > 0$. Old ice has air bubbles which increase in concentration with ice age; hence older ice has a higher surface albedo (Section 10.5) and $\partial \alpha_i/\partial y_i > 0$. Since ice thickness generally increases with ice age, $\partial y_i/\partial h_i > 0$. Therefore, $A_i \, d\alpha_i/dT_0 < 0$. In (13.42), the term $d\alpha_l/dT_0 = 0$, since there is no reason for the albedo of open water to vary with surface temperature. The albedo of melt ponds decreases with increasing melt pond depth. Thicker ice can support deeper melt ponds. Hence we can write

$$\frac{d\alpha_p}{dT_0} = \frac{\partial \alpha_i}{\partial h_p} \frac{\partial h_p}{\partial h_i} \frac{dh_i}{dT_0} \tag{13.44}$$

Since pond albedo decreases with increasing pond depth, we have $\partial \alpha_i / \partial h_p < 0$. We have also seen that $dh_i / dT_0 < 0$. However, the variation of pond depth with ice thickness is not straightforward. On one hand, thicker ice can support deeper melt ponds. On the other hand, as ponds deepen in thin ice, further deepening of the pond may be accelerated as the pond albedo lessens due to the influence of the underlying ocean, becoming "melt holes" as they melt through the ice, and the pond depth becomes undefined. The albedo of snow depends on snow depth and snow age. Deeper snow and more frequent snowfalls are associated with a higher value of surface albedo. If a warmer climate (associated with higher T_0) is also associated with higher snowfall amount in the polar regions, then $d\alpha_s / dT_0 > 0$, but the sign of this term must be regarded as uncertain, since the characteristics of snowfall in an altered climate are not known.

To summarize the preceding analysis of snow/ice-albedo feedback, a negative value of $d\alpha_0 / dT_0 = \alpha_j \, dA_j / dT_0 + A_j \, d\alpha_j / dT_0$ gives a positive value of the snow/ice-albedo feedback in (13.38), since $\partial F_{Q0}^{rad} / \partial \alpha_0 < 0$. The sign of $\alpha_j \, dA_j dT_0$ is unambiguously negative, although the sign of $A_j \, d\alpha_j / dT_0$ is less certain and may depend critically on whether snowfall over sea ice increases in a warmer climate. While the overall sign of the snow/ice-albedo feedback is not in doubt, its magnitude in climate models depends on the details of the snow and sea ice model parameterizations, such as snow albedo, melt ponds, sea ice dynamics, etc.

An additional factor to consider in the context of the snow/ice-albedo feedback is the influence of clouds on the surface albedo of snow and ice. As described in Section 10.5, the broadband surface albedo of snow and ice can be significantly higher under cloudy skies than under clear skies, because clouds deplete the incoming solar radiation in the infrared portion of the spectrum and change the direct beam radiation to diffuse radiation. If the local cloud-radiation feedback is nonzero, then an additional component to the snow/ice-albedo feedback must be considered.

13.6 Thermodynamic Control of the Tropical Ocean Warm Pool

In Section 11.3.1, the tropical ocean warm pool was discussed. Skin temperature is observed to be less than 34°C, while ocean temperatures measured at a depth of 0.5 m typically do not exceed 32°C. Geochemical studies of paleoclimatic data suggest that the maximum annually averaged equatorial sea surface temperatures have not exceeded about 30°C during warm climatic episodes. However, there is controversy about the paleoclimatic situation during the last ice age, where the warm pool may have been as cold as 24°C. Nevertheless, it appears that the equatorial sea surface temperature is remarkably insensitive to global climatic forcing, in contrast to the pronounced sensitivity of mid and high latitudes (e.g., ice ages). It appears that negative feedbacks are acting to stabilize the tropical ocean surface temperature. The nature of the negative feedbacks in this region continues to be hotly debated.

The simple planetary energy balance model described in Section 13.2 is predicated upon the principle that when averaged over the entire Earth and over an annual cycle, the net incoming solar radiation at the top of the atmosphere is equal to the outgoing longwave radiation balance at the top of the atmosphere. This assumption results in elimination of the term ΔF_{TOA}^{rad} in (13.9). To consider the energy balance at the top of the atmosphere for a region, the term ΔF_{TOA}^{rad} cannot be neglected since advection of heat into or out of the region will change the local energy balance at the top of the atmosphere.

Hence, in examining regional climate feedbacks, it is more fruitful to conduct the feedback analysis using a surface energy balance model, whereby the surface energy balance is written from (9.1) as

$$F_{Q0}^{net} - F_{Q0}^{adv} - F_{Q0}^{ent} = F_{Q0}^{rad} + F_{Q0}^{SH} + F_{Q0}^{LH} + F_{Q0}^{PR}$$

The sea surface temperature is determined by a balance between ocean heat transports and surface energy fluxes. For simplicity in the following discussion, the distinction between the skin sea surface temperature, T_0, and the ocean mixed layer temperature, T_m, is ignored, and it is assumed that $T_0 = T_m$ (see section 11.2 to recall the distinction between T_0 and T_m). From (13.8), we can write the feedback for a surface energy balance model as

$$f = \mathcal{G}_o \sum_j \frac{\partial F_{Q0}^{net}}{\partial I_i} \frac{dI_i}{dT_0} \tag{13.45}$$

Incorporating (9.1) into (13.45), we can write

$$f = \mathcal{G}_o \sum_j^{\infty} \left(\frac{\partial F_{Q0}^{rad}}{\partial I_j} \frac{dI_j}{dT_0} + \frac{\partial F_{Q0}^{SH}}{\partial I_j} \frac{dI_j}{dT_0} + \frac{\partial F_{Q0}^{LH}}{\partial I_j} \frac{dI_j}{dT_0} \right.$$
$$\left. + \frac{\partial F_{Q0}^{PR}}{\partial I_j} \frac{dI_j}{dT_0} + \frac{\partial F_{Q0}^{adv}}{\partial I_j} \frac{dI_j}{dT_0} + \frac{\partial F_{Q0}^{ent}}{\partial I_j} \frac{dI_j}{dT_0} \right) \tag{13.46}$$

The surface radiative flux feedback

$$f = \mathcal{G}_0 \sum_j \frac{\partial F_{Q0}^{rad}}{\partial I_j} \frac{dI_j}{dT_0} \tag{13.47}$$

depends upon all of the internal variables discussed in Sections 13.3 and 13.4: water vapor, lapse rate, cloud fractional area, cloud temperature, and cloud optical depth. Hence we can write (13.47) as

$$f = \mathcal{G}_0 \left(\frac{\partial F_{Q0}^{rad}}{\partial W_v} \frac{dW_v}{dT_0} + \frac{\partial F_{Q0}^{rad}}{\partial \Gamma} \frac{d\Gamma}{dT_0} + \frac{\partial F_{Q0}^{rad}}{\partial A_c} \frac{dA_c}{dT_0} \right.$$
$$\left. + \frac{\partial F_{Q0}^{rad}}{\partial T_c} \frac{dT_c}{dT_0} + \frac{\partial F_{Q0}^{rad}}{\partial \tau_c} \frac{d\tau_c}{dT_0} \right) \tag{13.48}$$

In contrast to the tropical water vapor feedback on the planetary energy balance (section 13.3.1), the tropical water vapor feedback on the surface energy balance is quite straightforward. Because the water vapor content of the tropical atmospheric boundary layer is so high, the longwave flux at the surface shows little variation with the amount of water vapor, so that F_{Q0}^{rad} is relatively insensitive to variation in W_v. The value of F_{Q0}^{rad} is, however, quite sensitive to variations cloud properties. Again, because the water vapor content of the tropical atmospheric boundary layer is so high, the longwave flux at the surface F_{Q0}^{LW} is insensitive to variations in cloud characteristics. However, the shortwave flux at the surface F_{Q0}^{SW} is quite sensitive to variations in cloud fraction where $\partial F_{Q0}^{rad}/\partial A_c \approx \partial F_{Q0}^{SW}/\partial A_c < 0$. The term $\partial F_{Q0}^{rad}/\partial \tau_c$ is also positive. As discussed in Section 13.5, determination of dA_c/dT_0 and $d\tau_c/dT_0$ depends not only on local thermodynamic processes but also on large-scale dynamical processes.

Since the surface sensible heat flux over the tropical ocean is an order of magnitude smaller than the latent heat flux, here we consider only the latent heat flux feedback (although the arguments are easily extended to include the sensible heat flux). The feedback associated with the latent heat flux is written from (13.8) as

$$f = \mathcal{G}_0 \sum_j \frac{\partial F_{Q0}^{LH}}{\partial I_j} \frac{dI_j}{dT_0} \tag{13.49}$$

From Section 9.1.2, the surface latent heat flux is determined to be

$$F_{Q0}^{LH} = \rho L_{iv} C_{DE} u_a \left(q_{va} - q_{v0} \right)$$

and hence $F_{Q0}^{LH}(u_a, \Delta q, C_{DE})$ where $\Delta q = q_{v0} - q_{va}$. We can therefore expand (13.49)

$$f = G_0 \left(\frac{\partial F_{Q0}^{LH}}{\partial u_a} \frac{du_a}{dT_0} + \frac{\partial F_{Q0}^{LH}}{\partial \Delta q} \frac{d\Delta q}{dT_0} + \frac{\partial F_{Q0}^{LH}}{\partial C_{DE}} \frac{dC_{DE}}{dT_0} \right)$$

$$\approx G_0 \rho L_{iv} \left(C_{DE} \Delta q \frac{du_a}{dT_0} + C_{DE} u_a \frac{d\Delta q}{dT_0} \right) \tag{13.50}$$

where the term dC_{DE}/dT_0 is ignored since it is estimated to be much smaller than the other terms. The individual terms in (13.50) have been evaluated using surface observations in the tropical Pacific Ocean (Zhang and McFadden, 1995). Typical values in the warm pool are $u_a = 5$ m s^{-1}, $\Delta q = 6$ g kg^{-1}, and $C_{DE} = 1.1 \times 10^{-3}$. It was shown that $d\Delta q/dT_0 > 0$, while $du_a/dT_0 < 0$. Wind speed dependence of the surface latent heat flux dominates for $T_0 > 301$ K, while the humidity dependence dominates for $T_0 < 301$ K. The decrease in surface latent heat flux at high surface temperatures is hypothesized to arise from the following mechanism: high surface temperature $\rightarrow$ increased instability and convection $\rightarrow$ increased large-scale low-level convergence $\rightarrow$ weaker surface wind $\rightarrow$ lower latent heat flux. In interpreting the magnitudes and signs of these derivatives, it should be kept in mind that an apparent empirical relationship between wind speed and humidity with T_0 is no guarantee that the primary factor giving rise to changes in wind speed or humidity is T_0. Wind speed and T_0 may appear to be related because both fields are related to a third and much more dominant factor, such as the large-scale coupled atmosphere–ocean circulation, which is controlled largely by the horizontal gradient in T_0 rather than the value of T_0 itself.

The feedback associated with the sensible heat flux of rain

$$f = G_0 \sum_j \frac{\partial F_{Q0}^{PR}}{\partial I_j} \frac{dI_j}{dT_0} \tag{13.51}$$

is believed to be smaller than the other terms in (13.46), since the long-term average magnitude of F_{Q0}^{PR} is about 2% of the value of F_{Q0}^{LH}. A change of F_{Q0}^{PR} with surface temperature might arise from a change in the amount of precipitation in an altered climate.

From (11.4), the heat flux from entrainment at the base of the ocean mixed layer, F_{Q0}^{ent}, can be written as

$$F_{Q0}^{ent} = \rho c_p u_{ent} \Delta T$$

where u_{ent} is the entrainment velocity at the base of the mixed layer and the term ΔT represents the jump in temperature across the base of the mixed layer. Recall from (11.6b) that the entrainment velocity is given by

$$u_{ent} = \frac{c_1 u_*^3 - c_2 \left(F_{B0}/\rho\right) h_m}{h_m \left(\alpha g \Delta T - \beta g \Delta s\right)}$$

The feedback associated with entrainment can therefore be written as

$$f = \mathcal{G}_0 \left(\frac{\partial F_{Q0}^{ent}}{\partial \Delta T} \frac{d\Delta T}{dT_0} + \frac{\partial F_{Q0}^{ent}}{\partial u_*^3} \frac{du_*^3}{dT_0} + \frac{\partial F_{Q0}^{ent}}{\partial F_{B0}} \frac{dF_{B0}}{dT_0} + \frac{\partial F_{Q0}^{ent}}{\partial h_m} \frac{dh_m}{dT_0} + \frac{\partial F_{Q0}^{ent}}{\partial \Delta s} \frac{d\Delta s}{dT_0} \right)$$

$$\approx \frac{\rho c_p}{h_m \alpha g \Delta T} \left(\mathcal{G}_0 \left(c_1 u_*^3 - c_2 F_{B0} h_m\right) \frac{d\Delta T}{dT_0} + \Delta T c_1 \frac{du_*^3}{dT_0} \right.$$

$$\left. - \Delta T c_2 h_m \frac{dF_{B0}}{dT_0} - \Delta T c_1 u_*^3 \frac{dh_m}{dT} \right)$$

(13.52)

The term Δs has been ignored, since there is not a strong halocline in the tropics, and $\Delta T > 0$ in the tropics. The feedback associated with entrainment depends on a complex interplay between surface momentum and buoyancy fluxes. Accumulation of buoyancy in the warm pool region alters the sensitivity of the sea surface temperature to wind forcing. When the mixed layer is shallow, entrainment cooling is more easily initiated. As heat and fresh water accumulate in the warm pool the threshold wind speed and duration of entrainment cooling increases, therefore rendering the mixed layer less sensitive to wind mixing.

The significance of the term F_{Q0}^{adv} can be explained as follows. If T_0 in the warm pool increases, both the east–west and meridional temperature gradients will increase. These gradients in upper ocean temperature generate increased transport of heat away from the warm pool. The meridional transport of heat away from the equatorial warm pool induces increased equatorial upwelling that cools the warm pool mixed layer. Heat that accumulates in the tropical western Pacific is exported in El Niño conditions towards the east, reducing upwelling in the central and eastern Pacific. Hence, the tropical ocean circulation moves heat to where the ocean more readily loses it to the atmosphere.

In summary, the largest terms involved in the tropical ocean warm pool feedback appear to be associated with solar radiation, the surface latent heat flux, mixed-layer entrainment, and oceanic advection processes. The separation of these effects based on empirical studies alone is extremely difficult because all contributing factors are operating simultaneously and only total changes are observed. To determine whether the relationships inferred from the short-term variability may be extrapolated to longer-term climate changes or used to assess model feedbacks operating in climate change experiments, one has to estimate the dynamical dependence of relationships between the sea surface temperature, clouds, radiation, and processes that control the depth and heat content of the ocean mixed layer. Feedback hypotheses for the warm pool can only be tested fully by models of the coupled ocean–atmosphere system, once these models have demonstrated sufficient realism.

13.7 High-latitude Ocean Feedbacks

The global ocean thermohaline circulation was described in Section 11.6, whereby large-scale overturning is driven by both buoyancy and mechanical forces generated in high-latitude oceans. The response of the ocean thermohaline circulation to perturbations is determined by four major feedbacks between the thermohaline circulation and the high-latitude temperature and salinity fields (Figure 13.9).

Consider an equilibrium situation where a positive perturbation to the freshwater flux (e.g., excess precipitation or sea ice melt) is imposed at high latitudes. A decrease in salinity corresponds to a decrease in density, which diminishes the sinking motion and the thermohaline circulation. The weakening of the thermohaline circulation reduces the poleward transport of relatively salty water from lower latitudes, which further decreases the polar salinity (loop 1 in Figure 13.9), which is a positive feedback.

At the same time, the decreased strength of the thermohaline circulation also reduces the northward heat transport, increasing the high-latitude surface density, which in turn intensifies the high-latitude convection and hence the overturning circulation (loop 2 in Figure 13.9), and hence is a negative feedback. This negative feedback partially compensates the positive feedback associated with salinity, but the compensation is not complete and the positive salinity feedback dominates.

Diminished overturning leads to lower surface temperatures, which results in reduced evaporation (loop 3 in Figure 13.9). If all of the evaporated water returns to the ocean locally in the form of precipitation, then there is no net effect and loop 3 is inactive. If precipitation falls outside the region of evaporation, then the feedback in loop 3 is positive. Lower surface temperature also results in increased sea ice formation, which increases the density and hence the thermohaline circulation.

A further result of decreased high-latitude surface temperature is that the meridional atmospheric circulation is enhanced by the stronger meridional surface temperature gradient, resulting in increased northward transport of atmospheric heat and moisture, which increases precipitation and decreases the high-latitude surface ocean density (loop 4 in Figure 13.9).

Given these feedbacks, what is the stability and variability of the thermohaline circulation? Climate stability can be analyzed reliably only if all important feedbacks

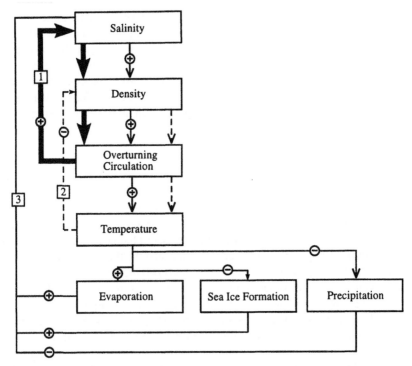

Figure 13.9 Signal flow graph of feedbacks affecting the thermohaline circulation. (Adapted from Willebrand, 1993.)

are represented accurately in a climate model. In spite of deficiencies in current climate models, useful sensitivity studies can be conducted even with relatively simple models. In models where the haline feedback (loop 1) dominates, multiple equilibrium states are possible. Alternate equilibrium states that have been found include: a conveyor-belt where the respective roles of Atlantic and Pacific oceans are reversed; a greatly diminished Atlantic circulation with less significant changes in the Pacific circulation; and symmetric circulations in both the Pacific and Atlantic. The nature and magnitude of the perturbation required to turn the system into a different state have been hypothesized to occur from perturbations to either the surface fresh-water budget or surface temperature (e.g., greenhouse warming). Modeled thermohaline circulations have collapsed on time scales of about 20 years. Positive feedback mechanisms can trigger instabilities of the circulation resulting in oscillatory phenomena. Observations and models suggest internal oscillations on time scales ranging from decades to millennia. The possibility that the ocean may switch from one state to another within a few decades is intriguing and indicates the importance of the interaction between the thermohaline circulation with the hydrological cycle.

Notes

An accessible treatment of feedback and control theory is given in *Feedback and Control Systems* (1967) by DiStefano *et al.*

Overviews of climate sensitivity and feedback analysis are given by Schlesinger (1986) and Hansen (1984).

An overview of the cloud-radiation feedback is given by Arking (1991).

An overview of feedbacks involving the ocean thermohaline circulation is given by Willebrand (1993).

Chapter 14 | Planetary Atmospheres

Our understanding of processes occurring on Earth is enhanced by examining thermodynamic processes occurring on other planets having different gravitational fields, solar illumination, orbital geometries, and atmospheric compositions.

Since the late 1950s, there have been hundreds of space exploration missions that have provided data on the structure of planetary surfaces and the composition and circulations of planetary atmospheres. Whereas the majority of the missions have been directed at the Earth's moon, a substantial number have been directed at Venus, Mars, Jupiter, and Saturn, with fewer for Uranus, Neptune, and Mercury. Except for the manned landings on the Earth's moon, these missions have been combinations of "fly-by" flights, orbiters, and unmanned landings. In addition to exploration of the solar system using spacecraft, measurements have been obtained using Earth-based spectral radiometers and radars. The totality of these data allows us to piece together descriptions of the planets and their atmospheres and to speculate on how each planetary atmosphere may have evolved since the formation of the solar system.

Of the myriad of topics that could be examined in the context of planetary science, we focus here on the thermodynamics of planetary atmospheres and surfaces. Specifically, we examine the mass, composition, and vertical structure of planetary atmospheres, and phase changes occurring in clouds and planetary ice caps. These thermodynamic characteristics and processes are examined in the context of the various gravitational fields, distances from the sun and orbital geometries, and varying atmospheric compositions characteristic of the different planets in the solar system.

14.1 Atmospheric Composition and Mass

Table 14.1 lists physical data for each of the planets, including size, orbital parameters, and surface pressure and temperature. These data provide a context for interpreting the composition and mass of the planetary atmospheres, as well as thermodynamic processes occurring within them.

Table 14.1 Physical data for the planets: m_P is the mass of the planet, r_P its radius, g_0 the surface gravity, R_P the average orbital radius from the sun, P the length of the planetary year (Earth days), ϕ the inclination of the axis of rotation, Ω is the length of the planetary day (Earth days), p_0 is surface pressure, and T_0 is surface temperature. Dagger indicates uncertainty in the data. An asterisk signifies that the surface of the planet is difficult to define and the surface temperature corresponds to an altitude where the pressure is equal to one Earth atmosphere.

Planet	m_P	r_P	g_0	R_P	P	ϕ	Ω	p_0	T_0
Mercury	3.3	2,439	3.76	58	88	$0^\dagger$	58.7	2×10^{-9}	452
Venus	48.7	6,051	8.88	108	225	< 3	−243	88.000	735
Earth	59.7	6,378	9.81	150	365	23.5	1.0	1,013	288
Mars	6.4	3,396	3.73	228	687	25.2	1.03	6	210
Jupiter	18,988	70,850	26.20	778	4,330	3.1	0.41	high	170*
Saturn	5,684	60,330	11.20	1,430	10,800	26.8	0.43	high	130*
Uranus	869	25,400	9.75	2,870	30,700	98.0	$-0.9^\dagger$	high	59*
Neptune	1,028	24,300	11.34	4,500	60,200	28.8	$0.7^\dagger$	high	59*
Pluto	0.13	1,150	0.07	5,900	90,700	?	$6.4^\dagger$	?	$40^\dagger$

The planets fall into two main classes: the inner and the outer planets. The inner or *terrestrial planets* (Mercury, Venus, Earth, and Mars) are similar in that all possess rocky and metallic cores. The outer planets (Jupiter, Saturn, Uranus, Neptune, and Pluto) are principally gaseous and, if they possess solid or liquid cores, they are made up of the lighter elements such as hydrogen and helium. Overall, the outer planets constitute 99.6% of the total planetary mass of the solar system. Jupiter possesses 71.2% of the total mass. By comparison, Earth constitutes only 0.22%.

The general differences in composition between the inner and outer planets can explained in the context of the evolution of the solar system, which is described briefly here. The *interstellar medium* that we call "space" contains clouds of hydrogen (78%) and helium (20%) plus trace quantities of heavier elements. These clouds are referred to as *nebula*. Stars are produced when the nebula collapses under the influence of its own gravitational field. If conditions are favorable, the gravitational collapse produces a core of sufficiently high temperature to produce fusion and the formation of a star. The similarity of the composition of the outer planets to the composition of the sun suggests that the outer planets were centers of gravitational collapse that did not produce fusion; hence, Jupiter and Saturn may be thought of as "failed" stars. The inner planets are aggregates of residuals of the heavier elements that coagulated during the early stages of formation of the solar system. Physical characteristics of the individual planets are summarized on the following page.

Mercury is the second smallest planet by mass in the entire solar system and has a very tenuous atmosphere. The surface of the planet is heavily cratered and appears to be similar to Earth's moon. There is evidence from Earth-based radar exploration indicating surface ice pockets on Mercury, in the permanently shadowed crater bottoms close to the poles. Similar pockets have been discovered on the Moon.

Venus, the planet most similar to the Earth in mass, is extremely hot, desert-like, and very dry, with evidence of surface weathering and erosion. Its atmosphere has a mass of almost 100 times that of the Earth's atmosphere and a surface temperature of about 740 K. The planet is completely covered with clouds.

Mars is half the size of Earth by radius and one ninth the planetary mass. The atmosphere is thin, with a mass that is roughly 1% of the Earth's atmosphere. The surface of the planet is colder than either Venus or Earth. Both clouds and polar ice caps are observed on Mars. With the exception of Mercury, Mars has the smallest atmosphere of all of the planets. Earth has the next smallest atmosphere.

The outer planets, with the exception of Pluto, are far greater in mass than the inner planets, are totally covered with clouds, and have massive atmospheres. The atmosphere of Pluto is very tenuous. The satellites of the outer planets are similar to the inner planets in the sense that they possess rocky cores. Surface water ice is evident on the Jovian moons Callisto, Ganymede, and Europa.

14.1.1 Mass

The mass and composition of an atmosphere depends on the planetary mass and geological structure, the temperature of the atmosphere, the strength of the planetary magnetic field, and the chemical composition of the gases.

The temperature of the gas establishes the speed distribution of the molecules of a given gas (1.7a). The mass of the planet determines the velocity of a molecule required to escape the planet's gravitational field. To escape from a planet, an object of mass m must have a kinetic energy ($mu^2/2$) that is greater than its gravitational potential energy. *Newton's law of gravitation* states that the force of gravity is given by

$$\mathscr{F}_g = \frac{G\, m_p\, m}{r^2} \tag{14.1}$$

where $G = 6.67 \times 10^{-11}$ N m^2 kg^{-1} is the gravitational constant, m_p is the mass of the planet, m is the mass of the smaller body, and r is the distance between the center of masses. The *escape velocity*, u_{esc}, of the smaller body is defined so that the kinetic energy of the smaller body is equal to its gravitational potential energy, $r\mathscr{F}_g$,

$$\frac{G\, m_p\, m}{r} = \frac{m\, u_{esc}^2}{2} \tag{14.2}$$

or

$$u_{esc} = \sqrt{\frac{2\,G\,m_P}{r}} \qquad (14.3)$$

The escape velocity does not depend on the mass of the escaping body but only on the mass of the planet and the distance from the center of mass of the planet. Since the mass of a planetary atmosphere is contained within heights above the surface which are much smaller than the radii of the planets, we can take r in (14.3) to be the radius of the planet, r_p. Table 14.2 lists the escape velocities for each planet.

No molecule with a speed less than the escape velocity can escape the gravitational pull of the planet. In Section 1.6, we derived the equation of state for an ideal gas utilizing only the average kinetic energy of the molecules, which was determined from a root-mean-square molecular speed. However, to examine the possibility of escape of molecules from a planetary gravitational field, we require not only the mean, but also the distribution, of molecular speeds. For an ideal gas at temperature T, the three-dimensional distribution of molecular speeds, $f(u)$, is given by the *Maxwell–Boltzmann distribution function*

$$f(u) = A\,4\,\pi u^2 \exp\left(-\frac{mu^2}{2kT}\right) \qquad (14.4)$$

which is a Gaussian distribution. The term A,

$$A = \left(\frac{m}{2\pi kT}\right)^{3/2}$$

Table 14.2 Escape velocities of the planets calculated from (14.3).

Planet	Escape Velocity $(\mathrm{ms^{-1}})$
Mercury	4,200
Venus	10,360
Earth	11,190
Mars	5,010
Jupiter	59,790
Saturn	35,450
Uranus	21,370
Neptune	23,760
Pluto	1,228

is defined such that f is normalized as

$$\int_0^\infty f(u)\,du = 1$$

The *most probable speed*, u_o, is defined to be that speed for which f is a maximum. By differentiating f with respect to u and setting the result equal to zero, we obtain

$$u_o = \sqrt{\frac{2kT}{m}} \tag{14.5}$$

The most probable speed increases with temperature and decreasing atomic mass. Table 14.3 provides values of u_o for atomic hydrogen, helium, nitrogen, and oxygen as a function of temperature.

Comparison of the values in Table 14.3 with those in Table 14.2, and also with planetary temperatures shown in Table 14.1, shows that an atom moving with the average speed within the planetary range of temperatures does not have a speed sufficient to escape any of the planets. From the Maxwell–Boltzmann distribution, the fraction f_{esc} of molecules with speeds greater than the escape velocity is written as

$$f_{esc} = A\,4\pi \int_{u_{esc}}^\infty u^2 \exp\left(-\frac{u^2}{u_o^2}\right) du$$

Table 14.3 Most probable speed, u_o, for atomic hydrogen, helium, nitrogen, and oxygen, at different temperatures.

Gas	Atomic Weight (amu)	Temperature			
		100	300	600	900
H	1	1290	2234	3160	3870
He	4	650	1125	1593	1950
N	14	390	675	956	1170
O	16	320	554	784	969

For a given atmospheric constituent, *thermal escape* is probably not important if

$$u_{esc} \leq b u_o \qquad (14.6)$$

where the factor b is between 5 and 6 (Lewis and Prinn, 1984). This means that at least one molecule in 10^{10} must exceed u_{esc}. If this is not the case, then the time needed to deplete the atmosphere of a constituent is longer than the age of the solar system (see Problem 14.1). Molecules that are most likely to escape a given planet have the smallest molecular weight and thus are in atomic form. Atomic species are more likely to occur high in the atmosphere where molecules encounter photons having sufficient energy to break molecular bonds. The chances of a molecule escaping from the planet are also greatest from the high levels of the atmosphere where a molecule has the least chance of colliding with another.

In an atmosphere with convective mixing, the various species of gases are well mixed, allowing determination of a specific gas constant for the mixture. However, at much greater heights in an atmosphere, mixing typically becomes less efficient and diffusive processes dominate. The height at which this basic change in the structure of a planetary atmosphere occurs is called the *exopause,* and the region above the exopause is referred to as the *exosphere.* In the exosphere, gases become stratified by mass.

The scale height for a constituent j of a planetary atmosphere can be written using (1.39) as

$$H_j = \frac{R_j T}{g}$$

Since $R_j = R^*/M_j$, where M_j is the molecular weight of the jth constituent, it is clear that the scale heights will be greater for lighter gases than for heavier gases. Therefore in the exosphere, atomic hydrogen overlies mixtures of hydrogen and helium and successively the heavier gases. As a result, the lightest gases find themselves in the highest regions of the atmosphere which have the largest molecular mean free paths. Therefore if the exosphere is sufficiently warm, the lighter constituents have an opportunity to escape the planetary gravitational field.

Gases can also escape from a planet via sputtering escape or solar wind sweeping. *Sputtering escape* refers to the actual collision of solar wind particles with atmospheric constituents in which momentum is transferred and the atom or molecule accelerated to speeds greater than the escape velocity of the planet. The *solar wind* consists of a stream of protons and electrons that move in the interplanetary magnetic field spiralling out from the sun. The effect of the solar wind on a planetary atmosphere depends on the strength of the planetary magnetic field. A sufficiently strong planetary magnetic field will divert the solar wind. Sputtering escape favors the loss of the lighter elements as these are more likely to achieve speeds greater than u_{esc} than heavier elements.

Primitive atmospheres of the inner planets were swept away by an intense solar wind shortly after the formation of the solar system, during which an outpouring of extra energy from the sun occurred. Secondary atmospheres of the inner planets formed through outgassing by volcanoes, fissures, and fumaroles. Mercury, Earth, Jupiter, and Saturn have strong enough magnetic fields to protect their atmospheres from being subjected to loss by solar wind sweeping. Venus and Mars, each with a weak magnetic field that is induced in the atmosphere by the solar wind itself, may be subject to some solar wind sweeping. On the Earth's moon, where there is no magnetic field, the solar wind continues to sweep away any gases that have been captured from meteors or through crustal processes.

Examination of the surface pressure values in Table 14.1 shows a general relationship between the mass of a planet and its surface pressure. Mercury and Mars, two planets with relatively small masses, have tenuous atmospheres, particularly Mercury with its relatively high planetary temperature. While surface pressure is undefined for the *Jovian planets* (Jupiter, Saturn, Neptune, Uranus) because there is no absolute planetary surface, the atmospheres on these planets are massive. Of all the planets, Earth does not follow the mass–pressure relationship well. It has the third smallest of all of the solar system atmospheres but the fifth smallest planetary mass.

A short discussion of Mercury's almost total lack of an atmosphere is warranted here. Table 14.2 gives an escape velocity for Mercury of 4200 m s^{-1}, the lowest of the planets except for Pluto, and a surface temperature in excess of 400 K. Using data from Tables 14.2 and 14.3 and (14.5), we obtain values of b of 1.7, 3.7, and 6.2 for H, He, and N, respectively, for $T = 400$ K. Thus, thermal escape would appear to be probable mechanism for explaining the lack of an atmosphere on Mercury. Also, the solar flux reaching Mercury is almost seven times larger than that reaching Earth, so that the dissociation potential of molecules is very high.

14.1.2 Atmospheric composition

Since the most abundant species in the interstellar medium is hydrogen, primitive planetary atmospheres consisted of hydrogen and helium, and simple compounds of hydrogen such as CH_4 and NH_3. Table 14.4 describes the present chemical composition of the planetary atmospheres, including some of the Jovian moons. The massive Jovian planets, far from the sun, appear to have maintained their primitive atmosphere. Since the outer planets do not contain a core of heavier elements, volcanic activity has not modified their atmospheres. By contrast, the gases in the present terrestrial atmospheres are dominated by effluents from volcanoes, primarily water vapor and carbon dioxide.

The atmosphere of Mercury is very tenuous and transitory, composed of helium and hydrogen from the solar wind plus traces of other gases such as oxygen. The atmosphere of Venus is primarily CO_2 and includes clouds thought to be composed of liquid H_2SO_4. The gaseous composition of the Martian atmosphere is similar to the Venusian atmosphere, and both water-ice and CO_2-ice clouds have been observed.

Table 14.4 Major components in the chemical composition of solar system bodies with substantial atmospheres. (Data from Wayne, 1991.)

Body	H_2	He	H_2O	CH_4	NH_3	CO_2	N_2	O_2	Ar
Venus		0.00002	0.00002			0.965	0.035		0.00007
Earth			0–0.04			0.0003	0.781	0.209	0.0093
Mars			0.00003			0.953	0.027	0.0013	0.016
Jupiter	0.90	0.10		0.0024	0.0002				
Saturn	0.96	0.04		0.002	0.0002				
Uranus	0.85	0.15							
Neptune	0.85	0.15		0.00003					
Titan	0.002			0.03			0.82		0.12

Although the geologies of the terrestrial planets are similar, leading one to anticipate that their atmospheres derived from outgassing would be the same, there are substantial differences in atmospheric composition among these planets. Perhaps the most striking difference is the abundance of water on Earth. Carbon dioxide in the Earth's atmosphere, in great contrast to Venus and Mars, is a trace gas. However, an inventory of total CO_2 on the planets, including that bound up in rock, shows that total CO_2 on Venus has a remarkably similar mass to that on Earth, although the majority of the CO_2 on Venus exists in the atmosphere, while on Earth it is concentrated in the planetary crust. CO_2 cannot exist on Venus in crustal form because of the extremely high surface temperatures, whereas on cooler planets such as Mars and Earth, these crustal compounds are stable. But perhaps the most important distinction between the atmosphere of the Earth and those of the other terrestrial planets is that the Earth's biosphere has resulted in the large concentration of molecular oxygen and nitrogen.

The outer planets (Jupiter, Saturn, Uranus, Neptune, and Pluto) have atmospheres that are markedly different in composition to atmospheres of the inner planets. Each of the outer planets is composed of hydrogen and helium. Except for Pluto, the outer planets are predominately cloud covered. Pluto has a very tenuous but extended atmosphere, possibly consisting of N_2, CO, and traces of CH_4. Several of the Jovian moons have atmospheres. Titan, the largest of Saturn's moons, has a massive cloudy atmosphere composed primarily of N_2. Titan's atmosphere has a red haze that suggests the existence of a photochemical hydrocarbon smog similar to that found in large cities of Earth.

14.2 Vertical Structure of Planetary Atmospheres

The characteristics of the vertical structure of planetary atmospheres are described here with reference to the Earth's atmosphere (Figure 14.1). The variation of temperature with height depends on the composition of the planetary atmosphere and

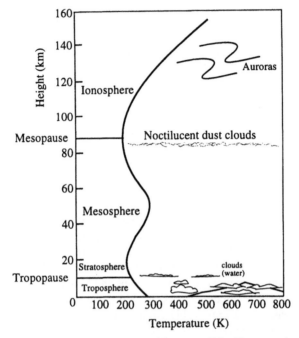

Figure 14.1 Vertical structure and features of Earth's atmosphere.

radiative properties of the gases. These properties determine the manner in which the atmosphere is heated. Generally, the highest temperatures are near the surface, with temperatures decreasing with height. This lowest layer of decreasing temperatures is the troposphere. There are several reasons why a troposphere exists on most of the planets. If the atmosphere is fairly transparent to solar radiation, then the planetary surface is heated and turbulent fluxes transport heat vertically, typically resulting in an adiabatic lapse rate. Also, if the atmosphere absorbs strongly in the infrared it can also change the vertical temperature profile. Above the troposphere the temperature gradient often increases with height. This stable region of increasing temperatures is referred to as the stratosphere and is usually caused by the absorption of incoming solar radiation. For example, the Earth's stratosphere occurs because of an abundance of ozone, which absorbs strongly in the ultraviolet wavelengths of the solar spectrum. Above the level of the maximum temperature in the stratosphere, the atmospheric temperature decreases; this region is referred to as the *mesosphere*, in which the atmospheric temperature resumes its decrease with height. Above the mesosphere is the *thermosphere*, where the temperature increases with height. The density of the atmosphere at these heights is low, and the atmosphere consists largely of the ions produced and heated by intense solar radiation. Collectively, the region where ions are abundant is referred to as the *ionosphere.*

14.2.1 Venus

A vertical temperature profile of the atmosphere of Venus is shown in Figure 14.2. The atmosphere is marked by a thick cloud deck between about 48 to 58 km above the surface, which are thought to be composed of sulfuric acid droplets. The temperature at the top of the cloud has been determined to be about 240 K, decreasing by a small amount towards the poles. Above the cloud is a deep haze layer of sulfuric acid droplets.

Beneath the cloud layer the temperature of the atmosphere increases steadily to about 740 K at the surface. The lapse rate between the surface and about 50 km has been measured to be very close to 7.7°C km^{-1}. The surface temperature has been measured directly at a number of latitudes and show a remarkable lack of dependency on latitude such that there is very little surface temperature difference between the equator and the poles.

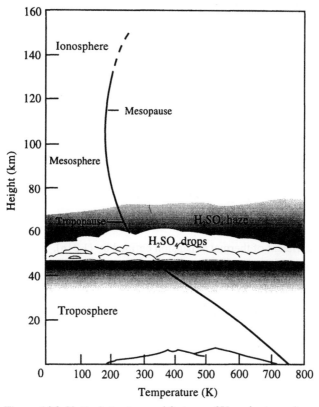

Figure 14.2 Vertical structure and features of Venus's atmosphere.

To assess whether the Venusian atmosphere is statically stable requires calculation of the adiabatic lapse rate. Recall that the dry adiabatic lapse rate for an ideal gas was derived in Section 2.4 from the first law of thermodynamics in enthalpy form for an adiabatic process (2.19):

$$c_p \, dT = v \, dp$$

Using the hydrostatic equation (1.33), the dry adiabatic lapse rate was written in (2.68) as

$$\Gamma_d = \frac{g}{c_p}$$

In principle, this expression for the dry adiabatic lapse rate can be applied to any of the planets, as long as the atmosphere is composed of an ideal gas and is hydrostatic. The critical point for CO_2 is at $T_{crit} = 304.2$ K and $p_{crit} = 72.9$ atm, clearly within the parameter range of the Venusian atmosphere. Because of the high temperatures and pressures of the Venusian atmosphere that exceed the critical point for CO_2, it is not obvious that (2.68) can be applied without incurring significant error.

The combined first and second law written following (2.71) is appropriate for considering a non-ideal gas

$$T \, d\eta = c_p \, dT - T\left(\frac{\partial v}{\partial T}\right)_p dp$$

Following (2.73), we can determine the dry adiabatic lapse rate from (2.71) to be

$$\Gamma_d = -\frac{T}{v}\frac{\partial v}{\partial T}\frac{g}{c_p} = \frac{\alpha T g}{c_p}$$

where α is the coefficient of thermal expansion defined by (1.31a).

From (2.17b), the value of c_p for a non-ideal gas is written as

$$c_p - c_v = \left[\left(\frac{\partial u}{\partial v}\right)_T + p\right]\left(\frac{\partial v}{\partial T}\right)_p$$

From the combined first and second law (2.31), we have

$$\frac{du}{dv} = T\frac{d\eta}{dv} - p$$

upon division by an incremental volume, dv. Using Maxwell's relation (2.51), we can write

$$\frac{du}{dv} = T\frac{dp}{dT} - p$$

so (2.17b) becomes

$$c_p = c_v + T\frac{\partial p}{\partial T}\frac{\partial v}{\partial T} \tag{14.7}$$

At the high temperatures that characterize the Venusian atmosphere, c_p shows a strong temperature dependence because of filling of the vibrational energy levels, and also because the high atmospheric pressure influences the equation of state. For linear molecules such as carbon dioxide, the specific heat at constant volume for an ideal gas, c_{vi}, is written as (following Callen, 1960)

$$c_{vi} = \frac{5}{2}R + \sum_{T_V}\left[R\left(\frac{T_V}{2T}\right)^2 \sinh^{-2}\left(\frac{T_V}{2T}\right)\right] \tag{14.8}$$

where T_V are the vibrational temperatures which for CO_2 have values $T_V = 960, 960, 200, 3380$ K. The equation of state for a gas at high pressure is often written in

$$\frac{pv}{RT} = 1 + \frac{B(T)}{v} + \frac{C(T)}{v^2} + \ldots = Z$$

where $B, C, \ldots$ are the virial coefficients and Z is called the *compressibility factor*. The relationship of real and ideal specific heats at constant volume is given by Callen (1960) as

$$c_v = c_{vi} - \frac{R}{v}\frac{d}{dT}\left(T^2\frac{dB}{dT}\right) - \frac{R}{2v^2}\frac{d}{dT}\left(T^2\frac{dC}{dT}\right) \tag{14.9}$$

Table 14.5 shows computations of c_p, c_{pi} using (14.7)–(14.9), $\Gamma_i = g/c_{pi}$, and Γ_d/Γ_i, where Γ_d is calculated using (2.71) for a CO_2 atmosphere over the temperature range 250–750 K and pressure range 1–100 atm. The value of c_{pi} increases by 30% over the temperature range of the Venusian atmosphere, with a corresponding decrease in Γ_i. Values of c_p show increasing departures from c_{pi} as the pressure increases, with the difference being relatively larger at the cooler temperatures. As atmospheric pressure increases beyond 1 atm, the value of the real dry adiabatic lapse

Table 14.5 Numerical results of adiabatic lapse rate computations for the Venusian atmosphere (100% CO_2) at various temperatures and pressures. (Data from Staley, 1970.)

T (K)	c_{pi} (J kg^{-1})	Γ_i (°C km^{-1})	c_p (J K^{-1}) and Γ_d/Γ_i for p (atm)				
			1	10	40	70	100
250	791	11.0	805	1090			
			1.013	1.073			
350	891	9.76	899	937	1088	1309	1638
			1.002	1.056	1.276	1.584	2.09
450	972	8.95	980	998	1060	1125	1186
			0.996	1.024	1.110	1.202	1.305
550	1040	8.37	1047	1057	1091	1127	1163
			0.996	1.010	1.005	1.097	1.134
650	1098	7.92	1103	1108	1129	1151	1175
			0.997	1.006	1.030	1.047	1.063
750	1142	7.62	1149	1153	1166	1181	1195
			0.995	1.000	1.012	1.021	1.030

rate exceeds the ideal value, by as much as a factor of two at $T = 350$ K and $p = 100$ atm. The discrepancy between the real and ideal adiabatic lapse rate values is greatest at the highest pressures and lowest temperatures. The variable dry adiabatic lapse rate must be accounted for in evaluating the static stability of the Venusian atmosphere. Lapse rates have been measured in the lowest 50 km to be 7.7 K km^{-1} which is very similar to the adiabatic lapse rate calculated in the lower atmosphere in Table 14.5. Thus, the temperature profile is dry adiabatic, indicating rising or sinking motions associated with either large-scale motions or vertical mixing.

Between about 40 km above the surface and the base of the cloud layer, the lapse rate decreases considerably to values far less than the dry adiabatic lapse rate. The cloud base is heated radiatively from below, and the sub-base region in turn is cooled. Given enough time, this results in a subadiabatic region just below cloud base, with a convective region in the lowest cloud layers. The observed temperature inversion near 64 km is suggestive of cloud-top radiative cooling, similar to that seen on Earth. The temperature at the top of the upper cloud layer is 230 K. Above the cloud layer the lapse rate continues to decrease to a nearly isothermal state between 95–110 km with temperature values of about 170 K. Temperatures then increase through the thermosphere to about 300 K, so that molecular escape from the exosphere by thermal escape processes is difficult. There is no stratosphere on Venus, indicating an absence of solar radiation absorption such as occurs on Earth by ozone.

14.2.2 Mars

Given the low pressures of the Martian atmosphere and the remoteness from critical points, we need not be concerned with any complications regarding the adiabatic lapse rate. The dry adiabatic lapse rate on Mars calculated for an ideal gas is 4.5 K km^{-1}. Temperature profiles in Figure 14.3 indicate that the observed Martian lapse rate is less than dry adiabatic. It is thought that the stable lapse rate is caused by absorption of solar radiation by dust suspended in the atmosphere. Whereas there always appears to be a residual concentration of dust, major dust storms are frequent

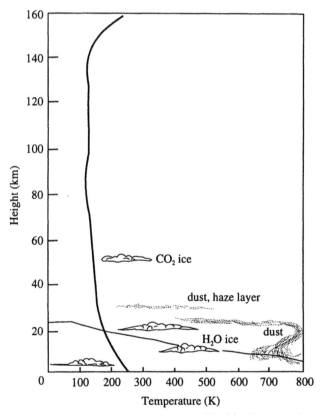

Figure 14.3 Vertical structure and features of the Martian atmosphere.

on Mars, sometimes covering the entire planet. Given the size of the dust particles, 75 m s^{-1} winds are necessary to lift the dust into the air. Dust storms originate in the southern hemisphere summer when the planet is closest to the sun. The increased solar insolation occurring at that time of the year causes rapid sublimation of the southern hemisphere CO_2 ice cap. The strong winds generated adjacent to the ice cap create local dust storms that lift dust into the atmosphere to heights greater than 35 km. Dust storms have been observed to obtain global dimensions within a few weeks. A Martian dust storm appears to be a self-limiting process. The absorption of solar radiation high in the atmosphere reduces solar heating of the ground, cooling the surface. This increases the stability of the boundary layer and decreases the wind speed, resulting in reduced lofting of surface dust. Once the source of new dust is cut off, the dust eventually settles out of the atmosphere.

One of the unique characteristics of Mars is its pronounced orographic features. For example, Mons Olympus, a broad volcanic cordillera, has a height of 25 km, nearly a factor of three larger than the highest peaks on Earth. Martian temperature profiles have shown that temperature profiles over high topography (+4 km above a reference level) and profiles over a valley region (−4 km below a reference level) at the same local time of day were virtually the same when superimposed over each other at a common surface reference level. The implication is that for the same values of insolation, the surface temperature is the same irrespective of the height of the topography. As a result, large horizontal temperature gradients occur between the mountains and the plains (Figure 14.4), forcing thermal circulations that reverse over the diurnal cycle. The topographically induced thermal circulations are important in determining the disposition of water and the formation of water clouds on the planet.

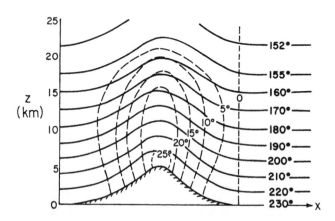

Figure 14.4 Schematic profile of the temperature anomaly (°C, dashed lines) associated with topography on Mars. The temperature gradients induce strong vertical circulations on diurnal time scales. (From Webster, 1976.)

14.2.3 Jovian Planets

Figures 14.5 and 14.6 show the vertical structures of the atmospheres of Jupiter and Saturn, respectively. The atmosphere of Jupiter can be divided into four major layers: troposphere, stratosphere, mesosphere, and thermosphere. The troposphere extends from an indeterminate depth within the planet to about 100 hPa. Above the tropopause is a weak stratosphere. Like Earth, the stratosphere is caused by absorption of solar radiation. However, instead of molecular oxygen and ozone being responsible, the absorption is thought to occur by methane and other absorbing gases. The temperature is nearly isothermal in the mesosphere but increases rapidly into the ionosphere which extends to great heights. Temperatures of 800–1000 K have been estimated by Voyager probes. The lapse rate in the upper Jovian troposphere is estimated to be about $2°C \ km^{-1}$, which is close to the adiabatic lapse rate. However, below

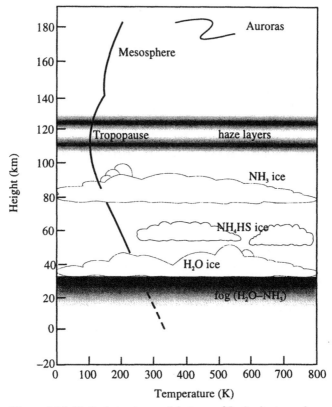

Figure 14.5 Vertical structure and features of Jupiter's atmosphere.

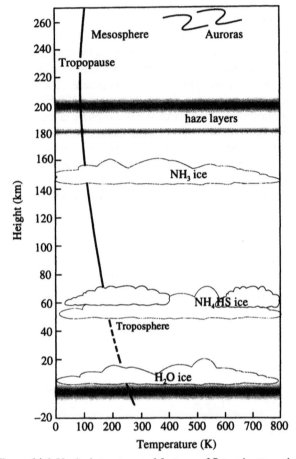

Figure 14.6 Vertical structure and features of Saturn's atmosphere.

about 1000 km into the interior of the Jovian atmosphere, it is necessary to compute real gas adiabatic lapse rates as we have done for Venus in Section 14.2.1. For hydrogen, the rotational transitions become important at temperatures greater than 100 K, and vibrational transitions begin around 700 K (Figure 14.7), causing a significant variation in c_p with atmospheric temperature.

There are many similarities between Jupiter and Saturn. Planetary size is very large compared to that of the Earth, they both rotate rapidly on their axes, and both have similar compositions made up mostly of H_2 and He. Both planets are cloud covered and have clouds that exits in distinctive layers through the depths of their atmospheres composed of NH_3, H_2O, and NH_4HS ice. Like Jupiter, the lapse rate of the upper troposphere of Saturn appears to be close to the ideal dry adiabatic lapse rate.

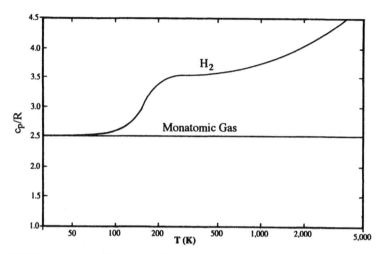

Figure 14.7 Heat capacity of hydrogen. The low-temperature heat capacity is that of a monatomic gas because temperatures are too low for intermolecular collisions to excite even the rotation of molecular hydrogen. Above 1000 K, the vibrational excitation of hydrogen begins to appear, but the energy of the vibrational fundamental is so close to the dissociation energy of the molecule that the molecule falls apart before the full vibrational contribution to the heat capacity can be realized. (After Lewis, 1995.)

Less is known about the structure and composition of the atmospheres of Uranus and Neptune. The Voyager mission has provided some indication that the atmosphere of Uranus contains a troposphere with a subadiabatic lapse rate and a warmer stratosphere and mesosphere. The stratosphere on Uranus appears to result from the photodissociation of CH_4.

14.3 Planetary Energy Balance

The planetary energy budget for the Earth was described in Section 12.1. There are five primary physical characteristics that control the radiation balance of a planet:

(i) The distance of the planet from the sun determines solar illumination.
(ii) The mass and composition of the planet. The mass determines the planetary escape velocity while the composition of the core determines the presence and strength of the planetary magnetic field.
(iii) The geological structure of the planet determines the outgassing of planetary effluents into the atmosphere.
(iv) Orbital parameters determine day and year length.
(v) Internal planetary heat sources.

Secondary controls on planetary energy budgets arise from:

(v) The evolving planetary albedo.
(vi) The infrared radiative properties of the planetary atmosphere.

14.3.1 The Planetary Greenhouse Effect

In Section 12.1, the radiative balance for a planet in radiation equilibrium was discussed, whereby the outgoing longwave radiation of the planet is equal to the net incoming solar radiation. The inner planets are essentially in radiation equilibrium. Jupiter, Saturn, Neptune, and possibly Uranus, however, are not in radiative equilibrium. Each of these planets actually radiates more energy to space that it receives from the sun: Jupiter, by a factor of 1.9; Saturn, by a factor of 2.2; and Neptune by a factor of 2.7. The generally close similarity in physical properties between Uranus and Neptune suggests that if Neptune has an internal heat source, then Uranus ought to also. The enhanced infrared emissions from the Jovian planets are apparently a product of the planet's own internal heat source. It has been theorized that these planets are evolutionary forms of a low mass star. Hence the present-day radiation energy may come from an internal reservoir of thermal energy generated by slow shrinkage of the entire planet..

For the inner planets, which are in planetary radiative equilibrium with the sun, the planetary radiation balance is written following (12.1) as

$$T_e^* = \left(\frac{S(1 - \alpha_P)}{4\sigma}\right)^{1/4}$$

The temperature, T_e^*, is defined as the equivalent black-body temperature at which the planet and its atmosphere radiates to space. Table 14.6 lists the equivalent black-body temperatures for the inner planets, along with values of the solar constant, S,

Table 14.6 Values of equivalent black-body planetary temperature, T_e^*, as a function of the planetary solar constant, S, and planetary albedo, α_P, and optical thickness of the atmosphere n.

Planet	α_p (%)	S (W m^{-2})	T_e^* (K)	T_0 (K)	n
Mercury	5.8	9200	442	452	0.09
Venus	71	2600	244	740	84
Earth	33	1370	254	288	0.65
Mars	17	600	217	223	0.12

and planetary albedo, α_P. Values of T_e^* decrease generally with distance of the planet from the sun, with the modulation of planetary albedo by clouds also providing an important control on T_e^*.

For a planet in radiative equilibrium, comparison of T_e^* with T_0 gives an indication of the planetary greenhouse effect. Note the especially large greenhouse effect of Venus. Mercury and Mars, with their tenuous atmospheres, have little greenhouse effect. Earth has a significant greenhouse effect, but nothing approaching that of Venus.

To examine the planetary greenhouse effect, consider a simple atmosphere characterized as follows (see Figure 14.8):

(i)   The atmosphere is made up of n slabs, each with temperature T_j.
(ii) Each slab contains a gas that absorbs longwave radiation. The thickness of each slab is such that it is just optically black with an emissivity of unity. That is, all longwave radiation entering the slab will just be absorbed.

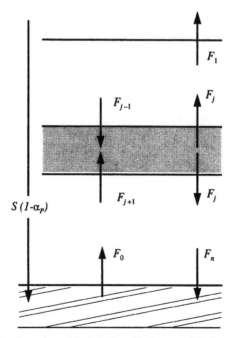

Figure 14.8 Schematic of an n-layer black-body radiation mondel. Arrows indicate the direction of the fluxes between the layers. It is assumed that the atmosphereis transparent to solar radiation and that the surface turbulent transfers of heat are ignored. The model relates the radiative surface temperature, T_0, to the effective temperature of the planet, T_e^*.

(iii) The longwave absorbing gas is equally mixed by mass throughout the atmospheric column. Thus, in order for each slab to be just optically black, it must contain the same mass of absorbing gas.

(iv) The slabs are assumed to be transparent to visible radiation. In this simplified atmosphere, only the surface reflects solar radiation.

(v) All other heating or transport processes are ignored.

Considering only radiative processes and assuming equilibrium, the energy balance in slab j is given by

$$2F_j = F_{j-1} + F_{j+1} \tag{14.10}$$

At the surface, the energy balance is

$$S(1 - \alpha_0) + F_n = \sigma T_0^4 \tag{14.11}$$

where the nth layer is the layer just above the surface, and the surface is assumed to emit as a black body. Since each of the layers is assumed to be black, we can use the Stephan–Boltzmann law to write $F_j = \sigma T_j^4$ and so on. Since the effective temperature is the radiating temperature of the uppermost black layer of the atmosphere, we can also write

$$F_1 = \sigma T_e^{*4} \tag{14.12}$$

By combining (14.10)–(14.12) and examining Figure 14.8, we can derive a relationship between the radiative surface temperature of a planet (T_0) and the effective temperature:

$$T_0 = (1 + n) T_e^* \tag{14.13}$$

where n is the number of optically black layers in the column or the optical depth.

From observations of T_0 and calculations of T_e^*, n can be estimated from (14.13). We see from Table 14.6 that $n \ll 1$ for Mercury and Mars. A very large value of $n = 84$ is obtained for Venus. The large greenhouse effect for Venus is attributable primarily to CO_2, which constitutes in excess of 95% of the atmospheric mass. Earth has a moderate value of $n = 0.65$.

At the very high temperature and pressure that characterize the Venusian surface, chemical reactions between the atmosphere and the surface become very important. It appears that the abundances of all of the reactive gases on Venus, including oxygen, carbon dioxide, and the sulfur compounds, are regulated primarily by buffer reactions with crustal minerals. A crustal source of CO_2 is of particular interest in the context of the greenhouse effect. The surface reactions appear to be dominated by the thermal reactions of sulfur- and carbon-containing species. The Venusian surface contains a considerable amount of calcium carbonate ($CaCO_3$), which provides a source of atmospheric CO_2 at high temperatures and pressures. As an increasing amount of CO_2 becomes incorporated into the atmosphere, the greenhouse effect is enhanced, which increases the planetary temperature further and increases the breakdown of the lithospheric carbonates. This positive feedback between increasing surface temperatures and increasing concentration of greenhouse gases is known as the *runaway greenhouse effect*.

14.3.2 Planetary Time Scales

If radiative transfer were the only process occurring in a planetary atmosphere, the surface temperatures of the planets would be determined solely by the net radiation at the top of the atmosphere. Thus, the equatorial regions would be warmest and the poles would be extremely cold. Whereas this is the case on Mars and to a lesser extent on Earth, other planets such as Venus and all of the Jovian planets have little or no equator-to-pole temperature gradients at the surface or deep in their interiors.

Simple arguments allow us to understand why there are differences in horizontal temperature gradients among the planets. There are two fundamental time scales that determine how a planetary atmosphere transfers heat. The *radiative time scale*, τ_{rad}, is the time it would take for an atmosphere above some pressure level p_0 to reduce its temperature by $1/e$ of its initial value via radiative cooling if solar radiation were turned off. The *dynamic time scale*, τ_{dyn}, is the time required to move a parcel over a characteristic distance in the atmosphere and, in so doing, transport heat from one location to another.

We can obtain a simple measure of a planetary radiative time scale as follows. Consider an atmosphere with a surface pressure of p_0. If the solar heating is turned off, the atmosphere will cool by thermal radiation to space. From (3.34), we can write the following expression for radiative heating:

$$\frac{\partial T}{\partial t} = \frac{g}{c_p}\frac{\partial F}{\partial p}$$

A simple estimate of the column radiative cooling for a black atmosphere can be obtained by using the slab model described in section 14.3.1. Assume that the entire column is initially in radiative equilibrium. At the instant the sun's heating is turned off, each of the n slabs remains in radiative equilibrium except for the top slab, $n = 1$. Once the top slab has cooled, the remainder of the slabs cool by coming into radiative equilibrium with the top slab. Hence the flux divergence at the top of the atmosphere controls the cooling of the column, and we can write

$$\frac{\partial T}{\partial t} = \frac{g}{c_p} \frac{\left(0 - \sigma T_e^{*\,4}\right)}{p_0} = -\frac{g}{c_p} \frac{\sigma T_e^{*\,4}}{p_0} \tag{14.14a}$$

where T_e^{*} is the equivalent black-body temperature of the planet. If the atmosphere is not black and $n < 1$, such as on Earth and Mars, then the atmosphere can cool more rapidly, since the entire atmosphere is cooling to space. In such a case, we increase the cooling rate relative to (14.11a) by scaling the surface pressure by n:

$$\frac{\partial T}{\partial t} = -\frac{g}{c_p} \frac{\sigma T_e^{*\,4}}{n\,p_0} \tag{14.14b}$$

The product np_0 can be considered as the width of a weighting function for the atmospheric transmission, which describes the pressure-depth of the atmosphere over which most of the emission to space occurs. To determine the radiative time scale, we integrate (14.14a) from T_e^{*} to $T_e^{*\prime} = T_e^{*}/e$, where $T_e^{*\prime}$ is an e-folding temperature for the cooling. Thus we obtain

$$\tau_{rad} = \frac{p_0 c_p}{g\,\sigma} \int_{T_e^{*\prime}}^{T_e^{*}} \frac{1}{T^4}\,dT = \frac{p_0 c_p}{4\,g\,\sigma T_e^{*\,3}\left(1 - \frac{1}{e^3}\right)} \approx \frac{p_0 c_p}{4\,g\,\sigma T_e^{*\,3}} \tag{14.15a}$$

or if $n < 1$, we have

$$\tau_{rad} = \frac{n\,p_0 c_p}{4\,g\,\sigma T_e^{*\,3}} \tag{14.15b}$$

From (14.15a,b) it is possible to identify the planetary properties that determine the radiative time scale. The value of τ_{rad} decreases with increasing mass of the planet through g, but increases with increasing mass of the atmosphere through p_0. For a given mass of atmosphere (p_0), τ_{rad} decreases as a function of the temperature of the radiating temperature, reflecting the fourth-power black body radiative flux dependency on temperature. Finally, τ_{rad} depends on the composition of the atmosphere through the heat capacity c_p. All other things being equal, the radiative time scale for a hydrogen–helium atmosphere ($c_p \approx 12,000$ J kg^{-1} K^{-1}) will be about an order of magnitude greater than for an atmosphere dominated by CO_2 or N_2 ($c_p \approx 1000$ J kg^{-1} K^{-1}).

If r_P is the planetary radius and u is a typical horizontal wind speed, the dynamic time scale may be defined as

$$\tau_{dyn} = \frac{r_P}{u} \qquad (14.16)$$

The ratio of the two time scales, ε, is given by

$$\varepsilon = \frac{4 \, r_P \, g \, \sigma \, T_e^{* \, 3}}{u \, p_0 \, c_p}, \qquad n \geq 1 \qquad (14.17a)$$

or

$$\varepsilon = \frac{4 \, r_P \, g \, \sigma \, T_e^{* \, 3}}{u \, n \, p_0 \, c_p}, \qquad n < 1 \qquad (14.17b)$$

which provides a measure of the relative importance of radiative and dynamic effects.

Values of the time scales for each of the planets with a significant atmosphere are given in Table 14.7. There are several different regimes for ε.

(i)   If $\varepsilon \gg 1$, the dynamic time scale is much greater than the radiative time scale, and hence radiative processes dominate. This regime is characteristic of Mars. This regime occurs if u and/or p_0 are very small, reflecting slow planetary motions and small heat capacity, respectively. Thus, the fluxes of heat by horizontal atmospheric

Table 14.7 Estimates of the planetary radiation and dynamic time scales and their ratio ε for planets in the solar system. Estimates are for the entire atmospheres of the planets except for Jupiter and Saturn where only the upper 1 atmosphere was considered. A value of c_p of 12,000 J K^{-1} kg^{-1} is used for Jupiter and Saturn.

Planet	T_e (K)	p_0 (atm)	n	τ_{rad}	u (ms^{-1})	τ_{dyn} (day)	ε
Venus	214	90	>1	36 yr	2	35	0.003
Earth	254	1	0.65	18 days	10	7	0.39
Mars	217	0.01	0.12	0.05 day	5	10	200
Jupiter	110	1	>1	60 yr	50	16	0.0007
Saturn	80	1	>1	299 yr	50	14	0.0001

motions would be small compared to the local cooling rate by radiative effects. With $\varepsilon \gg 1$, a parcel is in approximately local radiative equilibrium with outer space so that the temperature of the parcel during its motion would be determined by the local radiative fluxes. As a parcel is advected, it cools rapidly to space so that its initial temperature signature is rapidly lost. Consequently, on a planet with $\varepsilon \gg 1$, the equator-to-pole temperature gradient is very large, in response the latitudinal variations of solar radiation. Also, as the radiative cooling of the atmospheric parcel is so efficient, a very large diurnal variation of temperature is expected.

(ii) If $\varepsilon \ll 1$, the dynamic transports of heat dominate over radiative cooling. This regime, characteristic of Venus, Jupiter, and Saturn, occurs because u is very large, indicating rapid transports of heat by horizontal motions, and p_0 is large, indicating very large atmospheric heat storage. Consider again the motion of a parcel from one point on a planet to another. Since the atmospheric pressure is so high, the time to cool the parcel radiatively is very long. With a slow radiative cooling rate, the temperature of the parcel varies adiabatically, retaining its initial thermal signature. Thus, for $\varepsilon \ll 1$ a very small or negligible equator-to-pole temperature difference is expected. For similar reasons, the diurnal variation of temperature on such a planet would be negligibly small.

(iii) For $\varepsilon \sim 1$, there is parity between the two time scales. Such a situation occurs if the mass of the atmosphere is moderate and the velocities on the planet are relatively weak, which is the case for Earth. Under these conditions, a parcel moving from one location to another will radiatively cool but not at a rate fast enough to destroy the characteristics of its initial thermal properties. On such a planet, one would expect a pole-to-equator temperature difference but one where the temperature of a parcel at a particular latitude is not in radiative equilibrium with the net radiative fluxes at the top of the atmosphere. Furthermore, a moderate diurnal temperature is expected.

Venus has a planetary albedo of about 71% (Table 14.6) which is more than twice that of Earth. It is estimated that 24% of the incoming energy is absorbed in the cloud layer. The remaining energy penetrates the cloud layers but perhaps only 2% of the incident radiation at the top of the atmosphere reaches the surface. Observations indicate that the surface temperature of Venus is essentially the same at all latitudes and between the day and night sides of the planet, at about 740 K. In the lower atmosphere of Venus, $\varepsilon \ll 1$, due principally to the very large surface pressure. The vast heat capacity of the lower troposphere results in a radiative time scale that is orders of magnitude larger than the dynamic time scale (Table 14.7). Thus the motion of the atmosphere tends to homogenize the temperatures latitudinally and between the day and night sides of the planet.

For Earth, $\varepsilon \approx 0.4$. The mass of the Earth's atmosphere is sufficiently large that the radiative cooling time scale is nearly equivalent to the dynamic time scale. Whereas the strong pole-to-equator temperature gradient suggests that radiative effects are very important, dynamic transports of heat tend to predominate. For example, the latitudinal surface temperature gradient does not follow the net radiation at the top of the atmosphere exactly, suggesting that transports of heat continually modify the surface temperatures. Occasionally, dynamic transports of heat far exceed radiative effects. Each winter there are the familiar outbreaks of Arctic air masses that propagate to low latitudes before being modified by radiative heating and surface effects. Within the Earth's oceans, $\varepsilon \ll 1$ because of the high density and heat capacity of sea water. Since $\varepsilon \ll 1$, heat cannot be carried away in the ocean more quickly by radiative processes at the top of the column than can be transported by the deep ocean currents. As a result, the deep ocean is relatively homogeneous in temperature.

For Mars, $\varepsilon \gg 1$ principally because of the very low mass of the atmosphere, indicating that radiative processes dominate the dynamic transports of heat. As a result very large pole-to-equator temperature gradients are observed on the planet, as well as a strong diurnal cycle.

Estimates of the interior temperatures of Jupiter and Saturn indicate that there is little, if any, temperature gradient with latitude as one goes deeper into the interior of the atmosphere. This is a characteristic of all of the outer planets and is a further indication of extremely long radiative time scales compared to dynamic time scales. Thus, from (14.16a), $\varepsilon \ll 1$ for both Jupiter and Saturn.

Like Jupiter and Saturn, dynamic processes on Uranus and Neptune control temperature distributions deep within the atmospheres. The rotational axis of Uranus is in the solar plane so that one pole of the planet points towards the sun for over 50 years at a time. Yet, the temperature of the planet appears to be relatively homogeneous within the interior, indicating once again that in dense planetary atmospheres the radiative time scales are extremely long and the temperature distribution is dominated by dynamic processes.

14.4 Water on the Terrestrial Planets

Insight into the relative abundances and phases of water and carbon dioxide on Venus, Earth, and Mars can be gained by examining the states of the planetary atmospheres on CO_2 and H_2O phase diagrams (Figure 14.9). For CO_2, only the gas–ice phase transition is relevant for the current conditions on the terrestrial planets. However, as discussed in Section 14.2.1, the critical point of CO_2 is encountered in the Venusian atmosphere. On Venus, CO_2 exists in gaseous form along with H_2O. Water on Mars exists in both the solid or gaseous forms, but with the present vapor pressure in the Martian atmosphere, the liquid phase is precluded. On Earth, water exists in all three phases.

To understand how the terrestrial planets came to possess different abundances of H_2O and CO_2, it is instructive to consider the evolution of the planetary atmospheres. Figure 14.10 describes the evolution of the terrestrial atmospheres in the context of the water phase diagram, which is referred to as the *simple water theory* of atmospheric evolution. Immediately after the primitive atmospheres were swept away by the immense solar wind shortly after the formation of the solar system, each of the planets possessed essentially no atmosphere, with $n \approx 0$ and $T_0 \approx T_e$. Assuming a 17%

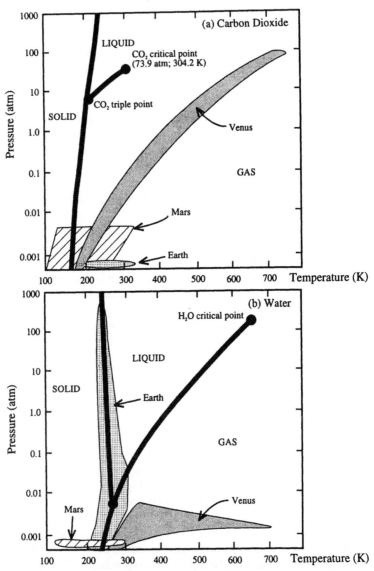

Figure 14.9 Phase diagrams for the planetary environments of Venus, Earth, and Mars for (a) CO_2 and (b) H_2O. Phase transition lines for CO_2 and H_2O are shown as heavy solid lines. The ranges in partial pressure–temperature space of the two substances on each planet are shown as shaded (Venus), dashed (Mars), and stippled (Earth) areas. (a) On Venus, CO_2 gas dominates the atmosphere from the surface to the outer reaches of the planet. A much lower concentration of CO_2 occurs on Mars. However, over the range of temperature, CO_2 can change phase between gas and solid form. CO_2 on Earth is only a trace constituent and is constrained in a gaseous form. (b) Water on Earth can exist in all three forms over a large temperature and pressure range. Venusian water is restricted to the vapor form over a large temperature range but at low pressure. Martian water occurs in relatively small concentrations but can exist either as a gas or as ice.

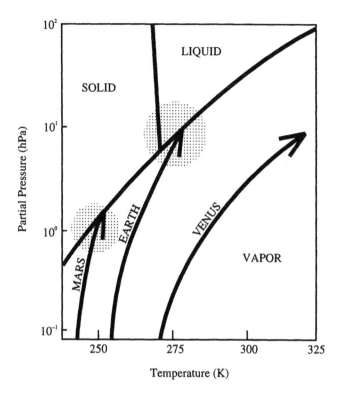

Figure 14.10 Water phase diagram illustrating the relative abundances and phases of water on Mars, Earth, and Venus.

surface albedo (the same albedo as Mars) for each of the primitive terrestrial planets, an initial equivalent black-body temperature for each planet is determined easily from (12.1). Assuming that the volcanic constituents are the same as those currently out-gassed by Earth (58% H_2O, 24% CO_2, 12% sulfuric gases, plus trace gases), water vapor and carbon dioxide would be the main constituents introduced into each planet's atmosphere. With continued outgassing, the optical depth would have increased so that gradually the surface temperature would start to exceed the equivalent black-body temperature of the planet.

Assuming the same outgassing on all of the planets, the trajectories of planetary atmospheres are seen in Figure 14.10 to intercept the water phase curves in different locations. Mars rapidly encounters the vapor–ice phase transition, so that all subse-quent outgassing of water vapor rapidly changes to ice and a greenhouse effect is retarded. The Earth trajectory intercepts the water phase curves in the vicinity of the triple point of water, allowing the formation of a complex hydrological cycle. Venus, on the other hand, starting with a considerably warmer primitive surface temperature, does not intercept any of the water phase transitions at all. In the following, the evolution of atmospheric water on each of these planets is examined in detail.

14.4.1 Mars

The simple water theory of Mars' evolution (Figure 14.10) places Mars in a state where H_2O remains locked in a perpetual ice phase. Thus H is inhibited from reaching the exosphere and escaping the planet. This picture is contrary to the geological evidence of liquid water on the Martian surface at some time in the planet's history, implying that at some point in time the surface temperature on Mars must have exceeded the freezing point of water with a high enough atmospheric vapor pressure to ensure the liquid phase. That is, Mars at some stage appears to have had a more massive atmosphere than at present and included greenhouse gases (e.g., CO_2 and H_2O). Quantities of CO_2 and H_2O are believed to be incorporated into polar ice caps and in the planetary crust where they exist as a permafrost or as carbonates. The existence of carbonates is suggestive of an earlier and wetter period on Mars, since carbonate-forming reactions require the presence of liquid water. Furthermore, as water cools, an increased amount of CO_2 gas is dissolved in the liquid, promoting the formation of carbonates. As water freezes, carbonate formation ceases. Curiously, the saturation vapor pressure with respect to ice is about equal to the currently observed partial pressure of CO_2. Either this is a coincidental value of the partial pressure of CO_2 or it suggests that a previously more massive atmosphere has collapsed, being in-gassed into liquid H_2O which eventually froze as the greenhouse of the atmosphere decreased.

Why might there have there been warm epochs on Mars? Consider the following hypothesis, which is consistent with the water theory of atmospheric evolution. From Figure 14.10 it is clear that, for primitive and early temperature ranges on Mars, the CO_2 mass in the atmosphere would continue to increase as a result of volcanic outgassing, even though the outgassed water would quickly become solid. A surface temperature below freezing would have precluded a sink of CO_2 via the formation of carbonates. As the CO_2 mass of the atmosphere increased and if the CO_2 growth trajectory was able to avoid the CO_2 gas–solid phase line (Figure 14.9), then an atmosphere of sufficient optical depth to raise the surface temperature above the freezing point of water might have developed.

14.4.2 Earth

Earth differs from the other terrestrial planets because it contains all three phases of water in abundant quantities, whose manifestations on Earth have been described in previous chapters of this text. The simple water theory of Earth's atmospheric evolution has the water trajectory arriving at the triple point of water, where a number of radiation and hydrological feedbacks have kept the climate of Earth fairly well constrained (Figure 14.10).

Temperatures in the outer portions of the Earth's exosphere exceed 400 K. Given the most probable speed for H (Table 14.3) and the escape velocity for Earth (Table 14.2), we arrive at a value of $b = 4$ in (14.5), which suggests that H should be an ephemeral species in the Earth's atmosphere. However, extremely large amounts of H have remained on the planet in the form of water. Thus, either the photodissociation rate in the upper atmosphere is not large enough or there are processes or phenomena endemic to the Earth that restrict the vertical diffusion of water (and hence hydrogen) to the upper levels of the atmosphere.

Evaporation from the warm tropical oceans provides the primary source of water vapor to the atmosphere. Water vapor is transported upwards into the atmosphere through convection, with considerable condensation occurring in the tropical troposphere and into the stratosphere, where global-scale circulations transport the water vapor to higher latitudes. The tropics are the only region on Earth where water vapor can reach the stratosphere through convection. Additional sources of stratospheric water arise from the oxidation of methane. The temperature of the tropical tropopause is the coldest anywhere on Earth, particularly over the tropical ocean warm pool, because of adiabatic cooling associated with the deep convection. Consider a rising parcel of air as it moves through the cold tropical tropopause, referred to as the *cold trap.* If the vapor pressure of the rising parcel exceeds the saturation vapor pressure at the tropopause temperature, then the excess water vapor will condense or deposit onto cloud particles which settle out into the upper troposphere. In this manner the cold trap restricts the upward flux of water. Measurements show that the water vapor pressure of the lower stratosphere is identical to the saturation vapor pressure with respect to ice at the temperature of the tropical tropopause. In the middle and upper stratosphere, observations show that the concentration of water vapor increases. This increase is due to the oxidation of methane which, being far away from phase transitions in the Earth environment, passes through the cold trap and into the stratosphere unimpeded.

The height of the cold trap is directly proportional to the sea surface temperature because of buoyancy effects. The warmer the sea surface temperature, the greater the depth of convection and the larger the adiabatic cooling of the parcels. Because of adiabatic cooling, the temperature of the cold trap is inversely proportional to the sea surface temperature (Figure 14.11). By extension, a warmer sea surface temperature implies a lower vertical flux of water vapor into the stratosphere, so that less water is available for dissociation, thus reducing the loss of hydrogen from the planet, allowing Earth to retain its water.

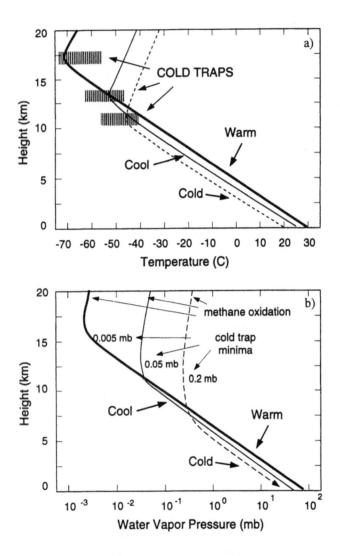

Figure 14.11 The impact of sea surface temperature on the moisture content of the stratosphere. Vertical profiles are shown for sea surface temperatures of 30°C (bold), 25°C (solid), and 20°C (dashed) for a) temperature and b) water vapor pressure. As the sea surface temperature decreases, the convective penetration decreases and the tropical tropopause is lower and warmer. Assuming that all water vapor in the stratosphere must be transported through the tropical tropopause, the tropopause acts as a cold trap and determines the moisture content of the stratosphere. Oxidation of methane in the stratosphere acts as a further source of water vapor, but one that is independent of sea surface temperature. (From Webster, 1994.)

14.4.3 Venus

As illustrated in Figure 14.10, water in the Venusian atmosphere has most likely existed only in the vapor phase throughout its evolution. During the initial period of outgassing after the primitive atmosphere was swept away, both water vapor and CO_2 continued to increase in the Venusian atmosphere. This increased the greenhouse effect of the planet and hence the planetary temperatures. As the atmospheric temperature increases, an increasingly greater water vapor pressure is required in order to cause condensation. However, the greenhouse effect caused by the increasing amount of water vapor was so large that condensation apparently never occurred. Once the surface temperature became sufficiently hot for a lithospheric source of CO_2 to become important, temperatures increased even more rapidly, further precluding condensation of water vapor.

Two processes may have contributed to the loss of hydrogen and, eventually, to the loss of water on the planet: thermal escape and solar wind sweeping. In the absence of water condensation, a cold trap is not possible. Thus, on Venus, there does not appear to be an intrinsic restriction to the vertical flux of water vapor. The temperature of the outer atmosphere of Venus is only about 200–300 K, so b in (14.5) would be between 5 and 10 for hydrogen. Assuming that the early exosphere temperature was close to the current values, thermal escape would be a marginal sink of hydrogen on Venus. However, we have noted that Venus lacks a magnetic field. Therefore, solar wind sweeping is a viable candidate for hydrogen loss on Venus.

14.5 Cloud Physics of the Terrestrial Planets

There is a great diversity of cloud forms among the four terrestrial planets. Earth's clouds were discussed in Chapters 5 and 8. Here we compare the types of clouds that are observed on the other inner planets with those observed on Earth and also compare the physical processes responsible for their formation.

Whereas the atmosphere of Mercury is nebulous, there appears to exist a haze that is observed as light whitish clouds. These are different from the clouds observed on Earth in the sense that they are not related to phase changes of constituents of the atmosphere. More likely, they are remnants of dust raised from the surface of the planet or from periodic outgassing that may occur on the planet.

Clouds cover the entire planet of Venus, although large-scale morphological cloud features are associated with the large-scale atmospheric dynamics. Cloud structure on Venus is quite complex, with three major cloud layers identified over the height range 45–70 km. There also appears to be lower haze regions, and a haze layer above 70 km. Three types of cloud particles have been identified. Small ubiquitous aerosols extend below the main cloud layers, in the altitude range 31–48 km, with $r \sim 0.1$ μm. Within each of the main three cloud layers are particles approximately 1–2 μm in diameter, which are composed of sulfuric acid (H_2SO_4). The production of cloud droplets by photochemical oxidation of SO_2 at the cloud tops and condensation of H_2SO_4 near the cloud base seems plausible. In the middle and lower cloud layers, larger solid crystals ($r > 30 \ \mu m$) are also present, apparently consisting of metal chlorides and elemental sulfur. It has been suggested that the Venusian clouds are precipitating. However, as the precipitate falls below the cloud, through the increasingly hot atmosphere, evaporation rapidly occurs. Observations of lightning discharges on Venus do not seem to indicate the presence of clouds associated with deep, vigorous convection, but rather that chemical processes are responsible for the buildup of electrical charge.

Clouds in the present-day Martian atmosphere are fairly simple. Both CO_2 and water condensates are evident in the form of frosts, fogs, and clouds. Surface-mixing fogs develop shortly after sunrise on valley bottoms and crater floors, as the warming of the sun evaporates some of the water from the soil and mixes with the colder air. Two types of frost have been observed on Mars: water ice frost and carbon dioxide frost. The carbon dioxide frost is less common because temperatures must drop to 148 K. Any CO_2 frost disappears shortly after sunrise. CO_2 frost is observed on the floor of craters less than 4 km elevation near the equator. The only possible explanation for such low temperatures in the equatorial region relates to the low thermal inertia of the material present in the crater bottoms, allowing rapid radiational cooling of the crater. Water frost formation occurs at temperatures between 191 and 196 K. Frosts consisting of water clathrate (water and CO_2) have been observed as well. Wave clouds form in the lee of topographical features. High-altitude clouds often form at night in the mid-latitudes of the winter hemisphere where temperatures are 120 K, but disappear within one hour after sunrise. The most significant and persistent of all clouds on Mars is the polar hood cloud, composed of ice crystals. The polar hood is a condensation phenomenon that occurs over the poles during the winter months. The hood disappears rapidly in the early spring. Dust storms appear to disrupt the hood cloud, occasionally resulting in its temporary disappearance. It is thought that at these times the dust particles act as condensation nuclei which cause CO_2 precipitation onto the polar ice cap.

14.6 Cloud Physics of the Jovian Planets

Clouds form on the Jovian planets due to rising motion, and occur in distinct bands, belts, and the Great Red Spot (and other related ovals). The global distribution of cloud regions and nomenclature are given in Figure 14.12. The EZ is the Equatorial Zone, NEB and SEB are respectively the North and South Equatorial Belts, the NTrZ and STrZ are respectively the North and South Tropical Zones, and the NTeZ and STeZ are North and South Temperate Zones. Poleward of about 50°, the zonal pattern is replaced by a highly varied structure. Cloud bands whose name end with the

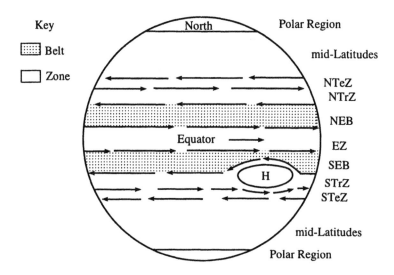

Figure 14.12 The belts and zones of Jupiter. These features do not extend into the disturbed mid-latitude polar regions. The arrows represent high-velocity winds located at the zone/belt boundaries. (After Barbato and Ayer, 1981.)

word "zone" appear to be regions where upward motions are sustained, characterized by bright white clouds. In contrast, cloud bands whose names end with the word "belt" are generally darker and more brown and thought to be areas where the upper clouds are absent due to subsidence. The Great Red Spot is located at 23°S, which is rotating counterclockwise and exhibits cold cloud top temperatures indicative of very high clouds.

The vertical structure of the Jovian clouds is quite complex. Although it is difficult to delineate the atmosphere on this planet, pressures at the 3000 K level are so high (50 kbar) that is probable that no condensed phases could be present. Consider a rising parcel of gas, cooling adiabatically from an initial temperature of 3000 K and pressure of 50 kbar (Figure 14.13). At a temperature of about 2700 K, iron (Fe) begins to condense. In sequence, MgO, Mg_2SiO_4, $MgSiO_3$, and SiO_2 condense, leaving each condensate as a distinct cloud layer. The metallic cloud particles do not travel along with the rising gas because they are too large; if they fall below the cloud base, they evaporate and their vapors are swept back upward to recondense. Thus discrete cloud layers are maintained.

After the condensation of silicone (SiO_2), then sodium oxide and sulfides of potassium, rubidium, and cesium condense. At temperatures near 400 K, ammonium halides, especially NH_4Cl, begin to condense. The only species remaining in the gas at the 300 K level are H_2, the rare gases, CH_4, NH_3, H_2O, and H_2S. The sequence of condensates is H_2O, NH_4SH, and HN_3; the temperature required for CH_4 condensation is too low to be reached on Jupiter (Figure 14.14).

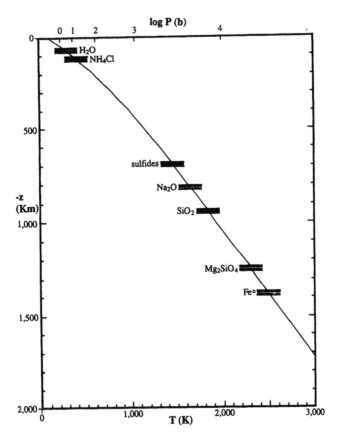

Figure 14.13 Cloud structure deep in Jupiter's atmosphere. The approximate condensation locations of a number of species in a solar-composition model of Jupiter's upper troposphere below the water clouds. The atmosphere is not transparent in the intercloud regions because of the opacity contributed by Rayleigh scattering and collision-induced absorption. (From Lewis, 1995.)

Condensation formed from H_2O, NH_3, and H_2S solutions is quite complex. First consider the two-component NH_3–H_2O system, which forms a solution. From (4.2), we have $f = 4 - \phi$. The condensates that are known to be possible at equilibrium in this system at pressures below 1 kbar include an aqueous NH_3 solution, solid H_2O ice, solid NH_3 ice, and the solid hydrates $NH_3 \cdot H_2O$ and $2NH_3 \cdot H_2O$. A phase diagram for specified pressure ($f = 3 - \phi$) for the NH_3–H_2O system is shown in Figure 14.15. For interpretation of Figure 14.15, it is useful to refer to Figure 4.8. The scalloped line in Figure 14.15 is the freezing curve at which a cooling begins to crystallize. Numerous eutectic points are seen from the minima in the freezing line. Each region

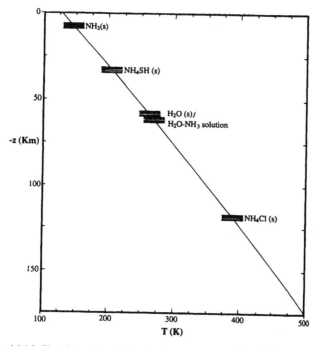

Figure 14.14 Cloud structures in Jupiter's upper troposphere. The saturation levels of ammonia, ammonium hydrosulfide, aqueous ammonia solution and water ice, and ammonium chloride are shown as calculated for a solar composition gas. (From Lewis, 1995.)

in Figure 14.15 labeled B, C, E, F, H, or I is a two-phase region containing one solid ice phase and liquid.

The vertical structure of clouds in the NH_3–H_2O–H_2S system has been calculated by assuming that the atmosphere of Jupiter has exactly the same composition as the sun (e.g. Lewis, 1995). Below the level of H_2O saturation, the concentrations of NH_3 and H_2O are constant. The ammonia concentration in the solution is small, less than $X(NH_3)$ = 0.2, so the first condensate is a dilute aqueous NH_3 solution. Only a few kilometers above the cloud base, the temperatures are low enough to freeze out pure H_2O ice.

Further complexity arises when we allow for condensation of hydrogen sulfide (H_2S). H_2S reacts readily with NH_3 to form the salt NH_4SH (ammonium hydrosulfide) and $(NH_4)_2S$ (ammonium sulfide). Condensation of ammonium hydrosulfide occurs at a temperature of 215 K. At this temperature, water vapor has already been depleted by about a factor of 300 by condensation. Above the NH_4SH condensation level, the H_2S and H_2O partial pressures drop off so rapidly that the cloud masses are negligible down to the point of saturation of ammonia, below 160 K. At this point, a third dense cloud layer condenses that is composed of crystals of solid ammonia.

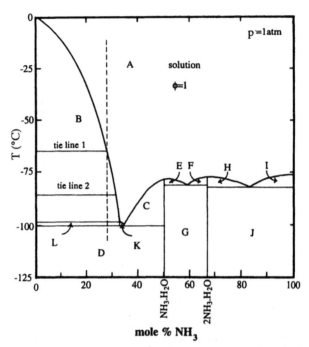

Figure 14.15 NH$_3$–H$_2$O phase diagram. In region A, a stable single-liquid solution of ammonia and water is present. B is a two-phase region containing water ice plus a solution of composition given by the illustrative tie lines (isotherms) 1 and 2. Cooling from 1 to 2 with constant total composition (vertical dashed line) causes ice to form and the concentration of ammonia in the residual liquid to increase. Regions C,E,F,H,I, and K are also two-phase regions ($\phi=2$) in which an ice phase coexists with a solution. Regions D,G,J, and L are two-phase regions occupied by two coexisting ices (such as ammonia hydrate and water ice in region D). Note the deep temperature minima (eutectic points) in the melting curve at which three phases, a solution and two ices, coexist ($\phi=3$). Above the scalloped line, only liquid is present; this curve is the liquidus curve. Below the lines atop regions D,G, and J, only solids are present; this is termed the solidus. (Lewis, 1995.)

14.7 Surface Ice

Besides Earth, a number of planets and moons have ice on part or all of their surfaces. Mars, for example, has a seasonal ice cap of CO$_2$ that forms over permanent polar H$_2$O ice deposits. Recent evidence suggests that permanent water ice exists on Mercury and possibly the Earth's moon in permanently shaded polar craters. The Jovian moons Europa, Callisto, and Ganymede possess water-ice surfaces. Io shows evidence of a solid SO$_2$ surface resulting from volcanic eruption effluents. Rhea, one of

Saturn's moons, appears to be composed principally of water ice. Finally, the surfaces of Neptune's moon Triton are covered with liquid nitrogen and methane ice. Overall, ice exists on planets because of the seasonal variation of solar radiation which promotes a deposition–sublimation or melting–freezing cycle, or because the surface is so cold as to reduce sublimation to negligible levels.

The largest polar cap on Mars occurs in the northern hemisphere and consists of a permanent H_2O base which is covered by frozen CO_2. The CO_2 ice cap is seasonal, disappearing in the northern summer. The southern hemisphere ice cap is composed primarily of CO_2. The seasonal variability of Mars is emphasized by two factors: the eccentricity of its orbit is much greater than either Venus or Earth; and its major atmospheric constituent, CO_2, undergoes a major deposition–sublimation cycle in the polar regions to produce the winter CO_2 ice caps. The eccentricity of the Martian orbit produces a deposition–sublimation process that is asymmetrical between the hemispheres. At the present time, southern hemisphere has a longer, cooler winter than the northern hemisphere but also a shorter, hotter summer. As a result, the winter southern hemisphere ice cap is more substantial than its northern hemisphere counterpart. CO_2 ice extends to about 40°S from the south pole and has an average thickness of about 23 cm. It has been estimated that 10–20% of the atmospheric mass exchanges take place in the seasonal deposition–sublimation process.

The surface of Jupiter's moons Callisto and Ganymede are heavily cratered. Ice on Ganymede appears in large parallel bands that occur between a global system of topographic rifts that reflect tectonic activity. Europa has a much smoother surface, consisting entirely of water ice. Similar to the Arctic Ocean sea ice on Earth, Europa exhibits ridges, faults, cracks, and evidence of flooding, suggesting a dynamic interface between an ice surface and an underlying ocean.

Surface temperature ranges on Mercury are very large, with temperatures during the day exceeding 700 K but at night falling to 100 K. The rotation rate of the planet is 59 days. All parts of the planet are illuminated by the sun during a Mercury year of 88 days. It is surprising then, that there appears to be some evidence from Earth-based radar exploration for the existence of permanent ice on Mercury. For ice to be a stable component, a temperature of about 100 K would be required. Such regions appear to occur on Mercury in the permanently shadowed crater bottoms close to the poles. Ice on the Earth's moon has been recently discovered in a deep basin that is largely in permanent darkness. The extent or the mass of either of these ice fields is not known. It appears that the source of the ice is impacts by comets. Both the Earth's moon and Mercury are subject to solar wind bombardment. After collisions have occurred, water molecules would either escape to space or find shelter from the solar wind. Deep, cold craters in perpetual shadow could act as collector regions.

Notes

The principal reference used for this chapter is *Physics and Chemistry of the Solar System* (1995) by Lewis. Additional references include *Atmospheres* (1981) by Barbato and Ayer, *Atmospheres* (1972) by Goody and Walker, and *Chemistry of Atmospheres* (1991) by Wayne.

Problems

1. Show that if $b > 7$ for a particular atmospheric species in (14.6), it is unlikely that the species will be lost to a planet by thermal escape processes.

2. Above the exopause, gases are stratified by density. Consider the Earth's atmosphere above the exopause. Assume that the temperature is isothermal at 500 K. Relative to their concentrations at the exopause, calculate the distributions of atomic hydrogen, helium, nitrogen, and oxygen in the exosphere. Using the concentrations of the gases given in Table 14.4 as the exopause values, calculate the height above the exopause at which the composition of the atmosphere will be 95% hydrogen.

3. Consider an atmosphere that is made up of five absorbing layers ($n = 5$), each of which is homogeneous and just absorbs any incident longwave radiation. Assume that the atmosphere is transparent to solar radiation but that the surface reflects fraction α_0 of solar radiation back to space.
i)   Show that the surface temperature is given by $T_0 = 1.57T_e$, where T_e is the effective temperature of the planet.
ii) Assume that $T_e = 220$ K. Determine the height and temperature of each of the pressure levels that delineate the n layers. Plot the vertical temperature profile as a function of height, using the temperature of each layer to represent the midpoint temperature of the layer.
iii) Assume that the atmosphere consists of an ideal CO_2 gas and that gravity is the same as on Earth. Using the graph in ii), delineate regions of static stability and instability.

4. On Venus the surface temperature of the day side of the planet has been measured to be the same as that on the night side. On Mars, on the other hand, extremely large temperature variations have been measured between day and night. Define a diurnal time scale for Venus and Mars and show that these measurements are reasonable.

5. Consider a 5 km mountain on Mars. What would the horizontal temperature difference be between a point 1 km above the mountain compared to an adjacent point 6 km above the plain? What would you expect the temperature differences to be at night? What is the critical feature of the Mars atmosphere that makes such a calculation possible? Why would this calculation be more difficult on Earth?

Appendix A | Notation

a	Helmholtz energy (intensive)
a_1	efficiency factor for accretion
A	area
$\mathcal{A}$	Helmholtz energy (extensive)
A_c	fractional area covered by clouds
A_i	fractional surface area covered by sea ice
A_{il}	affinity for freezing from liquid
A_{iv}	affinity for vaporization from ice
A_l	fractional surface area covered by open water
A_{lv}	affinity for vaporization from liquid
A_{lR}	autoconversion of cloud water into rain water
A_p	fractional surface area covered by melt ponds
A_s	fractional surface area covered by snow
$\mathcal{A}_\lambda$	monochromatic absorptivity
B_0	Bowen ratio
c	speed of light
c	specific heat capacity
c_l	specific heat capacity of liquid water
c_o	specific heat capacity of pure ice
c_p	specific heat capacity at constant pressure
c_{pd}	specific heat capacity of dry air at constant pressure
c_{pi}	specific heat capacity of ice at constant pressure
c_{pl}	specific heat capacity of water at constant pressure
c_{ps}	specific heat capacity of snow at constant pressure
c_{pv}	specific heat capacity of water vapor at constant pressure
c_{p0}	specific heat capacity of surface seawater at constant pressure
c_s	speed of sound
c_s	specific heat capacity of snow
c_v	specific heat capacity at constant volume
c_{vd}	specific heat capacity of dry air at constant volume
c_{vi}	specific heat capacity of an ideal gas at constant volume
c_{vv}	specific heat capacity of water vapor at constant volume
C	number of configurations

C	concentration of a scalar quantity
$\mathscr{C}$	capacitance
C_D	drag coefficient
C_{DE}	aerodynamic transfer coefficient for humidity
C_{DH}	aerodynamic transfer coefficient for temperature
C_{vl}	condensation rate of cloud water
$CAPE$	convective available potential energy
$CINE$	convection inhibition energy
CCN	cloud condensation nuclei
CF^{LW}	longwave cloud radiative effect
CF^{net}	cloud net radiative effect
CF^{SW}	shortwave cloud radiative effect
d	drop diameter
d	Earth-sun distance
D	dynamic depth
D	diffusion coefficient
$\mathscr{D}$	vapor diffusion term in the droplet growth rate equation
$\mathscr{D}$	cloud water dilution
D_s	diffusion coefficient for salt in water
D_v	diffusion coefficient for water vapor in air
e	partial pressure of water vapor
e_a	near-surface atmospheric vapor pressure
e_s	saturation vapor pressure with respect to liquid water
$e_\&$	vapor pressure with respect to snow
e_{si}	saturation vapor pressure with respect to ice
$e_{s,tr}$	saturation vapor pressure at the triple point
$\mathscr{E}$	efficiency
$\mathscr{E}$	collection efficiency
$\mathcal{E}_i$	quantities external to the climate system
E_g	internal energy of column of atmosphere
$\mathscr{E}_k$	translational kinetic energy
E_K	kinetic energy
E_0	energy of a single photon
$E_\mathscr{P}$	potential energy
E_{Rv}	evaporation of rain water
E_T	total potential energy
$\dot{E}_0$	surface moisture flux from evaporation

f	degrees of freedom
f	feedback factor
F	irradiance (adiant flux density)
$\mathcal{F}$	feedback
$\mathscr{F}$	force
F_{B0}	ocean surface buoyancy flux
F_c	conductive heat flux
F_{ci}	conductive heat flux in ice
F_{cs}	conductive heat flux in snow
$\mathscr{F}_g$	gravitational force
$\mathscr{F}_p$	vertical pressure gradient force
F_Q	flux density of heat or energy
$\mathscr{F}_R$	frictional force
F_s	salt flux
F_v	flux of water vapor
F_w	heat flux at base of ice
F_{s0}^{adv}	salt transport via fluid motions
F_{s0}^{ent}	salt transport via entrainment at the base of the ocean mixed layer
F_{s0}^{net}	ocean salinity storage term
F_{Q0}^{adv}	energy flux due to fluid motions
F_{Q0}^{ent}	energy flux due to entrainment at the base of the ocean mixed layer
F_{Q0}^{LH}	surface turbulent latent heat flux
F_{Q0}^{net}	ocean heat storage term
F_{Q0}^{PR}	surface sensible heat transfer due to precipitation
F^{rad}	net radiative flux
F_{Q0}^{rad}	net surface radiation flux
F_{Q0}^{SH}	surface turbulent sensible heat flux
F^{LW}	longwave flux
F_{TOA}^{LW}	upwelling longwave radiation at the top of the atmosphere
F_{TOA}^{rad}	net radiative flux at the top of the atmosphere
F^{SW}	shortwave flux
F_{TOA}^{SW}	solar flux at the top of the atmosphere
F_0^{LW}	downwelling infrared radiation at the surface
$F_0^{LW,clr}$	F_0^{LW} under cloudless conditions
F_0^{SW}	downwelling solar radiation at the surface
$F_0^{SW,clr}$	F_0^{SW} under cloudless conditions

F^*	black-body irradiance
F_λ	monochromatic irradiance
F_λ^*	black-body monochromatic irradiance
$F_{\rho v}$	diffusion of water vapor in air
g	Gibbs energy (intensive)
g	acceleration due to gravity
g_0	globally averaged acceleration due to gravity at Earth's surface
$g(h_i)$	ice thickness distribution
G	Gibbs energy (extensive)
G	Gravitational constant
G_f	gain, in the presence of feedback(s)
G_o	gain in the absence of feedback(s)
h	Planck's constant
h	moist static energy
h	enthalpy (intensive)
h_i	enthalpy of ice (intensive)
h_i	sea ice thickness
h_l	enthalpy of liquid (intensive)
h_m	ocean mixed layer depth
h_p	pond depth
h_s	snow depth
h_v	enthalpy of water vapor (intensive)
H	enthalpy (extensive)
H	scale height of the atmosphere
$\mathcal{H}$	relative humidity
$\mathcal{H}_i$	relative humidity with respect to ice
$\mathcal{H}_0$	surface air relative humidity
i	van't Hoff dissociation factor
I	radiance
IFN	ice-forming nuclei
I_j	quantities internal to the climate system
I^*	black-body radiance
I_λ	monochromatic radiance
I_λ^*	black-body monochromatic radiance
J	entropy flux
J_v	spectral entropy flux density
k	Boltzmann's constant

k	von Karman constant
k	Henry's law constant
k_1	efficiency factor for collection
k_λ^{abs}	mass absorption coefficient
k_λ^{ext}	extinction coefficient
$k_\lambda^{v,abs}$	volume absorption coefficient
K	eddy diffusion coefficient
$\mathcal{K}$	heat conduction term in the droplet growth rate equation
K_C	eddy diffusion coefficient for a scalar quantity
K_{lR}	collection of cloud water
K_T	mean bulk modulus
K_θ	eddy diffusion coefficient for potential temperature
K_1, K_2	horizontal eddy diffusion coefficients
l	mean free path
L	latent heat
L_{il}	latent heat of fusion
L_{iv}	latent heat of sublimation
L_{lv}	latent heat of vaporization
L_o	luminosity of the sun
LFC	level of free convection
LNB	level of neutral buoyancy
m	mass
m_d	mass of dry air
m_l	mass of liquid water
m_P	planetary mass
m_v	mass of water vapor
M	molecular weight
M_d	mean molecular weight of dry air
M_v	mean molecular weight of water vapor
MCL	mixing condensation level
n	number of moles
n	complex index of refraction
n_{im}	index of refraction (imaginary component)
n_{re}	index of refraction (real component)
N	Brunt-Vaisala frequency
N	number of particles (molecules, drops, etc.)
N_{CCN}	number of cloud condensation nuclei per unit volume

N_o	Avogadro's number
OLR	outgoing longwave radiation
p	pressure
p_a	pressure at reference level a
p_{crit}	critical pressure
p_d	pressure of dry air
p_l^o	vapor pressure over pure liquid water
p_s	saturation pressure
p_{tr}	triple point pressure
p_0	atmospheric pressure at the surface
p_o	vapor pressure over a pure solvent
P	probability
P	length of the planetary year
P_{ad}	adiabatic precipitation amount
$\dot{P}_{ad}$	adiabatic precipitation rate
$\dot{P}_r$	rainfall rate
$\dot{P}_s$	snowfall rate
q	heat (intensive)
q_p	constant-pressure heating
q_s	saturation specific humidity
q_v	constant-volume heating
q_v	specific humidity
q_{v0}	saturation specific humidity (at surface)
q_0	specific humidity at a surface
Q	heat (extensive); radiant energy
Q	climate forcing
Q_{abs}	absorption efficiency
Q_{ext}	extinction efficiency
Q_p	heating at constant pressure
Q_{sca}	scattering efficiency
r	radius
r^*	critical radius for droplet growth
r_e	effective radius
r_{ei}	effective radius for ice particle
r_{el}	effective radius for liquid particle
r_o	initial drop radius
r_P	planetary radius

r_δ	radius of snow element
R	collector drop radius
R	specific gas constant
R^*	universal gas constant
$\dot{R}$	river runoff rate
Re	Reynold's number
Ri_B	bulk Richardson number
R_d	specific gas constant for dry air
$\mathcal{R}_f$	feedback gain ratio
R_v	specific gas constant for water vapor
$\mathcal{R}_\lambda$	monochromatic reflectivity
R_ρ	buoyancy flux ratio
$\mathcal{R}^\circ$	reduced reflectivity
s	salinity
s_i	ice salinity
s_m	salinity of the mixed layer
s_0	ocean surface salinity
S	solar constant
$\mathcal{S}$	saturation ratio
$\mathcal{S}^*$	critical saturation ratio for droplet growth
$\mathcal{S}_i$	saturation ratio with respect to ice
Sc	spreading coefficient
S_C	body source term
S_s	source term for salinity
t^*	surface renewal time scale
T	temperature
T_a	near-surface atmospheric temperature
T_c	cloud temperature
T_{ci}	ice-cloud temperature
T_{cl}	liquid-cloud temperature
T_{crit}	critical temperature
T_D	dew-point temperature
T_{D0}	dew-point temperature at the surface
T_e	equivalent temperature
T_e^*	equivalent black-body temperature
T_f	freezing temperature
ΔT_f	freezing point depression

T_F	frost-point temperature
T_i	ice temperature
$\mathcal{T}_i$	transmissivity of sea ice to solar radiation
T_I	ice-bulb temperature
T_{Ia}	near-surface ice-bulb temperature
T_m	ocean mixed layer temperature
T_s	saturation temperature
T_{δ}	snow temperature
T_{sun}	sun temperature
T_{tr}	triple point temperature
T_v	virtual temperature
T_{vl}	liquid water virtual temperature
T_w	wet-bulb temperature
T_{Wa}	near-surface atmospheric wet-bulb temperature
T_{W0}	wet-bulb air temperature of the air at surface
$\mathcal{T}_\lambda$	monochromatic transmissivity
T_ρ	temperature at maximum density
T_v^*	monochromatic brightness temperature
T_0	surface temperature
$(\Delta T_0)^\wedge$	zero-feedback temperature change
T°	freezing point of a pure solvent
$\mathcal{T}^\circ$	reduced transmissivity
u	internal energy (intensive)
u	3-dimensional velocity
u_*	friction velocity
u_a	near-surface wind speed
u_{ent}	entrainment velocity
u_{esc}	escape velocity
u_o	most probably speed
u_x	x-component of the velocity
u_y	y-component of the velocity
u_z	z-component of the velocity
u_T	terminal velocity
u_0	ocean surface velocity
U	internal energy
v	specific volume

v_a	fractional volume of air bubbles
v_i	specific volume of ice
v_l	specific volume of liquid water
v_v	specific volume of water vapor
V	volume (extensive)
w	work (intensive)
w	mixing ratio
w_i	ice mixing ratio
w_l	liquid water mixing ratio
w_{lo}	threshold liquid water mixing ratio
w_l^{ad}	adiabatic liquid water mixing ratio
w_R	precipitation mixing ratio
w_s	saturation mixing ratio
w_{si}	saturation mixing ratio with respect to ice
w_t	total water mixing ratio
w_v	water vapor mixing ratio
w_0	water vapor mixing ratio at the surface
W	work (extensive)
$\mathcal{W}_i$	ice water path
$\mathcal{W}_l$	liquid water path
$\mathcal{W}_l^{ad}$	adiabatic liquid water path
W_{st}	surface tension work
$\mathcal{W}_v$	precipitable water, water vapor path
x	size parameter
X	thermometric property
X	mole fraction
X, X_i	climate variables
X_{H_2O}	mole fraction of water
X_{solt}	mole fraction of a solute
y_i	age of ice
z	height; depth
z_a	atmospheric reference level
z_s	lifting condensation level (LCL)
z_0	surface roughness
Z	geopotential height
Z	zenith angle; solar zenith angle
Z	compressibility factor

α	coefficient of thermal expansion
α	albedo
α_i	ice albedo
α_l	open water albedo
α_p	melt pond albedo
α_P	planetary albedo
α_s	snow albedo
$\alpha_{\lambda 0}$	surface spectral albedo
α_0	surface albedo
β	coefficient of saline contraction
β_{sca}	infrared diffusivity factor associated with scattering
γ	compressibility
γ	characteristic property
Γ	atmospheric temperature lapse rate
Γ_{ad}	adiabatic lapse rate in the ocean
Γ_d	dry adiabatic lapse rate
Γ_{env}	environmental lapse rate
Γ_m	lapse rate of a parcel subjected to entrainment
Γ_s	saturated adiabatic lapse rate
Γ_{si}	ice-saturation adiabatic lapse rate
δ	solar declination angle
δ	sensitivity (of a climate variable to change in another climate variable)
ε	ratio of the mass of water vapor to the mass of dry air (M_v/M_d)
ϵ	emissivity
ϵ_0	surface longwave emissivity
θ	potential temperature
θ_a	near-surface potential temperature
θ_e	equivalent potential temperature
θ_{il}	ice-liquid water potential temperature
θ_l	liquid water potential temperature
θ_m	potential temperature of the ocean mixed layer
θ_v	virtual potential temperature
θ_{va}	near-surface virtual potential temperature
θ_{v0}	virtual potential temperature at the surface
θ_W	wet-bulb potential temperature
θ_η	entropy potential temperature
θ_0	potential temperature (at surface)

κ	thermal conductivity
κ_i	thermal conductivity of sea ice
κ_o	thermal conductivity of pure ice
κ_s	thermal conductivity of snow
λ	wavelength
λ_{max}	wavelength of maximum emission
Λ	slope factor
Λ	available potential energy
μ	dynamic viscosity
μ	chemical potential
μ_d	chemical potential (dry air)
μ_i	chemical potential (ice)
μ_l	chemical potential (liquid)
μ_v	chemical potential (water vapor)
μ_{vs}	chemical potential of water vapor at saturation
μ_v^o	reference chemical potential for water vapor
ν	molecular viscosity
ν	frequency
η	entropy (intensive)
$\mathcal{H}$	entropy (extensive)
$\mathcal{H}_P$	planetary entropy
$\mathcal{H}_v^*$	monochromatic radiant entropy
ρ	density
ρ_a	air density
ρ_d	density of dry air
ρ_i	density of ice
ρ_l	density of liquid water
ρ_m	density of the ocean mixed layer water
ρ_o	density of the pure phase (pure water; pure ice)
ρ_v	density of water vapor
ρ_s	density of snow
ρ_θ	potential density (seawater)
ρ_0	surface density
σ	Stefan-Boltzmann constant
σ	surface tension between two phases
σ_{abs}	absorption cross section
σ_{ext}	extinction cross section

σ_{iv}	surface tension between ice and liquid
σ_{lv}	surface tension between the liquid and vapor phases
σ_{sca}	scattering cross section
$\sigma_{\delta l}$	surface tension between snow and liquid
$\sigma_{\delta v}$	surface tension between snow and vapor
σ_t	density of seawater (as departure from freshwater value, $\rho-\rho_o$)
τ_{abs}	absorption optical depth
τ_{ext}	extinction optical depth
τ_{sca}	scattering optical depth
τ_c	cloud optical depth
τ_g	Brunt-Vaisala period of oscillation
τ_{rad}	radiative timescale
τ_{dyn}	dynamic timescale
τ_λ	optical thickness, optical depth
ϕ	geopotential
ϕ	latitude
ϕ	inclination angle of the planetary axis of rotation
φ	number of phases in a system
χ	number of components in a system
ψ	hour angle
Ω, ω	solid angle
Ω	length of the planetary day
Ω_o	solid angle subtended by the sun at a point on the planet's sfc.

subscripts

a	adiabatic
a	near-surface reference level in atmosphere, usually 10 m
d	dry air
i	ice
l	liquid water
s	saturation
δ	snow
v	water vapor
λ	monochromatic; wavelength dependent
0	surface
abs	absorbtion
ext	extinction
sca	scattering

Appendix B | Physical Constants

Fundamental Constants

Universal gas constant (R^*)	8.314 J mol^{-1} K^{-1}
Boltzmann's constant (k)	1.38 x 10^{-23} J K^{-1}
Stefan–Boltzmann constant (σ)	5.67 x 10^{-8} W m^{-2} K^{-4}
Planck's constant (h)	6.63 x 10^{-34} J s
Speed of light (c^*)	2.998 x 10^8 m s^{-1}
Gravitational constant (G)	6.67 x 10^{-11} N m^2 kg^{-2}
Avogadro's number (N_A)	6.022 x 10^{26} kmol^{-1}

Sun

Luminosity (L_0)	3.92 x 10^{26} W
Mass (m)	1.99 x 10^{30} kg
Radius (r)	6.96 x 10^8 m

Earth

Average radius (a)	6.37 x 10^6 m
Equatorial radius	6.378 x 10^6 m
Polar radius	6.357 x 10^6 m
Standard gravity (g)	9.80665 m s^{-2}
Mass of Earth	5.983 x 10^{24} kg
Mass of ocean	1.4 x 10^{21} kg
Mass of atmosphere	5.3 x 10^{18} kg
Mean angular rotation rate (Ω)	7.292 x 10^{-5} rad s^{-1}
Solar constant (S)	1367±2 W m^{-2}
Mean distance from sun $(\overline{d})$	1.496 x 10^{11} m
Acceleration due to gravity at surface at earth (g_0)	9.81 m s^{-2}
Solar irradiance on a perpendicular plane at distance $\overline{d}$ from sun	1.37 x 10^3 W m^{-2}

Dry Air

Average molecular weight (M_d)	28.97 g mol^{-1}
Gas constant $(R) = R^*/M_d$	287 J K^{-1} kg^{-1}
Density at 0°C and 101.325 Pa	1.293 kg m^{-3}
Specific heat at constant pressure (c_p)	1004 J K^{-1} kg^{-1}
Specific heat at constant volume (c_v)	717 J K^{-1} kg^{-1}
Density at 0°C and 1000 mb pressure	287 J deg^{-1} kg^{-1}
Thermal conductivity at 0°C	2.40 x 10^{-2} J m^{-1} s^{-1} K^{-1}

Water

Molecular weight (M_v)	18.016 g mol^{-1}
Gas constant for vapor ($R_v = R^*/M_v$)	461 J K^{-1} kg^{-1}
Density of pure water at 0°C	1000 kg m^{-3}
Temperature of maximum density	3.99 °C
Density of ice 0°C	917 kg m^{-3}
Temperature of triple point	0.0100 °C
Specific heat of vapor at constant pressure	1952 J K^{-1} kg^{-1}
Specific heat of vapor at constant volume	1463 J K^{-1} kg^{-1}
Specific heat of liquid water at 0°C	4218 J K^{-1} kg^{-1}
Specific heat of ice at 0°C	2106 J K^{-1} kg^{-1}
Latent heat of vaporization at 0°C	2.5 x 10^6 J kg^{-1}
Latent heat of vaporization at 100°C	2.25 x 10^6 J kg^{-1}
Latent heat of fusion at 0°C	3.34 x 10^5 J kg^{-1}

Typical values for the ocean and atmosphere

Freezing point of seawater, $s = 24.7$ psu	T_f	−1.33	°C
Average surface temperature of the Earth	T_0	15	°C
Average atmospheric surface pressure	p_0	1.013 x 10^5	Pa
Average atmospheric surface density	ρ_0	1.225	kg m^{-3}
Atmospheric surface wind speed	u	5	m s^{-1}
Atmospheric drag coefficient (10 m) (neutral conditions)	C_D	0.95 x 10^{-3}	—
Atmospheric surface stress	τ_0	0.03	N m^{-2}
Specific heat of ocean—constant pressure and salinity: $T = 20$ °C, $s = 35$ psu	c_p	3994	J (kg K)$^{-1}$
Specific heat of ocean—constant volume and salinity: $T = 20$ °C, $s = 35$ psu	c_v	3939	J (kg K)$^{-1}$
Salinity of seawater	s	35	10^{-3} (psu)
Surface tension, clean water	σ	0.079	N m^{-1}
Molecular salinity diffusivity	D_s	1.5 x 10^{-9}	m^2 s^{-1}
Horizontal eddy diffusivity of salinity	$K_{s,1,2}$	1.5 − 3 x 10^3	m^2 s^{-1}
Vertical eddy diffusivity of salinity	$K_{s,3}$	3 − 7 x 10^5	m^2 s^{-1}
Horizontal eddy thermal diffusivity	$K_{Q,1,2}$	10 − 10^5	m^2 s^{-1}
Vertical eddy thermal diffusivity	$K_{Q,3}$	2 x 10^{-6} − 3 x 10^{-2}	m^2 s^{-1}
Molecular thermal conductivity	κ	0.596	W m^{-1} K^{-1}

Appendix C | Units and Their SI Equivalents

Table C.1 Prefixes for Decimal Multiples of SI Units

Multiple	Prefix	Symbol	Submultiple	Prefix	Symbol
10^{18}	exa	E	10^{-1}	deci	d
10^{15}	peta	P	10^{-2}	centi	c
10^{12}	tera	T	10^{-3}	milli	m
10^9	giga	G	10^{-6}	micro	μ
10^6	mega	M	10^{-9}	nano	n
10^3	kilo	k	10^{-12}	pico	p
10^2	hecto	h	10^{-15}	femto	f
10^1	deka	da	10^{-18}	atto	a

Table C.2 SI Units

Quantity	Name of Unit	Symbol	Definition
Length	meter	m	
Mass	kilogram	kg	
Time	second	s	
Temperature	Kelvin	K	
Force	newton	N	$kg\ m\ s^{-2}$
Pressure	pascal	Pa	$N\ m^{-2} = kg\ m^{-1}\ s^{-2}$
Energy	joule	J	$kg\ m^2\ s^{-2}$
Power	watt	W	$J\ s^{-1} = kg\ m^2\ s^{-3}$
Frequency	hertz	Hz	s^{-1}
Celsius temperature	degree Celsius	°C	$K-273.15$

Appendix D   Saturation Pressures over Pure Liquid Water and Pure Ice as a Function of Temperature

T (°C)	e_w (hPa)	e_i (hPa)	T	e_w	e_i	T	e_w	T	e_w
-50	0.0635	0.0393	-24	0.8826	0.6983	1	6.565	26	33.606
-49	0.0712	0.0445	-23	0.9647	0.7708	2	7.054	27	35.646
-48	0.0797	0.0502	-22	1.0536	0.8501	3	7.574	28	37.793
-47	0.0892	0.0567	-21	1.1498	0.9366	4	8.128	29	40.052
-46	0.0996	0.0639	-20	1.2538	1.032	5	8.718	30	42.427
-45	0.1111	0.0720	-19	1.3661	1.135	6	9.345	31	44.924
-44	0.1230	0.0810	-18	1.4874	1.248	7	10.012	32	47.548
-43	0.1379	0.0910	-17	1.6183	1.371	8	10.720	33	50.303
-42	0.1533	0.1021	-16	1.7594	1.505	9	11.473	34	53.197
-41	0.1704	0.1145	-15	1.9114	1.651	10	12.271	35	56.233
-40	0.1891	0.1283	-14	2.0751	1.810	11	13.118	36	59.418
-39	0.2097	0.1436	-13	2.2512	1.983	12	14.016	37	62.759
-38	0.2322	0.1606	-12	2.4405	2.171	13	14.967	38	66.260
-37	0.2570	0.1794	-11	2.6438	2.375	14	15.975	39	69.930

Appendix D Saturation Pressures over Pure Liquid Water and Pure Ice as a Function of Temperature

T (°C)	e_w (hPa)	e_i (hPa)	T	e_w	e_i	T	e_w	T	e_w
-36	0.2841	0.2002	-10	2.8622	2.597	15	17.042	40	73.773
-35	0.3138	0.2232	-9	3.0965	2.837	16	18.171	41	77.798
-34	0.3463	0.2487	-8	3.3478	3.097	17	19.365	42	82.011
-33	0.3817	0.2768	-7	3.6171	3.379	18	20.628	43	86.419
-32	0.4204	0.3078	-6	3.9055	3.684	19	21.962	44	91.029
-31	0.4627	0.3420	-5	4.2142	4.014	20	23.371	45	95.850
-30	0.5087	0.3797	-4	4.5444	4.371	21	24.858	46	100.89
-29	0.5588	0.4212	-3	4.8974	4.756	22	26.428	47	106.15
-28	0.6133	0.4668	-2	5.2745	5.173	23	28.083	48	111.65
-27	0.6726	0.5169	-1	5.6772	5.622	24	29.829	49	117.40
-26	0.7369	0.5719	0	6.1070	6.106	25	31.668	50	123.39
-25	0.8068	0.6322	-	-	-	-	-	-	-

Appendix E | Pseudo-adiabatic Chart

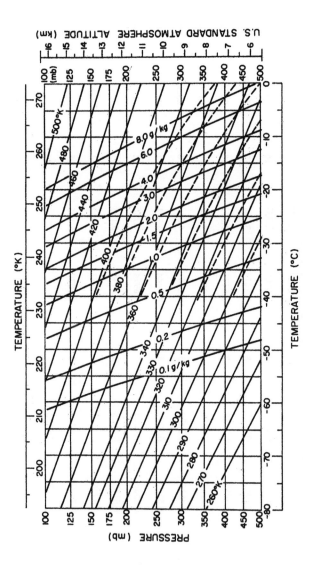

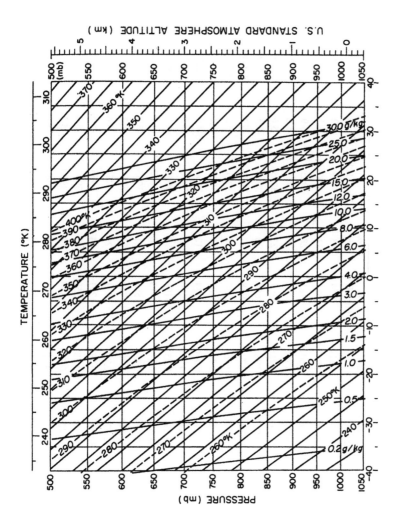

Appendix F Properties of Seawater

p (bar)	s	T (°C)	ρ − 1000 (kg m⁻³)	$\frac{\partial \rho}{\partial s}$	α (10⁻⁷ K⁻¹)	$\frac{\partial \alpha}{\partial s}$	c_p (J kg⁻¹ K⁻¹)	$\frac{\partial c_p}{\partial s}$	θ (10⁻³ °C)	$\frac{\partial \theta}{\partial s}$	c_s (m s⁻¹)	
0	35	−2	28.187	0.814	254	33	3989	−6.2	−2000	0	1439.7	1.37
0	35	0	28.106	0.808	526	31	3987	−6.1	0	0	1449.1	1.34
0	35	2	27.972	0.801	781	28	3985	−5.9	2000	0	1458.1	1.31
0	35	4	27.786	0.796	1021	26	3985	−5.8	4000	0	1466.6	1.29
0	35	7	27.419	0.788	1357	23	3985	−5.6	7000	0	1478.7	1.25
0	35	10	26.952	0.781	1668	20	3986	−5.5	10000	0	1489.8	1.22
0	35	13	26.394	0.775	1958	17	3988	−5.3	13000	0	1500.2	1.19
0	35	16	25.748	0.769	2230	15	3991	−5.2	16000	0	1509.8	1.16
0	35	19	25.022	0.764	2489	14	3993	−5.1	19000	0	1518.7	1.13
0	35	22	24.219	0.760	2734	12	3996	−4.9	22000	0	1526.8	1.10
0	35	25	23.343	0.756	2970	11	3998	−4.9	25000	0	1534.4	1.08
0	35	28	22.397	0.752	3196	9	4000	−4.8	28000	0	1541.3	1.06
0	35	31	21.384	0.749	3413	8	4002	−4.7	31000	0	1547.6	1.03
100	35	−2	32.958	0.805	552	31	3953	−5.8	−2029	−2	1465.1	1.38
100	35	0	32.818	0.799	799	28	3953	−5.7	−45	−2	1465.5	1.35
100	35	2	32.629	0.793	1031	26	3954	−5.6	1939	−2	1474.5	1.33
100	35	4	32.393	0.788	1251	24	3955	−5.5	3923	−2	1483.1	1.30
100	35	7	31.958	0.781	1559	21	3957	−5.3	6901	−1	1495.1	1.26
100	35	10	31.431	0.774	1844	18	3960	−5.2	9879	−1	1506.3	1.22
100	35	13	30.818	0.769	2111	16	3963	−5.1	12858	−1	1516.7	1.19
100	35	16	30.126	0.763	2363	14	3967	−5.0	15838	−1	1526.4	1.16
100	35	19	29.359	0.759	2603	13	3970	−4.9	18819	−1	1535.3	1.13

p (bar)	s	T (°C)	$\rho - 1000$ (kg m⁻³)	$\frac{\partial \rho}{\partial s}$	α (10⁻⁷ K⁻¹)	$\frac{\partial \alpha}{\partial s}$	c_p (J kg⁻¹ K⁻¹)	$\frac{\partial c_p}{\partial s}$	θ (10⁻³ °C)	$\frac{\partial \theta}{\partial s}$	c_s (m s⁻¹)	$\frac{\partial c_s}{\partial s}$
200	35	-2	37.626	0.797	834	28	3922	-5.5	-2076	-3	1472.8	1.39
200	35	0	37.429	0.791	1058	26	3923	-5.4	-107	-3	1482.3	1.36
200	35	2	37.187	0.786	1269	24	3925	-5.3	1862	-3	1491.2	1.33
200	35	4	36.903	0.781	1469	22	3927	-5.2	3832	-3	1499.8	1.30
200	35	7	36.402	0.774	1750	19	3931	-5.1	6789	-3	1511.8	1.26
300	35	-2	42.191	0.789	1101	26	3893	-5.2	-2140	-5	1489.9	1.39
300	35	0	41.941	0.783	1303	24	3896	-5.1	-186	-5	1499.3	1.36
300	35	2	41.649	0.778	1494	22	3899	-5.0	1771	-5	1508.2	1.33
300	35	4	41.319	0.774	1676	20	3903	-5.0	3728	-5	1516.6	1.30
400	35	-2	46.658	0.781	1351	24	3867	-4.9	-2221	-7	1507.2	1.39
400	35	0	46.356	0.776	1534	22	3871	-4.8	-279	-6	1516.5	1.36
400	35	2	46.017	0.771	1707	20	3876	-4.8	1665	-6	1525.3	1.33
400	35	4	45.643	0.767	1872	19	3880	-4.7	3610	-6	1533.7	1.30
500	35	-2	51.029	0.773	1587	22	3844	-4.7	-2316	-8	1524.8	1.38
500	35	0	50.678	0.769	1751	20	3849	-4.6	-386	-8	1534.0	1.35
500	35	2	50.293	0.764	1907	19	3854	-4.6	1546	-7	1542.7	1.32
600	35	-2	55.305	0.766	1807	20	3824	-4.4	-2426	-9	1542.6	1.37
600	35	0	54.908	0.762	1954	18	3829	-4.4	-506	-9	1551.6	1.34
600	35	2	54.841	0.758	2094	17	3835	-4.4	1416	-9	1560.2	1.31

Answers to Selected Problems

Chapter 1

1. 4×10^{-20} m^3

2.   a) 298 K; b) 0.61×10^{-20} J; c) 4.11×10^{-12} Pa

3.   a) 12 mg m^{-3}; b) 0.45 Pa

5.   555.3 kg

6.   a) 0.234 m^3; b) 925 hPa

7. $\Delta(z_2 - z_1) = 13$ m

8. $p = p_0 \left(\dfrac{T}{T_0} \right)^{\left(\frac{g}{R\Gamma} - 1 \right)}$

9.   6,092 m

10. a) $T(z) = \dfrac{g H^2}{2 R z} \left[1 + \left(\dfrac{z}{H} \right)^2 \right]$; b) 6.35 km

11. a) $p = \dfrac{T}{\Gamma} \left[1 - \left(\dfrac{p}{p_0} \right)^{\frac{R\Gamma}{g}} \right]$; b) 1.46 km; c) 50 m

Chapter 2

1.   a) $\Delta u = 0$; $\Delta h = 0$; $\Delta \eta = 18.53$ J kg^{-1} K^{-1} ; $\Delta \theta = 5.65$ K
b) $\Delta u = 14.4$ kJ kg^{-1} ; $\Delta h = 20.1$ kJ kg^{-1} ; $\Delta \eta = 73.6$ J kg^{-1} K^{-1} ; $\Delta \theta = 20$ K
c) $\Delta u = 3131.8$ J kg^{-1}; $\Delta h = 4384.6$ J kg^{-1}; $\Delta \eta = 0$; $\Delta \theta = 0$

2. $-1,102$ J kg^{-1}

3.   a) 0; b) 0; c) $\Delta \eta = N k \ln 2$; d) yes; e) no

4.   1,591 J K^{-1}

6. $\rho_\theta = \rho \left(\dfrac{p}{p_0} \right)^{c_v/c_p}$

8.   a) $\left(\dfrac{\partial u}{\partial v} \right)_T = \dfrac{1}{\gamma} \left[1 - \alpha T_0 - \gamma \rho_0 - \dfrac{v}{v_0} \right]$;   b) approximately 0.3%

Chapter 3

1.   299.5 K

2. $T_\lambda = \exp \left[-\dfrac{1}{2g} \, w \, k_{\lambda,abs} \left(p_0 \right) \sec Z \right]$

3. $T_\lambda = \mathcal{H} k_{\lambda,abs} \, \rho_0 \exp \left(-\dfrac{z}{\mathcal{H}} \right)$

Chapter 4

2.   93°C

3.   −12.3°C

4. $\Delta H = 3106$ J; $\Delta U = 3728$ J; $\Delta \eta = 9.32$ J K^{-1}

5.   a) $e = 21.2$ mb; b) $w_v = 0.0135$; c) $q_v = 0.0133$; d) $c_p = 1016.5$ J K^{-1} kg^{-1};
     e) $T_v = 305.6$ K

6. $w_v = \dfrac{w_0 \, p_0 \, H \, H_w}{H + H_w} \left[1 - \exp \left(\dfrac{H + H_w}{H \, H_w} z \right) \right]$

7. $w_v = \dfrac{\varepsilon \mathcal{H}_0}{2g} \, b \exp \left[a \left(T_0 - T_{tr} \right) \right]$

8.   a) $\dot{E} = 5.02 \times 10^{-5}$ kg m^{-2} s^{-1}; b) $\dot{E} = 4.65 \times 10^{-5}$ kg m^{-2} s^{-1}

9.   a) −0.94°C; b) −1.89°C; c) −1.64°C

10. $m_{NaCl} = 1.15 \times 10^5$ kg

Chapter 5

1. 234 J

2. Sc (olive oil)$=28 \times 10^{-3}$ N m^{-1}; Sc (parafin oil)$=5.6 \times 10^{-3}$ N m^{-1}

3. $r^* = 0.41$ μm, $S^*=0.0019$

4. $r > 2.16$ r^*

5. a) 2.15 s, 2.17×10^4 s, 2.17×10^6 s; b) 20.35 s, 2055 s, 2.06×10^5 s

7. a) $\mathcal{F} = 3/B$; b) $N=2A/B^3$; c) $B=3 \times 10^5$ m^{-1} , $A=2.7 \times 10^{24}$ m^{-6}, $w_l=0.65$ g kg^{-1}

Chapter 6

1. a) 3.4°C; b) 20 g kg^{-1}; c) 0.54; d) 0.73; e) 11.6°C; f) 25°C; g) 14°C; h) 25°C; i) –11.5°C

2. a) $e=40.3$ hPa; $w_v=26.1$ g kg^{-1}; b) 29.1°C; c) 0.0056 kg

3. a) 8°C; b) 10.72 hPa; c) 1 g kg^{-1}

4. 0.87 kg

5. a) 14°C; b) 5°C

6. $T=266$ K; $w_v=4.5$ g kg^{-1}; $w_l=10.2$ g kg^{-1}

Chapter 7

1. $z=81$ m; $u_z=2.12$ m s^{-1}

2. 333 s

3. a) 24.3 m s^{-1}; b) 25.8 m s^{-1}; c) 24.6 m s^{-1}

Chapter 8

1. $T=298$ K

2. $t=1412$ s

3. a) $r=177$ μm; b) 0.062 cm

4. $z = 107$ km

5. a) $N=443$ cm^{-1}, $r_e=6.4$ μm, $\sigma_{ext}=0.068$ μ^{-1}; b) $W_l=87$ g m^{-1}; c) $\tau_{ext}=20.4$

6. a) $\mathcal{A}=0.03$, $\mathcal{R}=0.4$, $\mathcal{T}=0.57$; b) $\mathcal{A}^\circ=0.24$, $\mathcal{R}^\circ=0.49$; c) $\varepsilon=0.98$;
    d) 8.2 K day^{-1}; e) -15.7 K day^{-1}

Chapter 9

1. a) $\dot{P} = 31$ mm hr^{-1}; b) $=145$ W m^{-2}

2. Noon: $F^{net}_{Q0} = 858$ W m^{-2}; 2 a.m.: $F^{net}_{Q0} = -171$ W m^{-2}

3. Noon: $F_{B0}= 6.3 \times 10^{-4}$ kg m^{-1} s^{-3}; 2 a.m.: $F_{B0}= -1.3$ kg m^{-1} s^{-3}

Chapter 10

1. 0.546

2. a) 1.4 W m^{-2}; b) 8.9 W m^{-2}; c) 22.8 W m^{-2}; d) 43.0 W m^{-2}; e) 69.7 W m^{-2};
    f) 115.2 W m^{-2}

3. -24°C

4. a) -17°C

5. 4.6×10^{-6} psu m s^{-1}

Chapter 11

3. Mechanical: 246 W m^{-2}; Buoyant: 188 W m^{-2}. 43% error if buoyancy effects
    are neglected.

Chapter 12

1. a) $Z = 90° - \delta$; b) $Z = |\phi - \delta|$

Chapter 14

4. a) 278.5 K; b) 1.55 W m^{-2} K^{-1}; c) 0.20 W m^{-2} K^{-1}

References

Anderson, E.A., 1976: A point energy and mass balance model of a snow cover, *NOAA Tech. Rep. NWS 19*, Washington, DC,. 150 pp.

Andrews, J.T., 1975: *Glacial systems: An approach to glaciers and their environments.* Duxbury Press, North Scituate, MA, 191 pp.

Ardanuy, P.E., L.L. Stowe, A. Gruber and M. Weiss, 1991: Shortwave, longwave and net cloud-radiative forcing as determined from Nimbus-7 observations. *J. Geophys. Res., 96*, 18537-18549.

Arking, A., 1991: The radiative effects of clouds and their impact on climate. *Bull. Amer. Meteorol. Soc., 71*, 795-813.

Assur, A., 1958: Composition of sea ice and its tensile strength. In Arctic Sea Ice, U.S. National Academy of Sciences, NRC Pub. 598, 106-138.

Bannon, P.R., C.H. Bishop and J.B. Kerr, 1997: Does the surface pressure equal the weight per unit area of a hydrostatic atmosphere? *Bull. Am. Meteorol. Soc., 78*, 2637-2642.

Barbato, J.P. and E.A. Ayer, 1981: *Atmospheres: A View of the Gaseous Envelopes Surrounding Members of our Solar System.* Pergamon Press, New York, 266 pp.

Baumgartner, A. and E. Reichel, 1975: *The World Water Balance: Mean Annual Global, Continental and Maritime Precipitation, Evaporation and Run-Off.* Elsevier Scientific Pub. Co., Amsterdam, 179 pp.

Berry, E. X., 1967: Cloud droplet growth by collection. *J. Atmos. Sci., 24*, 688-701.

Betts, A. K., 1973: Non-precipitating cumulus convection and its parameterization. *Quart. J. Roy. Meteorol. Soc., 99*, 178-196.

Bolton, D., 1980: The computation of equivalent potential temperature. *Mon. Weather Rev., 108*, 1046-1053.

Bromberg, J.P., 1984: *Physical Chemistry*, Allyn and Bacon, Inc., New York, 966 pp.

Bryden, H.L., 1973: New polynomials for thermal expansion, adiabatic temperature gradient and potential temperature gradient of sea water. *Deep-Sea Res., 20*, 401-408.

Byers, H.R., 1959: *General Meteorology*, McGraw-Hill, New York, 540 pp.

Byers, H.R., 1965: *Elements of Cloud Physics.* Chicago, University of Chicago Press, Chicago, 191 pp.

Callen, H.B., 1960: *Thermodynamics: An Introduction to the Physical Theories of Equilibrium Thermostatics and Irreversible Thermodynamics,* J. Wiley & Sons, Inc., New York, 376 pp.

Callies, U. and F. Herbert, 1988: Radiative processes and non-equilibrium thermodynamics. *J. Appl. Math. and Phys., 39*, 242-266.

Carmack, E.C., 1990: Large-scale physical oceanography of polar oceans. In *Polar Oceanography, Part A: Physical Science*, W.O. Smith, ed., Academic Press, San Diego, 171-222.

Cess, R.D. and G.L. Potter, 1987: Exploratory studies of cloud radiative forcing with a general circulation model. *Tellus, 39A*, 460-473.

Chahine, M.T., 1992: GEWEX: The Global Energy and Water Cycle Experiment. *EOS, Trans. Amer. Geophys. Union, 73*, 9-14.

Chu, P.C. and R.W. Garwood, Jr., 1988: Comment on "A coupled dynamic-thermodynamic model of an ice-ocean in the marginal ice zone," *J. Geophys. Res., 93*, 5155-5156.

Chu, P.C. and J.C. Gascard, 1991: *Deep Convection and Deep Water Formation in the Oceans.* Elsevier, Amsterdam, 382 pp.

Clayson, C.A., 1995: *Modelling the mixed layer in the tropical Pacific and air-sea interactions: Application to a westerly wind burst.* Ph.D thesis, Department of Aerospace Engineering Sciences, University of Colorado, Boulder, 174 pp.

Colbeck, S.C., 1982: An overview of seasonal snow metamorphism. *Rev. Geophys. Space Phys.*, **21**, 45-61.

Cotton, W. R. and R.A. Anthes, 1989: *Storm and cloud dynamics.* Academic Press, San Diego, 883 pp.

Curry, J.A., 1983: On the formation of continental polar air, *J. Atmos. Sci.*, **40**, 2278-2292.

Curry, J.A., J.L. Schramm, M.C. Serreze, and E.E. Ebert, 1995: Water vapor feedback over the Arctic Ocean. *J. Geophys. Res.*, **100**, 14,223-14,229.

Curry, J.A., W.B. Rossow, D. Randall and J.L. Schramm, 1996: Overview of arctic cloud and radiation properties. *J. Clim.*, **9**, 1731-1764.

DiStefano, J.J., A.R. Stubberud, and J. Williams, 1967: *Schaum's Outline of Theory and Problems of Feedback and Control Systems.* Schaum, New York, 371 pp.

Dutton, J.A., 1986: *The Ceaseless Wind: An Introduction to the Theory of Atmospheric Motion.* Dover Publications, Mineola, NY, 617 pp.

Ebert, E.E. and J.A. Curry, 1992: A parameterization of cirrus cloud optical properties for climate models. *J. Geophys. Res.*, **97**, 3831-3836.

Eisenberg, D.S. and W. Kauzmann, 1969: *The Structure and Properties of Water.* Oxford University Press, New York, 296 pp.

Emanuel, K.A., 1994: *Atmospheric Convection.* Oxford University Press, New York, 580 pp.

Flatau, P.J., R.L. Walko and W.R. Cotton, 1992: Polynomial fits to saturation vapor pressure, *J. App. Meteorol.*, **31**, 1507-1513.

Fletcher, N.H., 1962: *The Physics of Rainclouds.* Cambridge University Press, Cambridge, UK, 386 pp.

Foster, T.D. and E.C. Carmack, 1976: Temperature and salinity structure in the Weddell Sea. *J. Phys. Oceanogr.*, **6**, 36-44.

Franks, F., ed., 1972: *Water, a Comprehensive Treatise: Volume 1. The Physics and Physical Chemistry of Water.* Plenum Press, New York, 613 pp.

Geernaert , G.L. and W.L. Plant, 1990: *Surface waves and fluxes: Volume I - Current Theory.* Kluwer Academic Publishers, Amsterdam, 286 pp.

Gill, A.E., 1982: *Atmosphere-Ocean Dynamics.* Academic Press, New York, 662 pp.

Gokcen, N.A. and R.G. Reddy, 1996: *Thermodynamics*, Plenum Press, New York, 400 pp.

Goody, R.M. and J.C.G., Walker. 1972: *Atmospheres.* Prentice-Hall, Englewood Cliffs, NJ, 150 pp.

Goody, R.M., 1995: *Principles of Atmospheric Physics and Chemistry.* Oxford University Press, New York, 324 pp.

Goody, R.M. and Y.L. Yung., 1989: *Atmospheric Radiation: Theoretical Basis*, Oxford University Press, New York, 519 pp.

Gordon, A.L., 1982: Weddell deep water variability. J. Mar. Res., **40**, 199–217.

Gunn, R. and G.D. Kinzer, 1949: The terminal velocity of fall for water droplets in stagnant air, *J. Meteorol.*, **6**, 243-248.

Gunn, K.L.S. and J.S. Marshall, 1958: The distribution with size of aggregate snowflakes. *J. Meteorol.*, **15**, 452-461.

Hansen, J., A. Lacis, D. Rind. G. Russell, P. Stone, I. Fung, R. Ruedy and J. Lerner, 1984: Climate sensitivity: Analysis of feedback mechanisms. In *Climate Processs and Climate Sensitivity, Geophys. Monogr. Ser.,* **29**, 171-179.

Hartmann, W.K., 1983: *Moons and Planets,* Wadsworth Pub. Co., Belmont, CA, 509 pp.

Hartmann, D.L., 1994: *Global Physical Climatology.* Academic Press, San Diego, 411 pp.

Hauf, T., and H. Holler, 1987: Entropy and Potential Temperature. *J. Atmos. Sci.,* **20**, 2887-2901.

Hobbs, P.V., 1974: *Ice Physics.* Clarendon Press, Oxford, 837 pp.

Holland, M.M., J.L. Schramm, and J.A. Curry, 1997: Thermodynamic feedback processes in a single-column sea ice/ocean model. *Ann. Glaciol.,* **25**, 327-332.

Houghton, H.G., 1985: *Physical Meteorology.* MIT Press, Cambridge, MA, 442 pp.

Houze, Robert A., 1993: *Cloud Dynamics.* Academic Press, San Diego, CA, 573 pp.

Iribarne, J.V. and W.L. Godson, 1981: *Atmospheric Thermodynamics,* Kluwer Press, Boston, MA, 259 pp.

Kantha, L.H. and C.A. Clayson, 1999: *Small-Scale Processes in Geophysical Flows.* Academic Press, New York, in press.

Kessler, E., 1969: *On the Distribution and Continuity of Water Substance in Atmospheric Circulations.* Met. Monograph, 10, American Meteorological Society, Boston, MA, 84 pp.

Kiehl, J.T. and K.E. Trenberth, 1997: Earth's annual global mean energy budget. *Bull. Am. Meteorol. Soc.,* **78**, 197-208.

Killworth, P.D., 1983: Deep convection in the world ocean. *Rev. Geophys. Space Phys.,* **21**, 1-26.

Klett, J.D. and M.H. Davis, 1973: Theoretical collision efficiencies of cloud droplets at small Reynolds numbers, *J. Atmos. Sci.,* **30**, 107-117.

Kraus, E.B. and J.A. Businger, 1994: *Atmosphere-Ocean Interaction,* Oxford University Press, New York, 362 pp.

Lee, I.Y., 1989: Evaluation of cloud microphysics parameterizations for mesoscale simulations. *Atmos. Res.,* **24**, 209-220.

Levitus, S., 1982: *Climatological Atlas of the World Ocean,* Prof. Pap. 13, NOAA, U.S. Dept. of Commerce, Washington, D.C., 173 pp.

Lewis, J.S., 1995: *Physics and Chemistry of the Solar System.* Academic Press, San Diego, CA, 556 pp.

Lewis, J.S. and R.E. Prinn, 1984: *Planets and Their Atmospheres: Origins and Evolutions.* Academic Press, New York, 470 pp.

Lindzen, R.S., 1990: Some coolness concerning global warming. *Bull. Am. Meteorol. Soc.,* **71**, 288-299.

Liou, K.-N., 1980: *An Introduction to Atmospheric Radiation.* Academic Press, New York, 392 pp.

Liou, K.-N., 1992: *Radiation and Cloud Processes in the Atmosphere.* Oxford University Press, New York, 487 pp.

Liu, W.T., and J.A. Businger, 1975: Temperature profile in the molecular sublayer near the interface of a fluid in turbulent motion. *J. Geophys. Res.,* **80**, 403-420.

Lock, G.S.H., 1990: *The Growth and Decay of Ice.* Cambridge University Press, New York, 434 pp.

Louis, J-F., 1979: A parametric model of vertical eddy fluxes in the atmosphere. *Boundary-Layer Meteorol.,* **17**, 187-202.

Ludlam, F.H., 1980: *Clouds and Storms.* Pennsylvania State University Press, University Park, PA, 405 pp.

Malanotte-Rizzoli, P. and A.R. Robinson, eds., 1994: *Ocean Processes in Climate Dynamics: Global and Mediterranean Examples*. Kluwer Academic Press, Boston, 437 pp.

Manton, M.J. and W.R. Cotton, 1977: Parameterization of the atmospheric surface layer. *J. Atmos. Sci.*, **34**, 331-334.

Marshall, J.S. and W.M. Palmer, 1948: The distribution of raindrops with size. *J. Meteorol.*, **5**, 165-166.

Mason, B.J., 1971: *The Physics of Clouds*. Clarendon Press, Oxford, 671 pp.

Maykut, G.A., 1985: The ice environment. In *Sea Ice Biota*, R. Hornew, ed. CRC Press, Boca Raton, FL, 21-82.

Maykut, G.A., 1986: The surface heat and mass balance. In *The Geophysics of Sea Ice*, N. Untersteiner, ed., Plenum Press, New York, 395-463 pp.

McDonald, J.E., 1964: Cloud nucleation on insoluble particles. *J. Atmos. Sci.*, **21**, 109-116.

McDougall, T.J., 1987: Thermobaricity, cabbeling, and water-mass conversion, *J. Geophys. Res.*, **92**, 5448-5464.

Millero, F.J., 1978: Freezing pont of seawter. In *Eighth Report of the Joint Panel on Oceanographic Tables and Standards*, UNESCO Tech Pap. Mar Sci. No 28, Annex 6, UNESCO, Paris.

Millero, F.J., G. Perron, and J.E. Desnoyers, 1973: Heat capacity of seawater solutions from 5 to 35°C and 0.5 to 22% chlorinity. *J. Geophys. Res.*, **78**, 4499-4507.

Millero, F.J. and M.L. Sohn, 1992: *Chemical Oceanography*. CRC Press, Boca Raton, FL, 531 pp.

Mirinova, Z.F., 1973: Albedo of the Earth's surface and clouds, In *Radiation characteristics of the Atmosphere and the Earth's Surface*, I.A. Kondratev, ed. NASA, Springfield, VA, 580 pp.

Moore, W.J., 1972: *Physical Chemistry*, Prentice-Hall, Inc., Englewood Cliffs, NJ, 977 pp.

Morel, D. and D. Antoine, 1994: Heating rate within the upper ocean in relation to its bio-optical state. *J. Phys. Oceanog.*, **24**, 1652-1665.

Moritz, R.E., 1999: Sampling the temporal variability of the sea ice draft distribution. *J. Atmos. Ocean. Tech.*, submitted.

Nazintsev, Yu. L., 1964: Thermal balance of the surface of the perennial ice cover in the central Arctic (in Russian). *Tr. Ark. Antark. Nauchno Issled. Inst.*, **267**, 110-126.

Neshyba, S., 1987: *Oceanography: Perspectives on a Fluid Earth*. John Wiley and Sons, Inc., New York, 506 pp.

Neumann, G. and W.J. Pierson, 1966: *Principles of Physical Oceanography*, Prentice Hall, Inc., Englewood Cliffs, NJ, 545 pp.

Nicholls, S., 1984: The dynamics of stratocumulous: Aircraft observations and comparisons with a mixed layer model. *Quart. J. Roy. Meteorol. Soc.*, 110, 783-820.

Ono, N., 1967: Specific heat and heat of fusion of sea ice. In *Physics of snow and ice, 1*. H. Oura, ed., Institute of Low Temperature Science, Hokkaido University, Sapporo, Japan, 599-610 pp.

Oort, A.H., 1983: *Global Atmospheric Circulation Statistics 1958-1973*, NOAA Prof. Pap. 14, U.S. Government Printing Office, Washington, DC, 180 pp.

Oort, A.H. and J. Peixoto., 1983: Global angular momentum and energy balance requirements from observations. *Adv. Geophys.*, **25**, 355-490.

Open University, 1989: *Ocean Circulation*. Pergamon Press, New York, 238 pp.

Paterson, W.S.B., 1981: *The Physics of Glaciers*. Pergamon Press, New York, 380 pp.

Perovich, D.K., G.A. Maykut and T.C. Grenfell, 1986: Optical properties of ice and snow in the polar oceans, I. Observations. *Proc. SPIE Int. Soc. Opt. Eng.*, **637**, 232-241.

454　　　**References**

Pruppacher, H.R. and J.D. Klett, 1997: *Microphysics of Clouds and Precipitation*, Kluwer Academic Publishers, Boston, 954 pp.

Ramanathan, V., R.D. Cess, E.F. Harrison, P. Minnis, B.R. Barkstrom, E. Ahmad and D. Hartmann, 1989: Cloud-radiative forcing and climate: Results from the Earth Radiation Budget Experiment. *Science*, **243**, 57-63.

Raschke, E. and D. Jacob, eds., 1993: *Energy and Water Cycles in the Climate System*. Springer-Verlag, New York, 467 pp.

Reed, R.K., 1977: On estimating insolation over the ocean. *J. Phys. Ocean.*, **7**, 482-485.

Rogers, R.R., 1976: *A Short Course in Cloud Physics*. A. Wheaton & Co., Exeter, 266 pp.

Rogers, R.R. and M.K. Yau, 1989: *A Short Course in Cloud Physics*, Pergamon Press, New York, 293 pp.

Schlesinger, M.E., 1986: Equilibrium and transient climatic warming induced by increased atmospheric CO_2. *Clim. Dyn.*, **1**, 35-51.

Schmitz, W.J., 1995: On the interbasin-scale thermohaline circulation. *Rev. Geophys.*, **33**, 151-173.

Schmitz, W.J., 1996: *On the World Ocean Circulation: Vol. 1, Some Global Features/N. Atlantic Circulation*, Woods Hole Oceanographic Institution Tech Rep. WHOI-96-03, Woods Hole, MA.

Schramm, J.L., M.M. Holland, J.A. Curry, and E.E. Ebert, 1998: Modeling the thermodynamics of a distribution of sea ice thicknesses. Part I: Sensitivity to ice thickness resolution. *J. Geophys. Res.*, **102**, 23,079-23,092.

Schwerdtfeger, P., 1963: The thermal properties of sea ice, *J. Glaciology*, **4**, 789-807.

Singh, H.B., 1995: *Composition, Chemistry, and Climate of the Atmosphere*. Van Nostrand Reinhold, New York, 527 pp.

Sinha, N.K., 1977: Instruments and methods technique for studying structure of sea ice, *J. Glaciology*, **18**, 315-323.

Slingo, A.,1989: A GCM parameterization for the shortwave radiative properties of water clouds. *J. Atmos. Sci.*, **46**, 1419-1427.

Spencer, J.W., 1971: Fourier series representation of the position of the sun. *Search*, 2, 172 pp.

Spencer, R.W. and W.D. Braswell, 1997: How dry is the tropical free troposphere? Implications for global warming theory. *Bull. Am. Meteorol. Soc.*, **78**, 1097-1106.

Staley, D.O., 1970: The adiabatic lapse rate in the Venus Atmosphere. *J. Atmos. Sci.*, **26**, 219-223.

Stephens, G.L., 1978: Radiation profiles in extended water clouds. II: Parameterization schemes. *J. Atmos Sci.*, **35**, 2123-2132.

Stephens, G.L. and D.M. O'Brien, 1993: Entropy and climate. I: ERBE observations of the entropy production of the earth. *Quart. J .Roy. Meteorol. Soc.*, **119**, 121-152.

Stull, R.B., 1988: *An Introduction to Boundary Layer Meteorology*. Kluwer Academic Publishers, Boston, 666 pp.

Swinbank, W C., 1963: Long-wave radiation from clear skies. *Quart. J. Roy. Meteorol. Soc.*, **89**, 339-348

Tomczak, M. and J.S. Godfrey, 1994: *Regional Oceanography: An Introduction*. Pergamon Press, New York, 422 pp.

Tripoli, G.J. and W.R. Cotton, 1981: The use of ice-liquid water potential temperature as a thermodynamic variable in deep atmospheric models. *Mon. Wea. Rev.*, **109**, 1094-1102.

Trenberth, K.E., 1998: Atmospheric moisture residence times and cycling: Implications for rainfall rates with climate change. *Climatic Change*, **34**, (in press).

Turner, J.S., 1981: Small-scale mixing processes, In *Evolution in Physical Oceanography, B.* B.A. Warren and C. Wunsch, eds. MIT Press, Cambridge, MA, 236-262 pp.

Turton, J.D. and S. Nicholls, 1987: A study of the diurnal variation of stratocumulus using a multiple mixed layer model. *Quart. J. Roy. Meteorol. Soc.*, **113**, 969-1010.

UNESCO, 1981: *Tenth Report of the Joint Panel on Oceanographic Tables and Standards.* UNESCO Technical Papers in Marine Sci. No. 26, UNESCO, Paris.

Untersteiner, N., 1961: On the mass and heat budget of arctic sea ice. *Arch. Meteorol. Geophys. Biok.*, **12**, 151-182.

Untersteiner, N., ed., 1986: *The Geophysics of Sea Ice.* Plenum Press, New York, 1196 pp.

U.S. Standard Atmosphere Supplements, 1976: Environmental Science Services Administration, NASA, U.S. Air Force, U.S. Government Printing Office, Washington, DC, 289 pp.

Waldram, J.R., 1985: *The Theory of Thermodynamics.* Cambridge University Press, New York, 336 pp.

Warren, B.A., 1981: Deep circulation of the world ocean, In *Evolution of Physical Oceanography*, B. A. Warren and C. Wunsch, eds., M.I.T. Press, Cambridge, MA, pp. 6-41.

Wayne, R.P., 1991: *Chemistry of Atmospheres,* Oxford University Press, New York, 447 pp.

Webster, P.J., 1977: The Low-latitude circulation of mars. ICARUS, **30**, 626-664.

Webster, P.J., 1987: The variable and interactive monsoon. in *Monsoons*, J.S. Fein and P.L. Stephens, eds., John Wiley, New York, N.Y., 269-330.

Webster, P.J. and R. Lukas, 1992: TOGA COARE: The coupled ocean-atmosphere response experiment. *Bull. Amer. Meteorol. Soc.*, **73**, 1377-1416.

Webster, P.J., 1994: The role of hydrological processes in ocean-atmosphere interactions. *Rev. Geophys.*, **32**, 427-476.

Webster, P.J., V.O. Magana, T.N. Palmer, J. Shukla, R.A. Tomas, M. Yanai, and T. Yasunari, 1998: Monsoons: Processes, predictability, and prospects for prediction. *J. Geophys. Res.*, **103**, 14,451-14,510.

Weeks, W.F. and S.F. Ackley, 1986: The growth, structure and properties of sea ice. In *The Geophysics of Sea Ice*, Untersteiner, N., ed., Plenum Press, New York, 9-64.

Willebrand, J., 1993. Forcing the ocean by heat and freshwater fluxes. In *Energy and Water Cycles in the Climate System*, E. Raschke and D. Jacob, eds., Springer-Verlag, New York, 215-234.

Woods, J.D., 1984: The upper ocean and air-sea interaction in global climate. In *The Global Change*, J.T Houghton, ed., Cambridge University Press, London, 141-187.

World Meteorological Organization (WMO), 1987: *International Cloud Atlas, Vol. 2,* World Meteorological Organization, Geneva, Switzerland, 212 pp.

Yen, Q.-C., 1969: Recent studies on snow properties. *Adv. Hydrosci.*, 5, 173-214.

Young, K.C., 1993: *Microphysical Processes in Clouds.* Oxford University Press, New York, 427 pp.

Zhang, Y.C. and M.J. McPhaden, 1995: The relationship between sea surface temperature and latent heat flux in the equatorial Pacific. *J. Clim.*, **8**, 589-605.

Zhang, Y.C. and W.B. Rossow, 1997: Estimating meridional energy transports by the atmospheric and oceanic general circulations using boundary fluxes, *J. Clim.*, **10**, 2358-2373.

Zillman, J.W. 1972: A study of some aspects of the radiation and heat budgets of the Southern Hemisphere oceans. In *Meteorological Studies*, 26, Bureau of Meteorology, Department of the Interior, Canberra, Australia, 562 pp.

Index

International Geophysics Series

EDITED BY

JAMES R. HOLTON
Department of Atmospheric Sciences
University of Washington
Seattle, Washington

* Out of Print

Journal of Arid Environments

Emeritus Editor
J. L. Cloudsley-Thompson
London, United Kingdom

Executive Editors
C. F. Hutchinson
Office of Arid Land Studies, University of Arizona, Arizona, U.S.A.
W. G. Whitford
U.S.D.A - ARS Jornada Experimental Range, New Mexico State
University, New Mexico, U.S.A.

This monthly international journal publishes original scientific and technical research articles and reviews on climate, geomorphology, geology, geography, botany, zoology, anthropology, sociology, and technical development in arid, semi-arid, and desert environments. As a forum for multidisciplinary and interdisciplinary dialogue, it is addressed to research workers interested in all aspects of the desert environment and the problems they pose. It is an authoritative work of reference for administrators and government officials in arid and developing countries.

Database coverage includes Biological Abstracts (BIOSIS), Current Contents, Ecological Abstracts, EMBASE, Environment Abstracts, Environmental Abstracts, Geological Abstracts, and Research Alert.

Printed and bound by CPI Group (UK) Ltd, Croydon, CR0 4YY

03/10/2024

01040410-0013